MASS SPECTROMETRY IN SPORTS DRUG TESTING

WILEY-INTERSCIENCE SERIES IN MASS SPECTROMETRY

Series Editors

Dominic M. Desiderio
Departments of Neurology and Biochemistry
University of Tennessee Health Science Center

Nico M. M. Nibbering
Vrije Universiteit Amsterdam, The Netherlands

A complete list of the titles in this series appears at the end of this volume.

MASS SPECTROMETRY IN SPORTS DRUG TESTING

Characterization of Prohibited Substances and Doping Control Analytical Assays

Mario Thevis

German Sport University Cologne, Center for Preventive Doping Research—Institute of Biochemistry, Cologne, Germany

A JOHN WILEY & SONS, INC., PUBLICATION

Copyright © 2010 by John Wiley & Sons, Inc. All rights reserved

Published by John Wiley & Sons, Inc., Hoboken, New Jersey
Published simultaneously in Canada

No part of this publication may be reproduced, stored in a retrieval system, or transmitted in any form or by any means, electronic, mechanical, photocopying, recording, scanning, or otherwise, except as permitted under Section 107 or 108 of the 1976 United States Copyright Act, without either the prior written permission of the Publisher, or authorization through payment of the appropriate per-copy fee to the Copyright Clearance Center, Inc., 222 Rosewood Drive, Danvers, MA 01923, (978) 750-8400, fax (978) 750-4470, or on the web at www.copyright.com. Requests to the Publisher for permission should be addressed to the Permissions Department, John Wiley & Sons, Inc., 111 River Street, Hoboken, NJ 07030, (201) 748-6011, fax (201) 748-6008, or online at http://www.wiley.com/go/permission.

Limit of Liability/Disclaimer of Warranty: While the publisher and author have used their best efforts in preparing this book, they make no representations or warranties with respect to the accuracy or completeness of the contents of this book and specifically disclaim any implied warranties of merchantability or fitness for a particular purpose. No warranty may be created or extended by sales representatives or written sales materials. The advice and strategies contained herein may not be suitable for your situation. You should consult with a professional where appropriate. Neither the publisher nor author shall be liable for any loss of profit or any other commercial damages, including but not limited to special, incidental, consequential, or other damages.

For general information on our other products and services or for technical support, please contact our Customer Care Department within the United States at (800) 762-2974, outside the United States at (317) 572-3993 or fax (317) 572-4002.

Wiley also publishes its books in a variety of electronic formats. Some content that appears in print may not be available in electronic formats. For more information about Wiley products, visit our web site at www.wiley.com.

Library of Congress Cataloging-in-Publication Data:

Thevis, Mario, 1973–
Mass spectrometry in sports drug testing : characterization of prohibited substances and doping control analytical assays / Mario Thevis.
p. ; cm.
Includes bibliographical references and index.
ISBN 978-0-470-41327-2 (cloth)
1. Doping in sports–Diagnosis. 2. Mass spectrometry. 3. Drugs–Analysis. I. Title.
[DNLM: 1. Doping in Sports. 2. Spectrum Analysis–methods. 3. Substance Abuse Detection–methods. QT 261 T418m 2010]
RC1230.T54 2010
362.29–dc22

2009052128

Printed in Singapore

10 9 8 7 6 5 4 3 2 1

For Jan and Linus

CONTENTS

PREFACE

Mass spectrometry has become an essential aspect of sports drug testing. Since the early doping controls were implemented almost one century ago, considerable knowledge has been presented, published in various journal articles and book chapters. In this book, the evolution of drug testing in sports (predominantly focused on humans) is illustrated with the aim to inform the reader of the challenges, varieties, and capabilities of modern doping controls. These controls rely strongly on the specificity and sensitivity of chromatographic and mass spectrometric assays. Prohibited classes of substances are introduced and explained with an emphasis on the characterization of analytes being of paramount importance in the unequivocal identification of banned substances and doping administration. The mass spectrometric behavior of low and high molecular weight analytes is presented with descriptive examples outlining the principles, advantages, and limitations of the assays.

ACKNOWLEDGMENTS

I wish to express my deep gratitude to my mentors, in particular to Professor Wilhelm Schänzer, Dr. Hans Geyer, and Professor Joseph A. Loo, who gave me the opportunity to learn and the support to develop. In addition, I wish to thank all those colleagues and friends whose help was invaluable in preparing this book.

1 History of Sports Drug Testing

1.1 HISTORICAL ATTEMPTS OF ARTIFICIAL PERFORMANCE ENHANCEMENT

Sports competition is initiated by many different motivations, which include, but are not limited to, mankind's desire for excellence and perfection, the enjoyment of contests, and financial as well as social benefits associated with victory and success at sporting events.[1] In particular, the latter facts are frequently mentioned as the major reasons for attempts of sportsmen to artificially increase their physical performance and that this modern and contemporary issue was not present between 776 B.C. and 393 A.D. at the ancient Olympic Games. However, the belief that athletes participating at the ancient Panhellenic Games, which included the Olympic Games as the most prestigious sporting festival, were motivated only by the glory and appreciation that they might receive, is weakened even by the interpretation of the term "athlete." The origin of this expression, the Greek noun *athlon*, means "prize" or "reward;" its verbal form *athleuein* means "to compete for a prize." Consequently, the athlete is a person who competes for a prize, which was reportedly a great value even in ancient times,[2] and sport for sport's sake was not an ancient concept.[3] In fact, even thousands of years ago, athletes were seeking competitive advantage over their rivals in many different ways, which included manipulation of equipment and corruption of judges. Moreover, the consumption of certain mushrooms as psychogenic aids was reported,[4] and Philostratus (3rd–2nd century B.C.) and Pliny the younger (1st century A.D.) wrote notes about athletes consuming bread prepared with juice of the poppy plant (opium) and the use of a decoction of the hippuris plant, respectively.[5] All was done with considerable support by the doctors and the goal to enhance the athletes' performance, which would possibly be referred

Mass Spectrometry in Sports Drug Testing: Characterization of Prohibited Substances and Doping Control Analytical Assays, By Mario Thevis
Copyright © 2010 John Wiley & Sons, Inc.

to as "doping" in a modern context. Physicians acted as coaches and doctors simultaneously, and one of the most famous medical attendants was Herodicus, the indoctrinator of Hippocrates, who was particularly interested in athletes' nutrition and rehabilitation.[6] In addition to the human competitors, horses were also the subject of treatment to increase their endurance and stamina as reported in the ancient Rome where *hydromel*, a mixture of honey and water, was administered to horses in chariot races.[7] With the growing medical and pharmaceutical knowledge in the 19th and 20th century, more and more attempts of artificial performance enhancements were assumed and reported, which were conducted with humans and animals likewise. Consequently, various initiatives were started to counteract the misuse of drugs and methods to surreptitiously increase power, strength and athletic capabilities. Such approaches were first conducted with horse saliva in the early 20th century, and, approximately 50 years later, applied also to human urine and blood specimens.

1.2 BACKGROUND AND RATIONALE OF DOPING CONTROLS

Numerous reasons for banning drugs and methods of doping and manipulation from sports were defined by anti-doping authorities such as the medical commission of the International Olympic Committee (IOC) or the World Anti-Doping Agency (WADA) and first necessitated the recognition of the issue. In 1988, the IOC medical commission drafted a charter stating that, "The use of drugs and other substances and banned methods to enhance or accentuate athletic performance is a tragic reality that must be eliminated from modern sport."[8] These fundamental words still reflect the principles that are still seminal to anti-doping programs, which have been coined to preserve the "spirit of sport." According to the World Anti-Doping Code, the spirit of sport is characterized by various values including ethics, fair play and honesty, health, dedication and commitment, and respect for rules, laws, self, and other participants, etc.[9] Doping, however, contravenes to all of these aspects and, thus, modern doping controls are focused on substances and methods of doping that meet at least two of three criteria as defined by WADA: (1) A substance or method has the potential to enhance or enhances sport performance as evidenced by medical or scientific data, pharmacological effects, or experience; (2) a substance or method represents an actual or potential health risk to the athlete as evidenced by medical or scientific data, pharmacological effects, or experience; and (3) a substance or method violates the spirit of sport as defined in the World Anti-Doping Code.[9]

1.2.1 Cheating

Sportsmanship implements the idea of fair play and the integrity of all members of the sporting community. The ideology that only "eligible" persons should be allowed to compete was present also during the ancient Olympic Games and outlined in the facts that only athletes who were never convicted of a crime should participate. Moreover, sportsmen were requested to swear that they had trained for 10 months prior to coming to Olympia, and another 4 weeks on-site being supervised by the *Helenedonakai*—the judges.[4] Doping contravenes the most basic principles of fair play and results in beguilement of competitors and spectators. Both are hoodwinked, and in particular the deceived athlete might suffer from financial, social, and probably occupational disadvantages in addition to a personal disappointment, if he/she loses a competition against an athlete who artificially increased his/her performance. Consequently, doping must be regarded as cheating in numerous regards, and the rights of those athletes, who are devoted to clean and fair sports, must be protected.

1.2.2 Health Issues

Doping practices can compromise the short- and long-term physical and mental health of athletes; hence, health and safety concerns have been a major aspect of the fight against doping. Numerous articles were published dealing either with case reports about serious or even fatal consequences of drug abuse in professional and amateur sport as well as general undesirable effects observed and associated with doping,[10–20] which were supposedly the final trigger for international sport federations to establish anti-doping rules and test their athletes for drug abuse.[21]

1.2.3 Ethical Issues

According to the World Anti-Doping Code[9] and the common understanding of the intrinsic value of sport, doping categorically contradicts the spirit of sport. This issue has also ethically been evaluated and all values attributed to the spirit of sport have been subject of ethical considerations.[22] Fair play and honesty, character and education, and the virtue of athletes, are an integral part of sport pedagogy and pedagogic ethics, which is complemented by numerous additional aspects of ethics in sports concerning the health and the exploitation of the human (or animal) body as well as the respect for rules, laws, self, and other participants. The violation of ethical principles is not acceptable in sport, and noncompliance is regarded as a doping offence.

1.3 EARLY DETECTION METHODS: POSSIBILITIES AND LIMITATIONS OF ASSAYS WITHOUT MASS SPECTROMETRY

Doping of animals, primarily horse and hound, has been considered a major pacemaker of doping practices in modern human sports but also as a driving force of anti-doping activities. In 1666, the first decree was enacted in England, which prohibited the administration of substances to horses aiming to improve their performance in races at Worksop,[12] and severe consequences up to the death penalty were announced and executed as reported in the late 18th century in Cambridge (Great Britain) when horses were poisoned at Newmarket.[23] In light of such regulations and their strict enforcement, it was a logical consequence that the first successful attempts to detect doping agents using bioassays and analytical chemistry were introduced in horse racing rather than in human sports.

1.3.1 First Applications Using Chemical and Biological Approaches in Horse Doping Control

In 1910, the Austrian Jockey Club hired a Polish pharmacist named Alfons Bukowski[24] (who has, at some occasions, been referred to as a Russian scientist) to establish a method that allows the detection of alkaloids such as morphine and heroin in equine saliva. He reportedly succeeded in developing such a method but never disclosed any details and returned to his home country, which prompted the Austrian Jockey Club to call in Professor Sigmund Fränkel from the University of Vienna to install a new procedure enabling saliva drug testing. Although never published by Fränkel himself, the principle procedure was later described by G. Lander, the chief chemist for the Jockey Club of England,[25] who published a method that included various consecutive extraction and concentration steps followed by chemical reactions forcing precipitation and/or color reaction of alkaloids for visual inspection. In general, the employed approach was mainly a miniaturized application of the Stas-Otto process,[26] which gained public recognition as early as 1850 when its use helped reveal the murder of Gustave Fougnies and strongly influenced the newly born arena of forensic sciences.[27]

First, a comparably large volume of saliva was required, which was preferably obtained by washing the horse's mouth using a 0.16 M acetic acid solution over a period of up to 5 minutes. The obtained material was extracted using 90% pure ethanol and diluted acetic acid, followed by filtration and concentration of the extract *in vacuo* to an aqueous residue of approximately 5 mL. The solution was purified by ether extraction and its volume further reduced by evaporation to yield a viscous remainder, which was again extracted with small amounts of ethanol that was finally concentrated to dryness. The dry residue was extracted using 0.3 M aqueous hydrochloric acid, the aqueous layer was adjusted to alkaline pH followed by extraction using chloroform and benzene, the combined organic layers were concentrated to dryness, and the remaining residue reconstituted in 0.16 M acetic acid. The presence of an alkaloid (including cocaine, strychnine, quinine, morphine, and heroin) was visualized by formation of a precipitate or opalescence when placing approximately 5–10 μL of the saliva extract in a capillary test tube and adding different reagents as listed in Table 1.1.[25] Depending on the target analyte and the employed chemical, estimated detection limits between 0.05 and 20 μg were accomplished using diluted reference compounds. The applicability to authentic saliva specimens and, as such, the proof-of-principle, was provided using blank saliva samples of reportedly untreated animals as well as specimens derived from administration studies to outline the specificity of the method and the ability to "unambiguously" differentiate between positive and negative results.

TABLE 1.1: Estimated Detection Limits (μg) for Alkaloids Using Colorimetric Test Methods According to Lander (1930)[a]

Reagent	Alkaloid				
	Cocaine	Strychnine	Quinine	Morphine	Heroin
Iodine	0.05	0.05	0.05	10.0	0.50
Phosphomolybdic acid	0.05	0.10	0.20	1.00	0.50
Potassium mercuric iodide	0.05	0.20	0.40	10.0	2.00
Gold chloride	0.10	0.40	0.40	1.00	0.50
Tannic acid	—	0.50	0.40	20.0	20.0

[a] Ref. 25.

Chemical saliva tests underwent further developments that aimed for optimized extraction conditions of target analytes and more sensitive assays.[28] Using a defined array of tests, opiates were the first to screen for preferably by the Marquis reagent (formaldehyde and sulphuric acid), which yields a dark red-to-purple color in the presence of opiates, particularly morphine and heroin. Subsequently, analyses for strychnine (vanadic acid/sulphur-chromate test), quinine (bromine-ammonia test), cocaine, nikethamide, atropine, etc. (crystalline methods) were conducted. These and other compounds were used as mixtures, e.g. 1.5 g of heroin, 2.5 g of strychnine, 2 minims of nitroglycerine (accounting for 1/250 U.S. fluid ounces or 0.12 mL), 5 minims of *tinctura digitalis*, and 2 ounces of cola nut, and applied to race horses approximately 1 hour before a race.[23] Numerous additional concoctions made from stimulants, narcotics, herbal extracts, and organic as well as inorganic poisons were employed for horse doping purposes in the early 20th century,[29] and astonishing estimations about the prevalence of doping in equine sports were published mentioning more than 50% of doped horses in the United States in the early 1930s.[30] However, the numbers dropped in the following years possibly due to improved analytical procedures and certainly enforcement of severe punishments. Nevertheless, the ambitious efforts of pharmaceutical industries and the enormous numbers of continuously generated new drugs represented a great challenge for doping control chemists, and in particular the introduction of amphetamine and its derivatives initiated a new era of doping for human and veterinary sports. Consequently, more sensitive, comprehensive, and specific detection assays were required, which were developed and established for human and equine doping control specimens likewise (see section 1.3.2).

Besides chemical analytical options, bioassays based on mice and frogs were established by Munch[31–35] and Lucas.[28] The effect of selected drugs on the behavior of mice was observed for instance by Straub, who reported on the phenomenon that these animals carry their tail in an unmistakable S-shaped curve parallel to their backbone under the influence of minute amounts of morphine ("Straub tail response"),[36–39] which was due to a sustained contraction of the *sacrococcygeus dorsalis* muscles[40] (Fig. 1.1).[290] Moreover, Munch described the effects of barbiturates and sedatives in general as well as those resulting from stimulants such as strychnine and caffeine on mice, which were commonly recognized in a decrease in voluntary travel and an increased tendency to sleep or increased locomotion and muscular tremor, respectively.[32] In addition, the administration of strychnine and nikethamide to small frogs yielded characteristic convulsions, and movements were kymo-

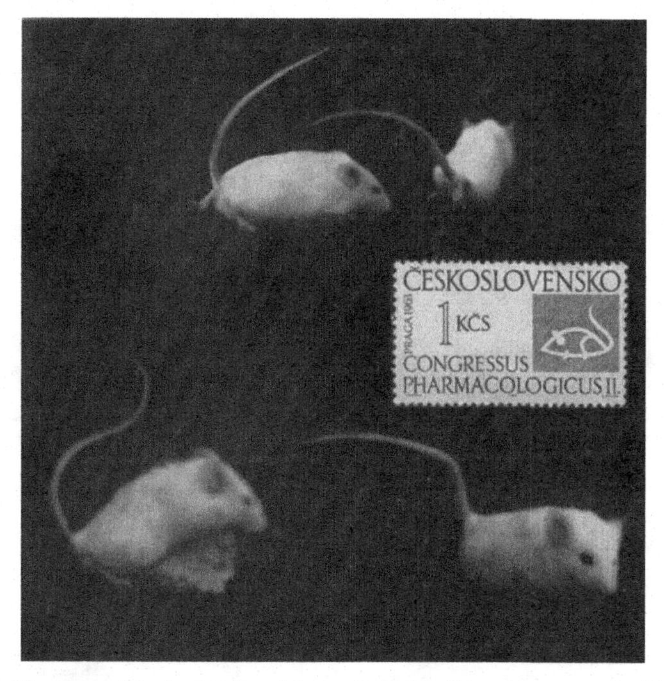

Figure 1.1: Photography of mice showing the Straub tail response to morphine applications; inset: stamp of 1963 ornamented with a Straub tail response mouse. From Klaus Starke, Die Geschichte des Pharmakologischen Instituts der Universität Freiburg, Figure 3, page 12. Reprinted by permission of Springer Science and Business Media. Copyright © Springer, Berlin 2004.

graphically recorded and evaluated.[29] Comprehensive studies were conducted on how these approaches would be useful and possibly "justiciable" for horse doping control purposes, and methods were established that required 2 mL of saliva or 1 mL of urine, aliquots of which were intraperitoneally administered to mice. Several thousand saliva and urine specimens derived from horses that did not receive any medication within a period of 24 hours were measured using the bioassay approach, and no changes in behavior, appearance, or attitude were observed within 15 minutes.[41] In contrast, samples collected from horses participating in administration studies caused positive reactions and were attributable to the applied drug(s), even when different substances were given simultaneously. Due to the fact that the injected samples were not sterilized, mice usually suffered from infections after 12–24 hours post-administration, but these reactions were differentiable from pharmacodynamic responses and not relevant for the test result. The mouse-based bioassay demonstrated reasonable sensitivity

TABLE 1.2: Estimated Detection Limits for Selected Drugs Using Mouse-based Bioassays According to Munch (1952)[a]

Drug	(μg / 20g mouse)
Amphetamine	80
Caffeine	600
Cocaine (HCl)	400
Codeine	60
Nikethamide	100
Pethidine	400
Desoxyephedrine	40
Dihydromorphinone	12
Ephedrine	200
Heroin	1
Morphine	60–80
Picrotoxin	40
Strychnine	1

[a] Ref. 32.

regarding a variety of analytes (Table 1.2), enabled fast delivery of test results, and was comparably cheap. Hence, it was considered as a rapid screening tool for horse drug testing, and the more time-consuming and laborious chemical analyses were conducted mainly when suspicious results were found. Several decades later, also the dilation of pupils of mice after subcutaneous injection of morphine and extracts derived from urine of morphine users was found to be a sensitive and specific test.[42]

1.3.2 First Applications Using Chemical Approaches in Human Doping Control

The history of drug abuse in human sports reaches back at least as far as 1865, when swimmers in the Amsterdam's canal races were competing under the influence of doping agents;[43] however, the administered drugs were never identified or disclosed. In contrast, the systematic misuse of stimulants, narcotics, and nitroglycerine by cyclists was reported particularly for participants of the infamous "Six-days" cycle races that began in 1879. Racing cyclists appeared to be the prime offenders, who used various drug cocktails consisting of coffee enriched with extra caffeine, which was further fortified with strychnine and cocaine as the race progressed. Alternatively or complementary, sugar cubes dipped in ether were consumed, capsules of nitroglycerine (to support and ease breathing) as well as mixtures of heroin and cocaine ("speedballs") were administered, and miracle drinks were provided by trainers that supposedly contained digitalis, atropine, and camphor.[21] Most drugs were applied without any supervision and, consequently, fatalities were not surprising. The most prominent and presumably first victim of doping practices was the British cyclist Arthur Linton, who was reported to have died in 1886 during a 600 km cycle race between Paris and Bordeaux. Numerous aspects, however, disproved this anecdote, including the fact that the first Paris-Bordeaux-race was in 1891 and Arthur Linton evidently died in 1896 of typhoid fever.[44] Nevertheless, numerous athletes died young for mostly unknown reasons in the late 19th and early 20th century, which might be attributed also to the misuse of drugs. Doping agents were applied in various sports such as cycling, running, boxing, soccer, etc. and even Olympic gold medal winners such as Thomas Hicks (Marathon, 1904) were reportedly and officially using drugs such as strychnine[45] without being sanctioned. Following World War II, the use of stimulants such as amphetamine and derivatives in sports increased enormously, and in particular cyclists and soccer players were found to be regular users.[46] The great health issues associated with the uncontrolled use of these drugs in sports prompted several European countries such as France, Belgium, and Italy to install anti-doping laws in 1965[21,47] and the need for doping control analytical assays was evident. Numerous approaches were established for a variety of stimulants and narcotics employing contemporary state-of-the-art technologies.

1.3.2.1 Methods without Chromatography Drug metabolism, disposition, elimination, and analysis have been of great interest for numerous clinical applications as well as doping control purposes, and both fields of research have frequently interacted with physical, chemical, biological, and biochemical sciences. Most prominent physicochemical properties of analytes were used to differentiate and identify target compounds in urine, blood, saliva, and sweat, which yielded a variety of analytical approaches primarily based on crystallographic, chromatographic, and/or colorimetric methods. Between 1930 and 1960, only few procedures were developed and published solely for sports drug testing purposes; most inventions were generally accepted for forensic, clinical, and doping control applications.

The early characterization of drugs, in particular of stimulants, was accomplished using microscopy, the determination of melting point and the corresponding refractive index of crystals (derived from various salts) as well as defined mixtures and their resulting melts.[48–50] In addition, microprecipitation, microchemcial tests with gold or platinic chloride and nephelometric analyses after complexation of alkaloids were reported,[51,52] but although "micro scale" methods were employed, most approaches were difficult to apply to biological matrices due to the large volumes required and the need of extensive sample preparation.

Colorimetric assays proved to be reasonably sensitive.[52] One of the first assays to determine benzedrine (the racemic mixture of d- and l-amphetamine) in human urine was reported in 1938,[53] which was based on an alkaline extraction of amines into ether, followed by a reaction with picric acid to yield a clear yellow solution, the color intensity of which was visually estimated against a calibration curve ranging from 0–20 µg/mL. The method was applied to a series of amines such as phenylethylamine, ephedrine, amphetamine, methamphetamine, mescaline, etc. in order to determine their excretion rates into urine after oral administration, and the author indicated a detection limit of 0.5 µg/mL.[54] However, the fact that various amines also occur naturally in urine specimens, which strongly influenced each individual blank value, as well as the entirely missing identification of a distinct compound made the test only moderately applicable for forensic and doping control purposes. More advanced methods with increased specificity were desired and established, commonly using extraction and re-extraction steps combined with preceding steam distillation of alkalinized urine (or other bodily specimens and extracts). Due to the volatility of amphetamines and various alkaloids as free bases, samples were usually buffered to pH 9–10 and distilled under reduced pressure into dilute acids. After re-adjusting the pH to 9–10, the distillate was extracted using organic solvents such as benzene or chloroform and the analytes were derivatized to colored compounds, for instance by means of diazotization, picric acid, or methyl orange. Following re-extraction into small volumes of dilute acid, quantitative results were obtained by colorimetry and photometry with detection limits of approximately 1 µg/mL.[51,55–60]

1.3.2.2 Methods Including Paper or Thin-layer Chromatography

Although the above-mentioned assays were comparably sensitive and optimized to ensure reduced susceptibility to interferences, the introduction of chromatography in general was considered a major breakthrough in drug testing. Various applications were established for numerous compounds, and special attention was first paid to the paper-chromatographic separation of alkaloids and stimulants[61–67] as well as diuretic agents,[68] which were, however, not prohibited in sports until

1988. Sample preparation and extraction techniques were commonly adapted from earlier methods based on colorimetry and photometry (*vide supra*), and extracts were subsequently subjected to paper chromatography followed by visualization of target analytes by means of different stains such as the Dragendorff's reagent (Bismuth nitrate and potassium iodide in dilute acetic or hydrochloric acid),[61,64] the Prussian blue reagent (also known as Berlin blue reagent, prepared of potassium ferricyanide and iron(III) chloride),[65] or an alcoholic solution of bromocresol green (3,3′,5,5′-tetrabromo-*m*-cresolsulfonphthalein).[67] Paper chromatography was rapidly complemented and/or substituted by thin-layer chromatography (TLC) in forensic and doping control analyses, in particular with regard to morphine-related narcotics and amphetamine and its derivatives,[69–85] as well as diuretics.[86] The principle sample preparation procedures remained and commonly required between 5 and 200 mL of urine (if available). After purification and extraction of target compounds, separation was accomplished by one-[75] or two-dimensional[74] TLC followed by visualization either on the plate using spray reagents or after elution of spots employing ultraviolet spectrophotometry.[73] Numerous spray reagents were tested such as ninhydrin, diphenylcarbazone/silver acetate/mercury(II) sulphate, iodoplatinate, iodine/potassium iodide, bromocresol green, and ammoniacal silver nitrate that exhibited either particular sensitivity or specificity for selected groups of compounds enabling the detection of 0.1–1.0 μg of target analytes per mL of urine.[87]

1.3.2.3 Methods Including Gas Chromatography The capability of gas chromatography (GC) to separate compounds relevant for doping controls was recognized in the late 1950s, and first proof-of-principle studies outlined the potential of GC systems to measure various classes of analytes.[88–92] Commonly, columns consisting of 1–3 m long stainless steel or glass tubing (0.3–0.6 cm outer diameter) filled with different stationary phases (e.g., 5% Carbowax 6000 and 5% KOH, 2% Carbowax 20M and 5% KOH, 2.5% SE-30, or 10% Apiezon L and 10% KOH) were employed to accomplish efficient separation of target substances of different polarities,[93] and analyzers such as flame ionization and nitrogen-phosphorus detectors (FID and NPD, respectively) as well as ionization β-ray (strontium[90]) or electron capture detectors were used to measure the eluting compounds. Initially, stimulants and alkaloids such as amphetamine, methamphetamine, caffeine, cocaine, ephedrine, strychnine, etc. and their hydroxylated metabolites isolated from biological matrices were the subject of research and routine analysis.[46,76,94–100] While sample extraction and concentration methodologies were mostly adapted from earlier procedures and only marginally altered, major aspects of gas chromatography to improve were soon identified, which included for instance the need for supporting information that increases confidence in the identification of a target compound, better chromatographic peak shapes, and reproducible separation of isomers. All of these issues were addressed by various derivatization strategies, which first improved chromatographic properties and, second, provided additional data to characterize a substance. A strategy to identify a compound by its retention times obtained from the native as well as derivatized analyte or two different derivatives was termed the "peak-shift technique"[101] and used as a common standard in confirmatory analyses. Trimethylsilylation (using e.g., hexamethyldisilazane or *N*-methyl-*N*-trimethylsilyltrifluoroacetamide [MSTFA][102]), acylation (using e.g., acetic, propionic, or heptafluorobutyric anhydride, bis[acylamide], etc.), alkylation, formation of several Schiff-bases (e.g., acetone-, propionaldehyde-, benzyl methyl ketone-Schiff-bases),[103] or preparation of mixed derivatives were utilized to modify the physicochemical nature of substances and, thus, enhance their traceability in (sports) drug testing. The most frequently used methods to chemically modify target analytes in doping controls were finally based on trimethylsilylation or acylation according to assays established by Donike and co-workers,[104–108] which enabled detection limits of approximately 1 µg/mL and represented a central element of the first comprehensive doping control program undertaken at the Olympic Games 1972 in Munich and subsequently conducted great sporting events.[109–111] Improved GC columns, in particular using capillary tubing, further increased the sensitivity and robustness of analyses, which has made GC an invaluable tool of past and present sports drug testing approaches,[112,113] especially as a separation unit for target compounds. However, the enormous complexity of biological matrices and the continuously increasing number of therapeutics have necessitated more specific and unequivocal analyzers than for instance NPD and FID, which resulted in the highly successful combination of GCs and mass spectrometers.

1.3.2.4 Methods Including Liquid Chromatography The analysis of substances by means of GC requires vaporization, and several compounds that gained relevance for doping controls in addition to "classical" doping agents are thermolabile or possess poor gas chromatographic properties even after derivatization and, thus, insufficient detection limits. Consequently, alternative approaches were necessary, and high performance liquid chromatography (HPLC), in particular employing reversed-phase columns, was found suitable for the detection of compounds that composed the continuously expanding list of prohibited compounds of doping such as diuretics,[114–119] (ring-hydroxylated) stimulants,[120–124] anabolic-androgenic steroids and corticosteroids,[125–138] analgesics,[139–142] etc. with and without derivatization. The commonly used detectors were ultraviolet (UV)- or fluorescence-based analyzers, depending on the nature of the analyte and/or the produced derivative.

Diuretics represent a heterogeneous class of compounds; however, most of them can be measured using conventional absorbance detectors such as UV-analyzers without further derivatization at reasonable detection limits of approximately 0.01–2.0 μg/mL.[116] Increased sensitivity and/or selectivity was reported for the loop diuretics furosemide and bumetanide as well as the potassium sparing diuretics amiloride and triamterene when using fluorescence detection due to the fact that numerous agents co-extracted with diuretics are visualized by means of UV but not fluorescence detection.[114]

Although stimulants related to amphetamine were determined using the above-mentioned GC-based approaches, options employing HPLC were evaluated and demonstrated great utility in particular regarding sensitivity and comprehensiveness. Target analytes were measured from urine either in native forms[122] or after pre- or post-column derivatization using *o*-phthalaldehyde, 4-chloro-7-nitro-benz-2,1,3-oxadiazole, sodium naphthaquinone-4-sulphonate, or bis(2,4,6-trichlorophenyl) oxalate, which allowed for detection limits ranging from 0.01–0.1 μg/mL.[120,121,124]

The analysis of steroidal agents such as androgens, corticosteroids, and anabolic androgenic steroids was of great interest in various fields including doping controls,[123] and the advantages of HPLC-based approaches over immunological methods in terms of specificity and reproducibility were major reasons for the developments of assays, enabling the detection of steroids in human and animal specimens.[135] Anabolic steroids such as metandienone were detected at concentrations of 10 ng/mL,[125] and other analytes lacking ultraviolet active functional groups were derivatized for instance by conversion into benzoates or *p*-nitrobenzoates.[130]

Few analgesics such as morphine and related opiates have been prohibited in human sports, and all members of the class of analgesic therapeutics are not allowed for the treatment of competing animals. Hence, various methods using HPLC were established to determine the presence of these compounds in urine and blood samples, and comprehensive assays enabled the detection of 10–100 ng/mL of analytes such as morphine, ketoprofen, naproxen, etc.[139,141,142]

Several of these assays are still in use in various fields of analytical chemistry but usually not in sports drug testing, where comprehensive, fast, and specific procedures are required. Although detection limits of numerous methods would fulfill so-called minimum required performance limits as established by WADA, the inferior specificity of UV-spectra compared to mass spectrometric information has led to several endeavors to combine the liquid chromatographic separation units via ionization interfaces to all kinds of mass spectrometers.

1.3.2.5 Methods Including Immunological Approaches (Radio) Immunoassays have been of great interest in endocrinology and related fields due to their capability to qualitatively and quantitatively determine trace amounts of peptide hormones.[143] Although most compounds were not prohibited in sports at the time of early detection method developments, various procedures would have been of benefit for doping controls also, e.g., assays for the detection of insulins,[144–146] corticotrophins,[147–149] human growth hormone (hGH),[150–156] human chorionic gonadotrophin (hCG),[157,158] luteinizing hormone (LH),[159,160] or erythropoietin (EPO).[161] However, their successors have been frequently used in sports drug testing and have become an important tool for the identification or purification of several drugs in doping control analyses.

Strategies for the immunological detection of substances that are prohibited in sports were introduced in 1975 by Brooks and colleagues[162] based on results first published by Sumner in 1974.[163] A major issue of the early anti-doping fight was, among others, the missing tool to uncover the assumed misuse of anabolic steroids, and the availability of radioimmunoassays represented a breakthrough in sports drug testing.[164–167] The extraction of steroids followed by acetylation and subsequent determination using radioimmunoassays targeting 17-alkylated steroids or 19-norsteroids provided a sensitive procedure enabling the detection of as low as 10 pg of metandienone per mL of urine or serum. These approaches were applied to anonymized samples collected at the 1974 Commonwealth Games (Christchurch, New Zealand) and European Games (Rome, Italy),[168] and further used officially to screen urine specimens at the Olympic Games of 1976 in Montreal (Canada).[167] Confirmatory analyses of samples that yielded adverse analytical findings were conducted by means of newly established GC-MS methods, and two samples of the 1974 Commonwealth Games as well as six specimens from the 1976 Olympic Games were reported "positive" with RIA and GC-MS. Although the quality of analytical results obtained by combinations of GC and MS considerably improved in the following years and became more and more the gold standard for doping controls, immunological assays were employed for various analytes such as stimulants,[169–172] opioids,[173–179] β_2-agonists,[180,181] diuretics,[182,183] benzoylecgonine, etc. in human and animal drug testing for decades.[184–186] Still, the importance of immunoassays for the detection of low molecular weight drugs dropped significantly with the constantly increasing performances of mass spectrometry-based methods and requirements of unequivocal identification of target analytes; however, the fact that high molecular weight compounds were hardly or not measurable from doping control specimens using conventional MS systems has made immunological approaches an invaluable tool in the past and present fight against doping. The most prominent examples for the application of immunoassays have been the detection of misuse regarding hCG and LH,[187–191] and, more recently, hGH[192–204] and EPO.[205–213] These methods have enabled the determination of prohibited compounds (e.g., hCG in urine samples of male athletes), the quantitation of banned substances such as hGH and its natural variants, or the quantitative analysis of parameters that indirectly indicate the potential misuse of drugs and methods of doping such as the administration of hGH and EPO or blood transfusions.

1.4 INTRODUCTION OF MASS SPECTROMETRY TO DOPING CONTROL ANALYSIS

Mass spectrometry of organic compounds and natural products has greatly influenced analytical approaches and possibilities of analyte characterization in complex matrices. Numerous seminal articles were published describing the fundamentals[214–219] and, thus, the instrumentation required for applied mass spectrometry in forensics, clinical, and doping control analysis. One of the earliest successful combinations of gas chromatography and mass spectrometry consisted of packed GC column interfaced to a Type 12-100 time-of-flight (TOF) MS (Bendix Aviation Corp.), which allowed to scan from m/z 1 to 6000 at an enormous scan rate of 2000/s with unit resolution up to m/z 200.[214] Alternative approaches included conventional magnetic field mass spectrometers such as the CEC Model 21-620 cycloidal path or the CEC Model 21-103B 180-degree MS (both Consolidated Electrodynamics Corp.) that covered a scan range from m/z 40 to 160 within a duty cycle time of 30s at a resolving power of 600.[215] Commonly, approximately 3–10% of the GC effluent was diverted from packed columns through a capillary to the ion source of the mass spectrometer, before capillary GC columns

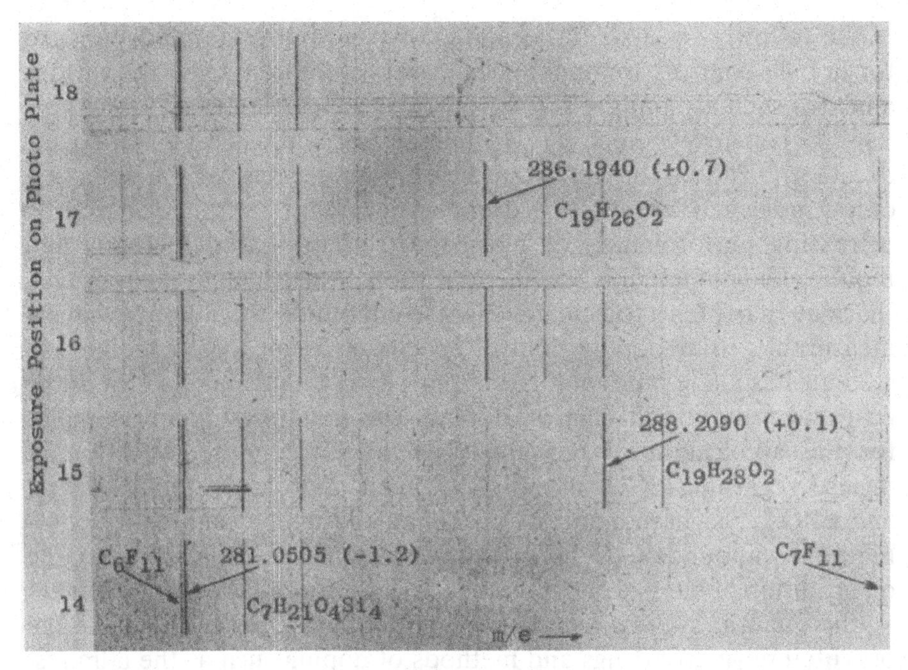

Figure 1.2: High-resolution mass spectra of androstane-3,17-dione (15) and androst-4-ene-3,17-dione (17), recorded on a GC-HRMS system (CEC Model 21-110) with a photographic plate detector. Reprinted with permission from J.T. Watson, K. Biemann, *Anal. Chem.*, 1965, 37 (7), p. 850, Figure 9. Copyright © 1965 American Chemical Society.

were directly interfaced to the analyzers.[217] Methods and instruments dramatically improved within a few years and enabled, for instance, the direct recording of high resolution mass spectra of steroids such as androstane-3,17-dione, and androst-4-ene-3,17-dione, which were injected into a GC that was interfaced to a double-focusing MS (Mattauch-Herzog design, CEC Model 21-110) with a photographic plate detector (Fig. 1.2).[219]

1.4.1 First Approaches and Adverse Analytical Findings

First official anti-doping rules as established by international federations became effective in 1966 (Fédération Internationale de Football Association, FIFA) and 1967 (Union Cycliste Internationale, UCI; Union Internationale de Pentathlon Moderne et Biathlon, UIPMB). Nevertheless, pilot studies with regard to human sports drug testing started as early as 1955,[46] and systematic controls at great sporting events were initiated at the FIFA World Cup in 1966. At that time, the

FIFA list of prohibited substances included narcotics such as morphine and heroin, stimulating agents (amphetamine and its derivatives, strychnine, micorene, phenmetrazine), diethyl ether, and trinitroglycerine. In 1967, UCI and UIPMB presented lists of banned compounds and drugs accordingly, and the IOC conducted first doping controls in Grenoble (86 samples) and Mexico City (667 samples) in 1968 following their own list of prohibited substances (Table 1.3).[47]

Early approaches of sports drug testing employing mass spectrometric techniques were reported in 1967 by Beckett and associates[93] who measured amphetamine and its derivatives from spiked urine specimens on a GC-MS system. A Perkin Elmer F11 GC equipped with a packed stainless steel column (2 m, i.d. 31 mm) and interfaced to a Hitachi-Perkin Elmer RMU-6E single focusing MS was used, which scanned from *m/z* 10 to 450 within 12 seconds and allowed the detection of less than 4 µg of analyte per mL of urine. Using three different sensitivity settings of the amplifier, informative electron ionization (EI) spectra were recorded (Fig. 1.3)[93] to serve as confirmatory data. Despite these successful couplings of GC and MS, also alternative assays using offline combinations, e.g., TLC and MS, were reported for the detection of methylamphetamine in doping controls.[220] Here, a Hitachi Perkin-Elmer RMU-6D MS was used, and although detection limits of the presented assay were not evaluated, the sensitivity was considered comparable to the earlier reported GC-MS-based method. A comprehensive use of GC-MS to substantiate suspicious test results in doping controls was present at the 1972 Olympic Games held in Munich, where seven adverse analytical findings with amphetamine (1), ephedrine (3), phenmetrazine (1), and nikethamide (2) were documented.[221–223] Besides common screening procedures employing GC with nitrogen-specific alkali-flame detector (N-FID),[109] an Atlas MAT CH-5 single sector MS interfaced to a temperature-programmed Hewlett-Packard GC Model 7600, which was equipped with a packed glass column (1.06 m, i.d. 2.5 mm, 2% Igepal CO-880 and 12.5% Apiezon L), was used to unambiguously confirm the presence or absence of prohibited compounds. An EI mass spectrum of an adverse analytical finding with nikethamide is depicted in Figure 1.4a (reconstructed). Compared to an EI mass spectrum recorded in 2008 (Fig. 1.4b), no significant difference is observed.

In order to enhance chromatographic properties of many target analytes with relevance to sports drug testing, numerous derivatives were prepared in the following years for instance from stimulants[224–226] and steroids.[168,227–230] The advantages of derivatization were manifold and various modifications were tested for more efficient GC separation

TABLE 1.3: First Lists of Prohibited Substances 1966–1968[a]

Federation	FIFA	UCI	UIMPB	IOC
Year	1966	1967	1967	1968
Occasion	World Cup (UK)	General	General	Olympic Games (Grenoble / Mexico City)
	Narcotics (including morphine, heroin, etc.)	Narcotics (including morphine, heroin, etc.)	Narcotics	Narcotic analgesics (including morphine, heroin, etc.)
	Substances related to amphetamine (including amphetamine and its methylated and hydroxylated derivatives)	Stimulants such as amphetamine, ephedrine, etc.	Amphetamine and its derivatives	Sympathomimetic amines (amphetamine, ephedrine, methylephedrine, etc.)
	Strychnine	Strychnine, ibokaine	Strychnine	CNS-stimulants (strychnine, analeptics, etc.)
	Phenmetrazine		Cocaine	
	Trinitroglycerine		Nitrate and related substances	
	Diethyl ether	Diethyl ether		
		Alcohol	Alcohol	Alcohol (on request by UIMPB after shooting competition)
		Antidepressants		Antidepressants (MAO-inhibitors, imipramine, etc.)
	Micorene		Camphor and pharmacologically related substances, analeptics	
			Hormones (natural and synthetic) if not used for more than one month)	
			Lobeline and related compounds	
			Vasodilators (peripheral)	
			Tranquilizer	Tranquilizer (e.g., phenotiazine)
			Purine bases	

[a] Ref. 47.

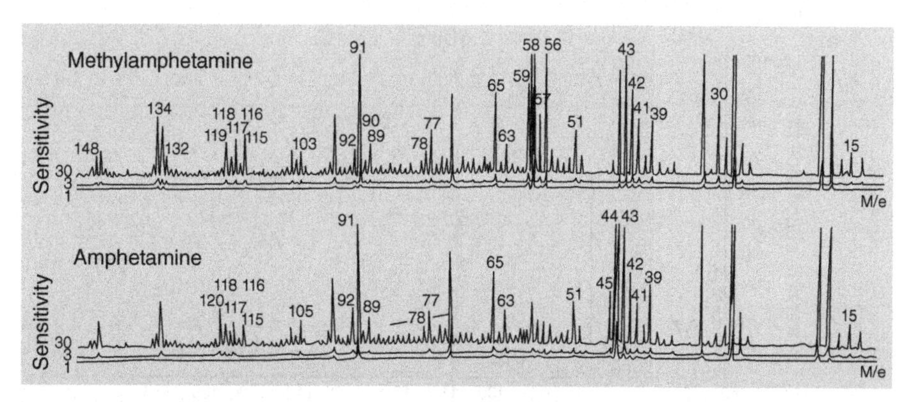

Figure 1.3: GC-EI-MS spectra obtained from urine samples containing methylamphetamine (top) and amphetamine, recorded on a Hitachi-Perkin Elmer RMU-6E single focusing MS. This figure has been reproduced with permission from Pharmaceutical Press, an imprint of RPS Publishing, the publishing organization of the Royal Pharmaceutical Society of Great Britain.

of analytes, improved peak shapes, and more characteristic mass spectra as obtained after EI. Stimulants were derivatized, e.g., to acetyl-, trifluoroacetyl-, pentafluoropropionyl-, heptafluorobutyryl-, perfluorooctanoyl-, trichloroacetyl-, or pentafluorobenzoyl-analogues[224,225] as well as mixed derivatives using trimethylsilylation and trifluoroacetylation.[102,105,226,231] Steroids were preferably acetylated, dimethylsilylated, trimethylsilylated,[168,227,230,232] or chloromethyldimethylsilylated,[228] and also mixed derivatives (methoxime/trimethylsilyl[233–235] or *N*-alkyl/trimethylsilyl[236] derivatives) were formed to yield adequate chromatographic peak shapes and informative EI mass spectra. However, also underivatized steroids were successfully measured using capillary GC and MS instruments.[237]

In 1975, the IOC list of prohibited substances was expanded by the class of anabolic androgenic steroids (AAS), and newly developed methods[168] were applied to samples collected at the 1976 Olympic Games in Montreal (Canada). From a total of 1786 specimens, which underwent conventional drug testing procedures, a selection of 275 was subjected to a special screening procedure based on GC-MS, which was dedicated to the detection of AAS.[238,239] After enzymatic hydrolysis of phase-II-metabolites, steroids such as metandienone (Dianabol), stanozolol (Stromba), 19-norethyltestosterone (Nivelar), 17α-ethyl-4-estren-17β-ol (Orabolin), 19-nortestosterone and its phenylpropionate ester (Durabolin) and respective phase-I-metabolites were fractionated on Sephadex LH-20 solid-phase extraction (SPE) columns. Aliquots were

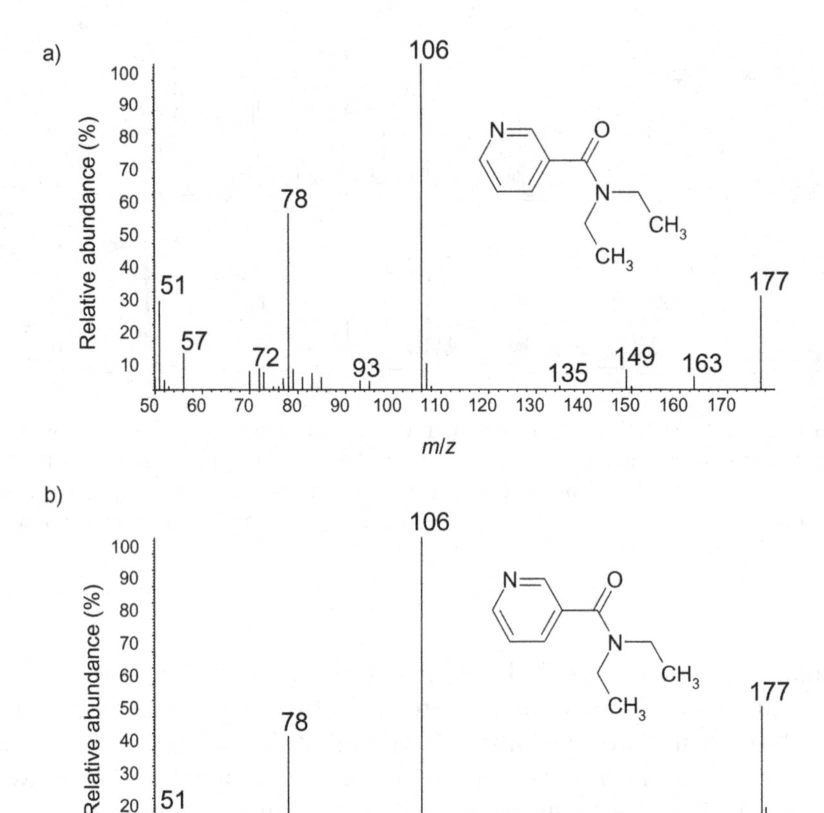

Figure 1.4: EI-mass spectra (a) of nikethamide found in a doping control specimen at the Olympic Games in Munich in 1972 (reconstructed), and (b) of a reference standard analyzed in 2008.

dried, and the dry residues were trimethylsilylated using pyridine/hexamethyldisilazane (HMDS)/trimethylchlorosilane (TMCS) followed by GC-MS analysis. The instrument used was a Pye 104 GC equipped with a 3% OV-1 column, which was connected to a Varian MAT 731 double-focusing high-resolution mass spectrometer (Mattauch-Herzog geometry), which allowed for repetitive scans every 8 seconds as well as selected ion monitoring (SIM). Using this approach, eight samples were tested positive for anabolic steroids, three of which were provided by gold- or silver medalists (all weightlifting).

Although invented as early as 1953,[240] the quadrupole mass spectrometer was not installed in doping control laboratories as detector in GC-MS systems until the late 1970s/early 1980s.[232,241–246] Providing fast scan speeds at sufficient resolution, benchtop GC-MS systems using quadrupole analyzers soon became state-of-the-art analytical tools in sports drug testing. Implemented for the drug testing programs at the 1982 Soccer World Cup in Spain, the 1983 World Championships in Athletics in Finland,[247] and the Pan American Games in 1983 in Venezuela (Caracas), GC-MS systems helped to uncover one of the biggest doping scandals in the latter event. Using two Hewlett-Packard 5996 GC-MS instruments with a capillary GC column (OV-1, 17 m, i.d. 0.2 mm, film thickness 0.11 μm) and EI source (heated to 250°C), SIM as well as full scan data were recorded in screening and confirmation methods. Earlier sample preparation methods were improved, and urine specimens underwent SPE, enzymatic hydrolysis, liquid-liquid extraction (LLE), and finally trimethylsilylation of extracted analytes before measurement.[184,244,245] A total of 19 athletes including 11 weightlifters, one cyclist, one fencer, one wrestler, one volleyball player, and four track and field athletes,[248–250] tested positive after competition. Four were sanctioned due to the use of stimulants (two ephedrine and two fencamfamine) and 15 were convicted of the use of AAS, in particular nortestosterone (12) was well as testosterone (1), metenolone (1), and metandienone (1).[110,251] In addition to these "clear" cases, reports and rumors on 12 further adverse analytical findings (out of 13 unlabeled urine specimens) prior to the start of the 1983 Pan American Games with regard to AAS misuse among the U.S. team, which were not sanctioned, became public. Moreover, the sudden departure of 12 American athletes when being informed about the efficiency of on-site drug tests[250] caused concern and confusion followed by endless speculations and few confessions.[252]

1.4.2 Progression of Analytical Methods

The detection methods being used in sports drug testing have continuously been improved, updated, and expanded in accordance with instrumental innovations and new drugs and challenges in doping controls. New standards and milestones in terms of analytical assays and available instruments were commonly set at great sporting events such as the Olympic Games.[111] In 1984 at the Los Angeles Olympic Games, for the first time all samples (1510 urine specimens) were screened for AAS using GC-MS, which was accomplished on Hewlett-Packard 5996 analyzers operated in SIM mode. Eleven adverse analytical findings were reported, including one stimulant (ephedrine) and ten anabolic steroids (eight nandrolone cases, one testosterone, and one metenolone case).[184,253]

Four years later at the Olympic Games in Seoul in 1988, the local doping control laboratory was equipped with 12 state-of-the-art Hewlett-Packard 5890/5970B GC-MS systems plus two HP5988A instruments interfaced to GC as well as LC (via thermospray).[185] The LC-MS version of the "Engine" (HP5988A) was used to measure corticosteroids from doping control specimens.[254] Revolutionary work regarding the determination of long-term metabolites of anabolic androgenic steroids such as stanozolol and the derivatization for sensitive detection using GC-MS systems[255] allowed the confirmation of three findings of stanozolol abuse, which represented one third of all positive specimens. The methodology on how to identify particular stanozolol metabolites in urine samples was installed as early as 1986 in Seoul by Schänzer and Donike, which was subsequently and successfully employed for the Olympic Games in 1988. In addition to the stanozolol findings, stimulants such as caffeine (1) and pemoline (1) as well as four cases of furosemide misuse were reported, but no further AAS.[253]

The doping control laboratory in Barcelona (Olympic Games 1992) used 13 GC-MS systems (HP5970, HP5971) and two LC-MS systems (HP5989) with particle beam interface (HP59980B). The latter was connected to an HP1090L Series II liquid chromatograph and was used to provide the first (and possibly last) adverse analytical finding (mesocarb) that has ever been confirmed using LC-particle beam-MS in sports drug testing.[256] In total, five positive doping control samples were reported with strychnine (1), norephedrine (1), clenbuterol (2), and mesocarb (1).

In 1994, the first GC-high resolution-MS system (HP5890/Finnigan MAT 95, double-focusing sector with reversed Nier-Johnson geometry) was installed for the Olympic Winter Games in Lillehammer (Norway),[257] in particular due to its superior sensitivity and selectivity for a specific subset of analytes such as stanozolol and metandienone metabolites as well as clenbuterol.[258] The GC-HRMS was complemented by seven HP5890/5970 and 5972 systems as well as one HP5890/Finnigan SSQ7000 low resolution GC-MS instruments to cope with the constantly increasing number of compounds to be analyzed.

The use of GC/combustion/isotope-ratio mass spectrometry (GC/C/IRMS) has demonstrated distinguished capabilities to reveal misuse of endogenous steroids such as testosterone, dihydrotestosterone, dehydroepiandrosterone (DHEA), etc. and was reported for the first time in 1994.[259] The analysis of carbon isotope ratios of steroids in doping control specimens progressively improved,[260–262] and in 1998 at the

Olympic Winter Games in Nagano (Japan), the first comprehensive analyses for synthetic endogenous steroids were conducted using an HP6890 GC coupled to an Isoprime IRMS (Micromass, Manchester, UK).

Polysaccharide-based plasma volume expanders were first analyzed in urine specimens during the Nordic Ski World Championships in Lahti (Finland, 2001) using a GC-MS-based method (HP 5890/5972 GC-MSD).[263–265] Seven adverse analytical findings were reported and gold, silver, and bronze medals were stripped, all won by Finnish male and female elite athletes.[266] Consequently, the established procedure was also employed at the subsequent Winter Olympic Games in 2002 (Salt Lake City, USA), but ever since, no additional positive case was found although methods were further expanded and improved using LC-MS/MS[267,268] and MALDI-TOFMS[269] approaches.

State-of-the-art GC-HRMS systems (Autospec Ultima) double-focusing sector instruments with EBE geometry, Micromass, Manchester, UK) were also used at the Athens Olympic Games and contributed significantly to the highest number of adverse analytical findings in the history of Olympic Games. Excluding those athletes who were sanctioned due to refusal/manipulation of doping control specimens (3) or "no-show" (2), 20 prohibited substances including AAS (stanozolol, testosterone, metandienone, nandrolone, methyltestosterone, oxandrolone), clenbuterol, cathine, ethamivan, heptaminol, and furosemide were reported. First approaches to use LC-MS/MS for the detection of peptide- and protein-based doping agents in human sports drug testing was described as early as 1993,[270] but it took approximately 10 years to become part of routine procedures[271–281] as for instance in Athens 2004, where LC-iontrap systems (Agilent 1100 LC/MSD SL) were employed. Major advances in sensitivity, accuracy, and resolution of mass spectrometers, which are described in more detail in Chapter 2, provided the basis for efficient method development and more comprehensive detection tools to cover a variety of prohibited peptide hormones such as insulins,[276,282–285] corticotrophins,[286] insulin-like growth factors,[287] luteinizing hormone releasing hormone,[288] as well as methods of manipulation using proteases.[289]

The need for complementary or new methods for doping control purposes is evident and a major task of sports drug testing related research. Continuously, new methods and approaches are developed to expand the portfolio of detection procedures for drugs and methods of manipulation, and currently applied MS-based methods are described in Chapter 6.

REFERENCES

1. Jokl, E. (1965) Sport and Culture, in *Doping* (eds A. de Schaepdryver and M. Hebbelink), Pergamon Press, Oxford, pp. 1–24.
2. Young, D.C. (1984) *The Olympic Myth of Greek Amateur Athletics*, Ares Publishers, Inc., Chicago.
3. Miller, S.G. (2004) *Ancient Greek Athletics*, Yale University Press, New Haven–London.
4. Toohey, K., and Veal, A.J. (2000) *The Olympic Games—A Social Science Perspective*, CABI Publishing, Oxon–New York.
5. Papagelopoulos, P.J., Mavrogenis, A.F., and Soucacos, P.N. (2004) Doping in ancient and modern Olympic Games. *Orthopedics*, **27**, 1226, 1231.
6. Arndt, K.H., and Arndt, C. (1980) Medicine at the Olympic Games. *Zeitschrift für Ärztliche Fortbildung (Jena)*, **74**, 389–392.
7. Morgan, C.E. (1957) Drug Administration to Racing Animals. *Journal of the American Veterinary Medical Association*, **130**, 240–243.
8. Wadler, G.I., and Hainline, B. (1989) *Drugs and the Athlete*, F.A. Davis Company, Philadelphia.
9. World Anti-Doping Agency (2003) *The World Anti-Doping Code*. Available at http://www.wada-ama.org/rtecontent/document/code_v3.pdf. Accessed 02-14-2008.
10. Kohler, M., Thevis, M., Schänzer W., and Püschel, K. (2008). Damage to health and fatalities from doping. *Rechtsmedizin*, **18**, 177–182.
11. Casavant, M.J., Blake, K., Griffith, J., *et al.* (2007) Consequences of use of anabolic androgenic steroids. *Pediatric Clinics of North-America*, **54**, 677–690.
12. Prokop, L. (1970) Zur Geschichte des Dopings und seiner Bekämpfung. *Sportarzt und Sportmedizin*, **21**, 125–132.
13. Neimann, J.L. (1985) Sudden death in athletes. *Annales de Cardiologie et d'Angéiologie*, **34**, 145–149.
14. Kennedy, M.C., and Lawrence, C. (1993) Anabolic steroid abuse and cardiac death. *Medical Journal of Australia*, **158**, 346–348.
15. Sein Anand, J., Chodorowski, Z., and Wisniewski, M. (2005) Multifactorial hypoglycaemic coma in female bodybuilder. *Przeglad lekarski*, **62**, 520–521.
16. Lage, J.M., Panizo, C., Masdeu, J., and Rocha, E. (2002) Cyclist's doping associated with cerebral sinus thrombosis. *Neurology*, **58**, 665.
17. Tentori, L., and Graziani, G. (2007) Doping with growth hormone/IGF-1, anabolic steroids or erythropoietin: Is there a cancer risk? *Pharmacological Research*, **55**, 359–369.
18. Hausmann, R., Hammer, S., and Betz, P. (1998) Performance enhancing drugs (doping agents) and sudden death—A case report and review of the literature. *International Journal of Legal Medicine*, **111**, 261–264.

19. Dhar, R., Stout, C.W., Link, M.S., *et al.* (2005) Cardiovascular toxicities of performance-enhancing substances in sports. *Mayo Clinic Proceedings*, **80**, 1307–1315.
20. Bernheim, P.J., and Cox, J.N. (1960) Heat stroke and amphetamine intoxication in a sportsman. *Schweizerische Medizinische Wochenschrift*, **90**, 322–331.
21. Donohoe, T., and Johnson, N. (1986) *Foul Play—Drug Abuse in Sports*, Basil Blackwell, Oxford.
22. Meinberg, E. (2007) *Dopingsport im Brennpunkt der Ethik*, Merus-Verlag, Hamburg.
23. Clarke, E.G. (1962) The doping of racehorses. *Medico-Legal Journal*, **30**, 180–195.
24. Roeske, W. (1968) Alfons Bukowski (1858–1921). *Archiwum Historii Medycyny*, **31**, 167–191.
25. Lander, G.D. (19300 The micro-detection of alkaloids. *Analyst*, **55**, 474–476.
26. Autenrieth, W. (1905) *The Detection of Poisons and Strong Drugs*. P. Blakiston's Son & Co., Philadelphia.
27. Thorwald, J. (1964) *Das Jahrhundert der Detektive*. Droemersche Verlagsanstalt A.G., Zurich.
28. Lucas, GHW. (1939) The Saliva Test. *Canadian Journal of Comparative Medicine*, **3**, 67–72.
29. Wilsdorf, G., and Graf, G. (1998) Historische Aspekte zur Entwicklung der Dopingforschung beim Pferd an der Veterinärmedizinischen Bildungsstätte in Berlin 1925–1945. *Berliner und Münchener Tierärztliche Wochenschrift*, **111**, 222–227.
30. Addis-Smith, L.F. (1961) The changing pattern of "doping" in horse racing and its control. *New Zealand Veterinary Journal*, **9**, 121–128.
31. Munch, J.C. (1931) *Bioassays, A Handbook of Quantitative Pharmacology*. Williams & Wilkins, Baltimore.
32. Munch, J.C., Sloane, A.B., and Latven, A.R. (1952) The value of bioassays in detecting narcotics. *Bulletin in Narcotics*, **4**, 23–26.
33. Munch, J.C. (1934) Saliva tests. I. Morphine. *Journal of the American Pharmaceutical Association*, **23**, 766–773.
34. Munch, J.C. (1935) Saliva tests. III. Detecting the administration of some opium derivatives to horses. *Journal of the American Pharmaceutical Association*, **24**, 557–560.
35. Munch, J.C. (1934) Saliva tests. II. Heroin. *Journal of the American Pharmaceutical Association*, **23**, 1185–1187.
36. Straub, W. (1911) Eine empfindliche biologische Reaktion auf Morphin. *Deutsche medizinische Wochenschrift*, **37**, 1462.
37. Straub, W. (1912) Die pharmakodynamische Wirkung des Narkotins im Opium. *Biochemische Zeitschrift*, **41**, 419–430.

38. Hermann, O. (1912) Eine biologische Nachweismethode des Morphins. *Biochemische Zeitschrift*, **39**, 216–231.
39. Maier, L. (1931) Quantitative Morphinbestimmung mit Hilfe des biologischen Verfahrens. *Archiv für Experimentelle Pathologie und Pharmakologie*, **161**, 163–172.
40. Pong, S.F., Sweetman, J.M., Pong, A.S., and Carpenter, J.F. (1987) Evaluation of oral skeletal muscle relaxants in the morphine-induced Straub tail test in mice. *Drug Development Research*, **11**, 53–57.
41. Keil, W., and Kluge, A. (1934) Über die Anwendung des Mäuseschwanzphänomens zur Auswertung von Morphin- und Skopolaminpräparaten. *Archiv für Experimentelle Pathologie und Pharmakologie*, **174**, 493–501.
42. Forst, A.W., and Deininger, R. (1949) Eine neue Methode zum spezifischen biologischen Nachweis von Morphin im Harn an der Mäusepupille. *Archiv für Experimentelle Pathologie und Pharmakologie*, **206**, 416–438.
43. Prokop, L. (1970) The struggle against doping and its history. *Journal of Sports Medicine and Physical Fitness*, **10**, 45–48.
44. Craig, S. (2000) Riding high. *History Today*, **50**, 18–19.
45. Hoberman, J. (2007) History and prevalence of doping in the marathon. *Sports Medicine*, **37**, 386–388.
46. Venerando, A. (1963) Doping: Pathology and ways to control it. *Medicina dello Sport*, **3**, 972–983.
47. Clasing, D. (2004) *Doping und seine Wirkstoffe*. Spitta Verlag GmbH, Köln.
48. Kofler, L. (1938) Mikroskopische Methoden zur Identifizierung organischer Substanzen. *Angewandte Chemie*, **51**, 703–714.
49. Kofler, L. (1945) Mikro-Methoden zur Kennzeichnung organischer Stoffe und Stoffgemische.Verlag Chemie, Berlin.
50. Dultz, G. (1940) Nachweis und Bestimmung von β-Phenylisopropylamin und β-Phenylisopropylmethylamin. *Zeitschrift für analytische Chemie*, **120**, 84–88.
51. McNally, W.D., Bergman, W.L., and Polli, J.F. (1947) The quantitative determination of amphetamine. *Journal of Laboratory and Clinical Medicine*, **32**, 913–917.
52. Deckert, W. (1936) Ein neues, leicht ausführbares Schnellverfahren zur quantitativen Bestimmung kleinster Mengen von Morphin in Harn, Blut und anderem biologischen Material. *Archiv für Experimentelle Pathologie und Pharmakologie*, **180**, 656–671.
53. Richter, D. (1938) A colour reaction for benzedrine. *Lancet*, **234**, 1275.
54. Richter, D. (1938) Elimination of amines in man. *Biochemical Journal*, **32**, 1763–1769.
55. Jacobsen, E. (1940) Gad I. Die Ausscheidung des β-Phenylisopropylamins bei Menschen. *Archiv für Experimentelle Pathologie und Pharmakologie*, **196**, 280–289.

56. Brodie, B.B., and Udenfriend, S. (1945) The estimation of basic organic compounds and a technique for the appraisal of specificity. *Journal of Biological Chemistry*, **158**, 705–714.
57. Beyer, K.H., and Skinner, J.T. (1940) The detoxication and excretion of beta phenylisopropylamine (benzedrine). *Journal of Pharmacology and Experimental Therapeutics*, **68**, 419–432.
58. Schoen, K. (1944) A rapid and simple method for the determination of ephedrine. *Journal of the American Pharmaceutical Association*, **33**, 116–118.
59. Keller, R.E., and Ellenbogen, W.C. (1952) The determination of d-amphetamine in body fluids. *Journal of Pharmacology and Experimental Therapeutics*, **106**, 77–82.
60. Axelrod, J. (1953) Studies on sympathomimetic amines. II. The biotransformation and physiological disposition of d-amphetamine, d-p-hydroxyamphetamine and d-methamphetamine. *Journal of Pharmacology and Experimental Therapeutics*, 315–326.
61. Munier, R., and Macheboeuf, M. Microchromatographie de partage des alcaloides et de diverses bases azotees biologiques. *Bulletin De La Societe De Chimie Biologique*, **31**, 1144–1162.
62. Munier, R., and Macheboeuf, M. (1951) Microchromatographie de partage sur papier des alcaloides et de diverses bases azotees biologiques. 3. Exemples de separations de divers alcaloides par la technique en phase solvante acide (familles de latropine, de la cocaine, de la nicotine, de la sparteine, de la strychnine et de la corynanthine). *Bulletin De La Societe De Chimie Biologique*, **33**, 846–856.
63. Mannering, G.J., Dixon, A.C., Carrol, N.V., and Cope, O.B. (1954) Paper chromatography applied to the detection of opium alkaloids in urine and tissues. *Journal of Laboratory and Clinical Medicine*, **44**, 292–300.
64. Jatzkewitz, H. (1953) Ein klinisches Verfahren zur Bestimmung von basischen Suchtmitteln um Harn. *Hoppe-Seyler's Zeitschrift für Physiologische Chemie*, **292**, 94–100.
65. Kaiser, H., and Jori, H. Beiträge zum toxikologischen Nachweis von Dromoran "Roche", Morphin, Dilaudid, Cardiazol Coramin und Atropin mit Hilfe der Papierchromatographie. *Archiv der Pharmazie und Berichte der Deutschen Pharmazeutischen Gesellschaft*, **287**, 224–242; contd.
66. Kaiser, H., and Jori, H. (1954) Beiträge zum toxikologischen Nachweis von Dromoran "Roche", Morphin, Dilaudid, Cardiazol Coramin und Atropin mit Hilfe der Papierchromatographie. *Archiv der Pharmazie und Berichte der Deutschen Pharmazeutischen Gesellschaft*, **287**, 253–258; concl.
67. Vidic, E. (1955) Eine Methode zur Identifizierung papierchromatographisch isolierter Arzneistoffe. *Archiv für Toxikologie. Fuehner Wielands Sammlung von Vergiftungsfällen*, **16**, 63–73.

68. Pilsbury, V.B., and Jackson, J.V. (1966) Identification of the thiazide diuretic drugs. *Journal of Pharmacy and Pharmacology*, **18**, 713–720.
69. Dole, V.P., Kim, W.K., and Eglitis, I. (1966) Detection of narcotic drugs, tranquilizers, amphetamines, and barbiturates in urine. *JAMA: The Journal of the American Medical Association*, **198**, 349–352.
70. Dole, V.P., Kim, W.K., Eglitis, I. (1966) Extraction of narcotic drugs, tranquilizers, and barbiturates by cation-exchange paper, and detection on a thin-layer chromatogram by a series of reagents. *Psychopharmacology Bulletin*, **3**: 45–48.
71. Schubert, B. (1967) Identification and metabolism of some doping substances in horses. *Acta Veterinaria Scandinavica*, **Suppl 21**, 21–101.
72. Mule, S.J. (1964) Determination of narcotic analgesics in human biological materials. *Analytical Chemistry*, **36**, 1907–1914.
73. Harms, D.R. (1965) Identification of morphine and codeine in the urine by thin-layer chromatography and ultraviolet spectrophotometry. *American Journal of Medical Technology*, **31**, 1–8.
74. Yoshimura, H., Oguri, K., and Tsukamoto, H. (1966) Detection of morphine in urine. II. An improved method by thin-layer chromatography utilizing potassium platinum iodide as the reagent for both coloration and fluorescence. *Chemical and Pharmaceutical Bulletin (Tokyo)*, **14**, 1286–1290.
75. Heyndrickx, A., and De Leenheer, A. (1967) Toxicological analysis of weckamines (amphetamine, pervitin, preludin and ritalin) in pharmaceutical compounds and urine of persons suspected from doping. *Journal de pharmacie de Belgique*, **22**, 109–126.
76. Moerman, E., and Vleeschnouwer, D. (1967) Detection of doping: Separation and identification of amphetamine and related compounds in urine. *Archives Belges de Médecine Sociale*, **25**, 455–461.
77. Debackere, M., and Laruelle, L. (1968) Isolation, detection and identification of some alkaloids or alkaloid-like substances in biological specimens from horses with special reference to doping. *Journal of Chromatography*, **35**, 234–247.
78. Parker, K.D., and Hine, C.H. (1967) Manual for the determination of narcotics and dangerous drugs in the urine. *Bulletin in Narcotics*, **19**, 51–57.
79. Heaton, A.M., and Blumberg, A.G. (1969) Thin-layer chromatographic detection of barbiturates, narcotics, and amphetamines in urine of patients receiving psychotropic drugs. *Journal of Chromatography*, **41**, 367–370.
80. Ono, M., and Engelke, B.F. (1969) Procedures for assured identification of morphine, dihydromorphinone, codeine, norcodeine, methadone, quinine, methamphetamine, etc., in human urine. *Bulletin in Narcotics*, **21**, 31–40.
81. Jansen, G.A., and Bickers I. (1971) Rapid method for simultaneous qualitative assay of narcotics, cocaine, quinine and propoxyphene in the urine. *Southern Medical Journal*, **64**, 1072–1074.

82. Imai, Y., Kawakubo, T., Otake, I., and Namekata, M. (1972) Studies on the detection of doping drugs. I. A thin–layer chromatographic screening procedure for detecting drugs from urine sample of race horses. Yakugaku zasshi. *Journal of the Pharmaceutical Society of Japan*, **92**, 1074–1081.
83. Gorodetzky, C.W. (1972) Sensitivity of thin-layer chromatography for detection of 16 opioids, cocaine and quinine. *Toxicology and Applied Pharmacology*, **23**, 511–518.
84. Machata, G. (1966) Der chemische Nachweis des Dopings beim Sport. *Deutsche Zeitschrift für die gesamte gerichtliche Medizin*, **57**, 335–341.
85. Bäumler, J., and Rippstein, S. (1961) Die Dünnschichtchromatographie als Schnellmethode zur Analyse von Arzneimitteln. *Pharmaceutica Acta Helvetiae*, **36**, 382–388.
86. Sohn, D., Simon, J., Hanna, M.A., *et al.* (1973) The detection of thiazide diuretics in urine. Column extraction and thin-layer chromatography. *Journal of Chromatography*, **87**, 570–575.
87. Kaistha, K.K., Tadrus, R., and Janda R. (1975) Simultaneous detection of a wide variety of commonly abused drugs in a urine screening program using thin-layer identification techniques. *Journal of Chromatography*, **107**, 359–379.
88. Lloyd, H.A., Fales, H.M., Highet, P.F., *et al.* (1960) Separation of alkaloids by gas chromatography. *Journal of the American Chemical Society*, **82**, 3791.
89. Parker, K.D., Fontan, C.R., Kirk, P.L. (1962) Separation and identification of some sympathomimetic amines by gas chromatography. *Analytical Chemistry*, **34**, 1345–1346.
90. Parker, K.D., Fontan, C.R., and Kirk, P.L. (1963) Rapid gas chromatographic method for screening of toxicological extracts for alkaloids, barbiturates, sympathomimetic amines, and tranquilizers. *Analytical Chemistry*, **35**, 356–359.
91. Brochmann-Hanssen, E., and Svendsen, A.B. (1962) Gas chromatography of sympathomimetic amines. *Journal of Pharmaceutical Sciences*, **51**, 393.
92. Brochmann-Hanssen, E., and Svendsen, A.B. Separation and identification of sympathomimetic amines by gas-liquid chromatography. *Journal of Pharmaceutical Sciences*, **51**, 938–941.
93. Beckett, A.H., Tucker, G.T., and Moffat, A.C. (1967) Routine detection and identification in urine of stimulants and other drugs, some of which may be used to modify performance in sport. *Journal of Pharmacy and Pharmacology*, **19**, 273–294.
94. Kolb, H., and Patt, P.W. (1965) Beitrag zum Arzneimittelnachweis in Körperflüssigkeiten durch Gaschromatographie. *Arzneimittel-Forschung*, **8**, 924–927.
95. Beckett, A.H., and Rowland, M. Urinary excretion kinetics of methylamphetamine in man. *Journal of Pharmacy and Pharmacology*, **17**, 109–114.
96. Beckett, A.H., and Rowland, M. (1965) Determination and identification of amphetamine in urine. *Journal of Pharmacy and Pharmacology*, **17**, 59–60.

97. Beckett, A.H., and Rowland, M. (1965) Urinary excretion kinetics of amphetamine in man. *Journal of Pharmacy and Pharmacology*, **17**, 628–639.
98. Beckett, A.H., and Wilkinson, G.R. (1965) Identification and determination of ephedrine and its congeners in urine by gas chromatography. *Journal of Pharmacy and Pharmacology*, **17**, 104–106.
99. Mule, S.J. (1971) Routine identification of drugs of abuse in human urine. I. Application of fluorometry, thin-layer and gas-liquid chromatography. *Journal of Chromatography*, **55**, 255–266.
100. Donike, M. (1966) Der Dopingnachweis mit Hilfe chromatographischer Methoden. *Sportarzt und Sportmedizin*, **17**, 81–84.
101. Langer, S.H., and Pantages, P. (1961) Peak-shift technique in gas-liquid chromatography: Trimethylsilyl ether derivatives of alcohols. *Nature*, **191**, 141–142.
102. Donike, M. (1969) N-Methyl-N-trimethylsilyl-trifluoracetamide, ein neues Silylierungsmittel aus der Reihe der silylierten Amide. *Journal of Chromatography*, **42**, 103–104.
103. Capella, P., and Horning, E.C. (1966) Separation and identification of derivatives of biologic amines by gas-liquid chromatography. *Analytical Chemistry*, **38**, 316–321.
104. Donike, M. (1970) Stickstoffdetektor und Temperaturprogrammierte Gas-Chromatographie, ein Fortschritt für die Routinemäßige Dopingkontrolle. *Sportarzt und Sportmedizin*, **21**, 27–30.
105. Donike, M. (1973) Acylierung mit Bis(Acylamiden); N-Methyl-Bis(Trifluoracetamid) und Bis(Trifluoracetamid), zwei neue Reagenzien zur Trifluoracetylierung. *Journal of Chromatography*, **78**, 273–279.
106. Donike, M., Jaenicke, L., Stratmann, D., and Hollmann, W. (1970) Gas chromatographic detection of nitrogen-containing drugs in aqueous solutions by means of the nitrogen detector. *Journal of Chromatography*, **52**, 237–250.
107. Donike, M., and Derenbach, J. (1976) Die Selektive Derivatisierung Unter Kontrollierten Bedingungen: Ein Weg zum Spurennachweis von Aminen. *Zeitschrift für analytische Chemie*, **279**, 128–129.
108. Donike, M. (1970) Temperature programmed gas chromatographic analysis of nitrogen-containing drugs: The reproducibility of retention times (I). *Chromatographia*, **3**, 422–424.
109. Donike, M., Stratmann, D. (1974) Temperaturprogrammierte gas-chromatographische Analyse stickstoffhaltiger Pharmaka: Die Reproduzierbarkeiten der Retentionszeiten und der Mengen bei automatischer Injektion (II) "Die Screeningprozedur für flüchtige Dopingmittel bei den Olympischen Spielen der XX. Olympiade München 1972." *Chromatographia*, **7**, 182–189.
110. Schänzer, W. (1999) Dem Doping keine Chance, in *25 Jahre Trainerausbildung* (ed J. Kozel), Sport&Buch Strauss, Cologne, pp. 59–94.

111. Hemmersbach, P. (2008) History of mass spectrometry at the Olympic Games. *Journal of Mass Spectrometry*, **43**, 839–853.
112. Mueller, R.K. (1995) Chromatographic techniques—The basis of doping control. *Journal of Chromatography B*, **674**, 1–11.
113. Müller, R.K., Grosse, J., Thieme, D., *et al.* (1999) Introduction to the application of capillary gas chromatography of performance-enhancing drugs in doping control. *Journal of Chromatography A*, **843**, 275–285.
114. Herraez-Hernandez, R., Campins-Falco, P., and Sevillano-Cabeza, A. (1992) Estimation of diuretic drugs in biological fluids by HPLC. *Chromatographia*, **33**, 177–185.
115. Tisdall, P.A., Moyer, T.P., and Anhalt, J.P. (1980) Liquid-chromatographic detection of thiazide diuretics in urine. *Clinical Chemistry*, **26**, 702–706.
116. Ventura, R., Nadal, T., Alcalde, P., *et al.* (1991) Fast screening method for diuretics, probenecid and other compounds of doping interest. *Journal of Chromatography A*, **655**, 233–242.
117. Fullinfaw, R.O., Bury, R.W., and Moulds, R.F. (1987) Liquid chromatographic screening of diuretics in urine. *Journal of Chromatography*, **415**, 347–356.
118. Cooper, S.F., Masse, R., and Dugal, R. (1989) Comprehensive screening procedure for diuretics in urine by high-performance liquid chromatography. *Journal of Chromatography*, **489**, 65–88.
119. Ventura, R., and Segura, J. (1996) Detection of diuretic agents in doping control. *Journal of Chromatography B*, **687**, 127–144.
120. Dye, D., East, T., and Bayne, W.F. (1984) High-performance liquid chromatographic method for post-column, in-line derivatization with o-phthaladehyde and fluorometric detection of phenylpropanolamine in human urine. *Journal of Chromatography*, **284**, 457–461.
121. Farrell, B.M., and Jefferies, T.M. (1983) An investigation of high-performance liquid chromatographic methods for the analysis of amphetamines. *Journal of Chromatography*, **272**, 111–128.
122. Slais, K., Nielen, M.W., Brinkman, U.A., and Frei, R.W. (1987) Screening of amphetamines by gradient microbore liquid chromatography and precolumn technology. *Journal of Chromatography*, **393**, 57–68.
123. Schänzer, W. (1984) Untersuchungen zum Nachweis und Metabolismus von Hormonen und Dopingmitteln, insbesondere mit Hilfe der Hochdruckflüssigkeitschromatographie. Institute of Biochemistry, German Sport University Cologne, Dissertation.
124. Hayakawa, K., Hasegawa, K., Imaizumi N., *et al.* (1989) Determination of amphetamine-related compounds by high-performance liquid chromatography with chemiluminescence and fluorescence detections. *Journal of Chromatography*, **464**, 343–352.
125. Frischkorn, C.G., and Frischkorn, H.E. (1978) Investigations of anabolic drug abuse in athletics and cattle feed. II. Specific determination of

methandienone (Dianabol) in urine in nanogram amounts. *Journal of Chromatography*, **151**, 331–338.

126. Heftmann, E., and Hunter, I.R. (1979) High-pressure liquid chromatography of steroids. *Journal of Chromatography*, **165**, 283–299.
127. Hunter, I.R., Walden, M.K., and Heftmann, E. (1979) High-pressure liquid chromatography of androgens. *Journal of Chromatography*, **176**, 485–487.
128. Darney, K.J., Jr., Wing, T.Y., and Ewing, L.L. (1983) Simultaneous measurement of four testicular delta 4-3-ketosteroids by isocratic high-performance liquid chromatography with on-line ultraviolet absorbance detection. *Journal of Chromatography*, **257**, 81–90.
129. Cochran, R.C., Darney, K.J., Jr., and Ewing, L.L. (1979) Measurement of testosterone with a high-performance liquid chromatograph equipped with a flow-through ultraviolet spectrophotometer. *Journal of Chromatography*, **173**, 349–355.
130. Fitzpatrick, F.A., and Siggia, S. (1973) High resolution liquid chromatography of derivatized non-ultraviolet absorbing hydroxy steroids. *Analytical Chemistry*, **45**, 2310–2314.
131. Siggia, S., and Dishman, R.A. (1970) Analysis of steroid hormones using high resolution liquid chromatography. *Analytical Chemistry*, **42**, 1223–1229.
132. Kawasaki, T., Maeda, M., and Tsuji, A. (1981) Determination of 17-oxosteroids in serum and urine by fluorescence high-performance liquid chromatography using dansyl hydrazine as a pre-labeling reagent. *Journal of Chromatography*, **226**, 1–12.
133. Kawasaki, T., Maeda, M., and Tsuji, A. (1982) Determination of 17-oxosteroid glucuronides and sulfates in urine and serum by fluorescence high-performance liquid chromatography using dansyl hydrazine as a prelabeling reagent. *Journal of Chromatography*, **233**, 61–68.
134. Butterfield, A.G., Lodge, B.A., Pound, N.J., and Sears, R.W. (1975) Combined assay, identification, and foreign related steroids test for methandrostenolone by high-speed liquid chromatography. *Journal of Pharmaceutical Sciences*, **64**, 441–443.
135. Rose, J.Q., and Jusko, W.J. (1979) Corticosteroid analysis in biological fluids by high-performance liquid chromatography. *Journal of Chromatography*, **162**, 273–280.
136. Culbreth, P.A., and Sampson, E.J. (1981) Liquid chromatography measurement of cortisol in methylene chloride extracts of aqueous solutions. *Journal of Chromatography*, **212**, 221–228.
137. Canalis, E., Reardon, G.E., and Caldarella, A.M. (1982) A more specific, liquid-chromatographic method for free cortisol in urine. *Clinical Chemistry*, **28**, 2418–2420.
138. Althaus, Z.R., Rowland, J.M., Freeman, J.P., and Slikker, W., Jr. (1982). Separation of some natural and synthetic corticosteroids in biological

fluids and tissues by high-performance liquid chromatography. *Journal of Chromatography*, **227**, 11–23.

139. Jane, I., and Taylor, J.F. (1975) Characterisation and quantitation of morphine in urine using high-pressure liquid chromatography with fluorescence detection. *Journal of Chromatography*, **109**, 37–42.
140. Duggin, G.G. (1976) Phenacetin estimation from biological samples by high-performance liquid chromatography. *Journal of Chromatography*, **121**, 156–160.
141. Jefferies, T.M., Thomas, W.O., and Parfitt, R.T. (1979) Determination of ketoprofen in plasma and urine by high-performance liquid chromatography. *Journal of Chromatography*, **162**, 122–124.
142. Upton, R.A., Buskin, J.N., Guentert, T.W., *et al.* (1980) Convenient and sensitive high-performance liquid chromatography assay for ketoprofen, naproxen and other allied drugs in plasma or urine. *Journal of Chromatography*, **190**, 119–128.
143. Berson, S.A., Yalow, R.S., Glick, S.M., and Roth, J. (1964) Immunoassay of protein and peptide hormones. *Metabolism*, **13**:Suppl, 1135–1153.
144. Yalow, R.S., and Berson, S.A. (1959) Radiobiology: Assay of plasma insulin in human subjects by immunological methods. *Nature*, **184**, 1648–1649.
145. Yalow, R.S., and Berson, S.A. (1960) Immunoassay of endogenous plasma insulin in man. *Journal of Clinical Investigation*, **39**, 1157–1175.
146. Samols, E., and Ryder, J.A. (1961) Studies on tissue uptake of insulin in man using a differential immunoassay for endogenous and exogenous insulin. *Journal of Clinical Investigation*, **40**, 2092–2102.
147. Felber, J.P. (1963) ACTH antibodies and their use for a radio-immunoassay for ACTH. *Experientia*, **19**, 227–229.
148. Yalow, R.S., Glick, S.M., Roth, J., and Berson, S.A. (1964) Radioimmunoassay of human plasma acth. *Journal of Clinical Endocrinology and Metabolism*, **24**, 1219–1225.
149. Felber, J.P., and Aubert, M.L. (1971) Radioimmunoassay for plasma ACTH. *Hormone and Metabolic Research*, **3**:Suppl, 3:73–77.
150. Ehrlich, R.M., and Randle, P.J. (1961) Immunoassay of growth hormone in human serum. *Lancet*, **2**, 230–233.
151. Ehrlich, R.M., and Randle, P.J. (1961) Immunoassay of growth hormone in human serum. *Proceedings of the Royal Society of Medicine*, **54**, 646–647.
152. Utiger, R.D., Parker, M.L., and Daughaday, W.H. (1962) Studies on human growth hormone. I. A radio-immunoassay for human growth hormone. *Journal of Clinical Investigation*, **41**: 254–261.
153. Parker, M.L., Utiger, R.D., and Daughaday, W.H. (1962) Studies on human growth hormone. II. The physiological disposition and metabolic

fate of human growth hormone in man. *Journal of Clinical Investigation*, **41**, 262–268.

154. Li, C.H., Moudgal, N.R., and Papkoff, H. (1960) Immunochemical investigations of human pituitary growth hormone. *Journal of Biological Chemistry*, **235**: 1038–1042.
155. Touber, J.L., and Maingay, D. (1963) Heterogeneity of human growth hormone. Its influence on a radio-immunoassay of the hormone in serum. *Lancet*, **1**, 1403–1405.
156. Glick, S.M., Roth, J., Yalow, R.S., and Berson, S.A. (1963) Immunoassay of human growth hormone in plasma. *Nature*, **199**, 784–787.
157. Brody, S., and Carlstroem, G. (1965) Human chorionic gonadotropin in abnormal pregnancy. Serum and urinary findings using various immunoassay techniques. *Acta Obstetricia et Gynecologica Scandinavica*, **44**, 32–44.
158. Benuzzi-Badoni, M., Lemarchand-Beraud, T., Gomez-Vuilleumier, J., *et al.* A quick radioimmunoassay for plasma HCG and Lh determination with preliminary results. *Helvetica Medica Acta*, **35**, 490–503.
159. Rizkallah, T., Taymor, M.L., Park, M., and Batt, R. (1965) An immunoassay method for human luteinizing hormone of pituitary origin. *Journal of Clinical Endocrinology and Metabolism*, **25**, 943–948.
160. Thomas, K., and Ferin, J. (1968) A new rapid radioimmunoassay for HCG (LH, ICSH) in plasma using dioxan. *Journal of Clinical Endocrinology and Metabolism*, **28**, 1667–1670.
161. Birgegard, G., Miller, O., Caro, J., and Erslev, A. (1982) Serum erythropoietin levels by radioimmunoassay in polycythaemia. *Scandinavian Journal of Haematology*, **29**, 161–167.
162. Brooks, R.V., Firth, R.G., and Sumner, N.A. (1975) Detection of anabolic steroids by radioimmunoassay. *British Journal of Sports Medicine*, **9**, 89–92.
163. Sumner, N.A. (1974) Measurement of anabolic steroids by radioimmunoassay. *Journal of Steroid Biochemistry*, **5**, 307.
164. Brooks, R.V., Jeremiah, G., Webb, W.A., and Wheeler, M. (1979) Detection of anabolic steroid administration to athletes. *Journal of Steroid Biochemistry*, **11**, 913–917.
165. Hampl, R., Picha, J., Chundela, B., and Starka, L. (1979) Radioimmunoassay of nortestosterone and related steroids. *Journal of Clinical Chemistry and Clinical Biochemistry*, **17**, 529–532.
166. Hampl, R., and Starka, L. Practical aspects of screening of anabolic steroids in doping control with particular accent to nortestosterone radioimmunoassay using mixed antisera. *Journal of Steroid Biochemistry*, **11**, 933–936.
167. Dugal, R., Dupuis, C., and Bertrand, M.J. (1977) Radioimmunoassay of anabolic steroids: An evaluation of three antisera for the detection of

anabolic steroids in biological fluids. *British Journal of Sports Medicine*, **11**, 162–169.

168. Ward, R.J., Shackleton, C.H., and Lawson, A.M. (1975) Gas chromatographic—Mass spectrometric methods for the detection and identification of anabolic steroid drugs. *British Journal of Sports Medicine*, **9**, 93–97.
169. de la Torre, R., Badia, R., Gonzalez, G., *et al.* (1996) Cross-reactivity of stimulants found in sports drug testing by two fluorescence polarization immunoassays. *Journal of Analytical Toxicology*, **20**, 165–170.
170. Colbert, D.L., Gallacher, G., and Mainwaring-Burton, R.W. (1985) Single-reagent polarization fluoroimmunoassay for amphetamine in urine. *Clinical Chemistry*, **31**, 1193–1195.
171. Turner, G.J., Colbert, D.L., and Chowdry, B.Z. (1991) A broad spectrum immunoassay using fluorescence polarization for the detection of amphetamines in urine. *Annals of Clinical Biochemistry*, **28**(Pt 6), 588–594.
172. Colbert, D. L., and Childerstone, M. (1987) Multiple drugs of abuse in urine detected with a single reagent and fluorescence polarization. *Clinical Chemistry*, **33**, 1921–1923.
173. Tobin, T., Watt, D.S., Kwiatkowski, S., *et al.* (1988) Non-isotopic immunoassay drug tests in racing horses: a review of their application to pre- and post-race testing, drug quantitation, and human drug testing. *Research Communications in Chemical and Pathology and Pharmacology*, **62**, 371–395.
174. Tobin, T., Kwiatkowski, S., Watt, D.S., *et al.* Immunoassay detection of drugs in racing horses. XI. ELISA and RIA detection of fentanyl, alfentanil, sufentanil and carfentanil in equine blood and urine. *Research Communications in Chemical and Pathology and Pharmacology*, **63**, 129–152.
175. Tobin, T., Tai, H.H., Tai, C.L., *et al.* (1988) Immunoassay detection of drugs in racing horses. IV. Detection of fentanyl and its congeners in equine blood and urine by a one step ELISA assay. *Research Communications in Chemical and Pathology and Pharmacology*, **60**, 97–115.
176. Woods, W.E., Tai, H.H., Tai, C., *et al.* (1986) High-sensitivity radioimmunoassay screening method for fentanyl. *American Journal of Veterinary Research*, **47**, 2180–2183.
177. Mcdonald, J., Gall, R., Wiedenbach, P., *et al.* Immunoassay detection of drugs in horses 1. Particle concentration fluoroimmunoassay detection of fentanyl and its congeners. *Research Communications in Chemical Pathology and Pharmacology*, **57**, 389–407.
178. Stanley, S., Jeganathan, A., Wood, T., *et al.* Morphine and etorphine. 14. Detection by ELISA in equine urine. *Journal of Analytical Toxicology*, **15**, 305–310.
179. Mcdonald, J., Gall, R., Wiedenbach, P., Bass, V.D., *et al.* (1988) Immunoassay detection of drugs in racing horses. 3. Detection of morphine in equine

blood and urine by a one-step ELISA assay. *Research Communications in Chemical Pathology and Pharmacology*, **59**, 259–278.

180. Ventura, R., Gonzalez, G., Smeyers, M.T., *et al.* (1998) Screening procedure for beta-adrenergic drugs in sports drug testing by immunological methods. *Journal of Analytical Toxicology*, **22**, 127–134.
181. Van Eenoo, P., and Delbeke, F.T. (2002) Detection of inhaled salbutamol in equine urine by ELISA and GC/MS2. *Biomedical Chromatography*, **16**, 513–516.
182. Woods, W.E., Wang, C.J., Houtz, P.K., *et al.* Immunoassay detection of drugs in racing horses. VI. Detection of furosemide (Lasix) in equine blood by a one step ELISA and PCFIA. *Research Communications in Chemical Pathology and Pharmacology*, **61**, 111–128.
183. Stanley, S., Wood, T., Goodman, J.P., *et al.* (1994) Immunoassay detection of drugs in racing horses: detection of ethacrynic acid and bumetanide in equine urine by ELISA. *Journal of Analytical Toxicology*, **18**, 95–100.
184. Catlin, D.H., Kammerer, R.C., Hatton, C.K., *et al.* (1984) Analytical chemistry at the Games of the XXIIIrd Olympiad in Los Angeles. *Clinical Chemistry*, **33**, 319–327.
185. Park, J., Park, S., Lho, D., *et al.* Drug testing at the 10th Asian Games and 24th Seoul Olympic Games. *Journal of Analytical Toxicology*, **14**, 66–72.
186. Park, J. (1991) Doping test report of 10th Asian Games in Seoul. *Journal of Sports Medicine and Physical Fitness*, **31**, 303–317.
187. Kicman, A.T., Brooks, R.V., and Cowan, D.A. (1991) Human chorionic gonadotrophin and sport. *British Journal of Sports Medicine*, **25**, 73–80.
188. Stenman, U.H., Unkila-Kallio, L., Korhonen, J., and Alfthan, H. (1997) Immunoprocedures for detecting human chorionic gonadotropin: Clinical aspects and doping control. *Clinical Chemistry*, **43**, 1293–1298.
189. Kicman, A.T., Brooks, R.V., Collyer, S.C., *et al.* (1990) Criteria to indicate testosterone administration. *British Journal of Sports Medicine*, **24**, 253–264.
190. Delbeke, F.T., Van Eenoo, P., and De Backer, P. (1998) Detection of human chorionic gonadotrophin misuse in sports. *International Journal of Sports Medicine*, **19**, 287–290.
191. de Boer, D., de Jong, E.G., van Rossum, J.M., and Maes, R.A. (1991) Doping control of testosterone and human chorionic gonadotrophin: A case study. *International Journal of Sports Medicine*, **12**, 46–51.
192. Wu, Z., Bidlingmaier, M., Dall, R., and Strasburger, C.J. (1999) Detection of doping with human growth hormone. *Lancet*, **353**, 895.
193. Bidlingmaier, M., and Strasburger, C.J. (2007) Technology insight: detecting growth hormone abuse in athletes. *Nature Clinical Practice. Endocrinology & Metabolism*, **3**, 769–777.
194. Bidlingmaier, M., Wu, Z., and Strasburger, C.J. (2000) Test method: GH. *Best Practice & Research Clinical Endocrinology & Metabolism*, **14**, 99–109.

195. Bidlingmaier, M., Wu, Z., and Strasburger, C.J. (2001) Doping with growth hormone. *Journal of Pediatric Endocrinology and Metabolism*, **14**, 1077–1084.

196. Bidlingmaier, M., Wu, Z., and Strasburger, C. (2003) Problems with GH doping in sports. *Journal of Endocrinological Investigation*, **26**, 924–931.

197. Holt, R.I. (2007) Meeting reports: Beyond reasonable doubt: Catching the growth hormone cheats. *Pediatric Endocrinology Reviews*, **4**: 228–232.

198. Holt, R.I., and Sonksen, P.H. (2008) Growth hormone, IGF-I and insulin and their abuse in sport. *British Journal of Pharmacology*, **154**, 542–556.

199. Abellan, R., Ventura, R., Palmi, I., *et al.* (2008) Immunoassays for the measurement of IGF-II, IGFBP-2 and -3, and ICTP as indirect biomarkers of recombinant human growth hormone misuse in sport values in selected population of athletes. *Journal of Pharmaceutical and Biomedical Analysis*, DOI: 10.1016/j.pba.2008.1005.1037.

200. Erotokritou-Mulligan, I., Bassett, E.E., Kniess, A., *et al.* (2007) Validation of the growth hormone (GH)-dependent marker method of detecting GH abuse in sport through the use of independent data sets. *Growth Hormone and IGF Research*, **17**, 416–423.

201. Healy, M.L., Dall, R., and Gibney, J. (2005) Toward the development of a test for growth hormone (GH) abuse: A study of extreme physiological ranges of GH-dependent markers in 813 elite athletes in the postcompetition setting. *Journal of Clinical Endocrinology and Metabolism*, **90**, 641–649.

202. Jenkins, P. (2001) Growth hormone and exercise: Physiology, use and abuse. *Growth Hormone and IGF Research*, **11**, S71–77.

203. Kicman, A., Miell, J., and Teale, J. (1997) Serum IGF-I and IGF binding proteins 2 and 3 as potential markers of doping with human GH. *Clinical Endocrinology (Oxf)*, **47**, 43–50.

204. Sartorio, A., Agosti, F., Marazzi, N., *et al.* (2004) Combined evaluation of resting IGF-I, N-terminal propeptide of type III procollagen (PIIINP) and C-terminal cross-linked telopeptide of type I collagen (ICTP) levels might be useful for detecting inappropriate GH administration in athletes: A preliminary report. *Clinical Endocrinology (Oxf)*, **61**, 487–493.

205. Lasne, F., and de Ceaurriz, J. (2000) Recombinant erythropoietin in urine. *Nature*, **405**, 635.

206. Lasne, F., Martin, L., Crepin, N., and de Ceaurriz J. (2002) Detection of isoelectric profiles of erythropoietin in urine: differentiation of natural and administered recombinant hormones. *Analytical Biochemistry*, **311**, 119–126.

207. Lasne, F., Thioulouse, J., Martin, L., and de Ceaurriz, J. (2007) Detection of recombinant human erythropoietin in urine for doping analysis: Interpretation of isoelectric profiles by discriminant analysis. *Electrophoresis*, **28**, 1875–1881.

208. Abellan, R., Ventura, R., Pichini, S., *et al.* (2004) Evaluation of immunoassays for the measurement of soluble transferrin receptor as an indirect biomarker of recombinant human erythropoietin misuse in sport. *Journal of Immunological Methods*, **295**, 89–99.
209. Abellan, R., Ventura, R., Pichini, S., *et al.* (2004) Evaluation of immunoassays for the measurement of erythropoietin (EPO) as an indirect biomarker of recombinant human EPO misuse in sport. *Journal of Pharmaceutical and Biomedical Analysis*, **35**, 1169–1177.
210. Nissen-Lie, G., Birkeland, K., Hemmersbach, P., and Skibeli, V. (2004) Serum sTfR levels may indicate charge profiling of urinary r-hEPO in doping control. *Medicine and Science in Sports and Exercise*, **36**, 588–593.
211. Parisotto, R., Gore, C., Emslie, K., *et al.* (2000) A novel method utilising markers of altered erythropoiesis for the detection of recombinant human erythropoietin abuse in athletes. *Haematologica*, **85**, 564–572.
212. Sharpe, K., Ashenden, M.J., and Schumacher, Y.O. (2006) A third generation approach to detect erythropoietin abuse in athletes. *Haematologica*, **91**, 356–363.
213. Segura, J., Pascual, J.A., and Gutiérrez-Gallego, R. (2007) Procedures for monitoring recombinant erythropoietin and analogues in doping control. *Analytical and Bioanalytical Chemistry.* **388**, 1521–1529.
214. Gohlke, R.S. (1959) Time-of-flight mass spectrometry and gas-liquid partition chromatography. *Analytical Chemistry*, **31**, 535–541.
215. Lindeman, L.P., and Annis, J.L. (1960) Use of a Conventional Mass Spectrometer as a Detector for Gas Chromatography. *Analytical Chemistry*, **32**, 1742–1749.
216. Ebert, A.A. (1961) Improved sampling and recording systems in gas chromatography-time-of-flight mass spectrometry. *Analytical Chemistry*, **33**, 1865–1870.
217. Gohlke, R.S. (1962) Time-of-flight mass spectrometry: Application to capillary column gas chromatography. *Analytical Chemistry*, **34**, 1332–1333.
218. Ryhage, R. (1964) Use of a mass spectrometer as a detector and analyzer for effluents emerging from high temperature gas liquid chromatography columns. *Analytical Chemistry*, **36**, 759–764.
219. Watson, J.T., and Biemann, K. (1965) Direct recording of high resolution mass spectra of gas chromatographic effluents. *Analytical Chemistry*, **37**, 844–851.
220. Alfes, H., and Clasing, D. (1969) Identifizierung geringer Mengen Methamphetamins nach Körperpassage durch Kopplung von Dünnschichtchromatographie und Massenspektrometrie. *Deutsche Zeitschrift für die gesamte gerichtliche Medizin*, **64**, 235–240.
221. Clasing, D., Donike, M., and Klümper, A. (1975) Dopingkontrollen bei den Spielen der XX. Olympiade München 1972—Teil 3. *Leistungssport*, **5**, 303–306.

222. Clasing, D., Donike, M., and Klümper, A. (1974) Dopingkontrollen bei den Spielen der XX. Olympiade München 1972—Teil 1. *Leistungssport*, **4**, 130–134.
223. Clasing, D., Donike, M., and Klümper, A. Dopingkontrollen bei den Spielen der XX. Olympiade München 1972—Teil 2. *Leistungssport*, **4**, 192–199.
224. Anggard, E., and Hankey, A. (1969) Derivatives of sympathomimetic amines for gas chromatography with electron capture detection and mass spectrometry. *Acta Chemica Scandinavica*, **23**, 3110–3119.
225. Wilkinson, G.R. (1970) The GLC separation of amphetamine and ephedrines as perfluorobenzamide derivatives and their determination by electron capture detection. *Analytical Letters*, **3**, 289–298.
226. Donike, M. (1975) N-Trifluoracetyl-O-trimethylsilyl-phenolalkylamine-Darstellung und massenspezifischer gaschromatographischer Nachweis. *Journal of Chromatography*, **103**, 91–112.
227. Sweeley, C.C., Elliot, W.H., Fries, I., and Ryhage, R. (1966) Mass spectrometric determination of unresolved components in gas chromatographic effluents. *Analytical Chemistry*, **38**, 1549–1553.
228. Brooks, C.J.W., and Middleditch B.S. (1971) The mass spectrometer as a gaschromatographic detector. *Clinica Chimica Acta*, **34**, 145–157.
229. Brooks, C.J.W., Harvey, D.J., Middleditch, B.S., and Vouros, P. (1973) Mass spectra of trimethylsilyl ethers of some δ^5-3β-hydroxy c_{19} steroids. *Organic Mass Spectrometry*, **7**, 925–948.
230. Donike, M. (1975) Zum Problem des Nachweises der anabolen Steroide: Gas-chromatographische und massenspezifische Möglichkeiten. *Sportarzt und Sportmedizin*, **26**, 1–6.
231. Donike, M. (1976) "Abfallprodukte" der Dopinganalytik. *Ärztliche Praxis*, **28**, 3935–3937.
232. Donike, M., and Zimmermann, J. (1980) Zur Darstellung von Trimethylsilyl-, Triethylsilyl- und tert.- Butyldimethylsilyl- enoläthern von Ketosteroiden für gas- chromatographische und massenspektrometrische Untersuchungen. *Journal of Chromatography*, **202**, 483–486.
233. Bjorkhem, I., and Ek, H. (1982) Detection and quantitation of 19-norandrosterone in urine by isotope dilution-mass spectrometry. *Journal of Steroid Biochemistry*, **17**, 447–451.
234. Bjorkhem, I., and Ek, H. (1983) Detection and quantitation of 3 alpha-hydroxy-1-methylen-5 alpha-androstan-17-one, the major urinary metabolite of methenolone acetate (Primobolan) by isotope dilution—Mass spectrometry. *Journal of Steroid Biochemistry*, **18**, 481–487.
235. Bjorkhem, I., Lantto, O., and Lof, A. (1980) Detection and quantitation of methandienone (Dianabol) in urine by isotope dilution—Mass fragmentography. *Journal of Steroid Biochemistry*, **13**, 169–175.
236. Lantto, O., Bjorkhem, I., Ek, H., and Johnston, D. (1981) Detection and quantitation of stanozolol (Stromba) in urine by isotope dilution-mass

fragmentography. *Journal of Steroid Biochemistry and Molecular Biology*, **14**, 721–727.

237. Dürbeck, H.W., and Büker, (1980) I. Studies on Anabolic Steroids. The Mass Spectra of 17α-Methyl-17β-hydroxy-1,4-androstadien-3-one (Dianabol) and its metabolites. *Biomedical Mass Spectrometry*, **7**, 437–445.
238. Laurin, C.A., and Letourneau, G. (1978) Medical report of the Montreal Olympic Games. *American Journal of Sports Medicine*, **6**, 54–61.
239. De Merode, A. (1979) Doping tests at the Olympic Games in 1976. *Journal of Sports Medicine and Physical Fitness*, **19**, 91–96.
240. Paul, W., and Steinwedel, H. (1953) Ein neues Massenspektrometer ohne Magnetfeld. *Zeitschrift für Naturforschung*, **8**, 448–450.
241. Donike, M. (1980) Dopinganalytik—Entwicklung und Entwicklungstendenzen. *Therapie-Woche*, **30**, 3156–3163.
242. Donike, M., Gielsdorf, W., and Schänzer, W. (1980) Der analytische Nachweis von Phenylbutazon in Plasma und Urin. *Der Praktische Tierarzt*, **61**, 884–887.
243. Donike, M. (1982) Fortschritte in der Dopinganalytik—Identifizierung und Quantifizierung von Dopingsubstanzen, in *Entwicklung und Fortschritte der Forensischen Chemie* (eds W. Arnold and K. Püschel), Verlag Dr. Dieter Helm, Hamburg. pp. 217–227.
244. Donike, M., Bärwald, K.R., and Klostermann, K. (1982) Detection of endogenous testosterone, in *Leistung und Gesundheit, Kongressbd, Dtsch. Sportärztekongress* (eds H. Heck, W. Hollmann, and H. Liesen), Deutscher Ärtze-Verlag, Köln, pp. 293–298.
245. Donike, M., Zimmermann, J., Bärwald, K.R., *et al.* (1984) Routinebestimmung von Anabolika in Harn. *Deutsche Zeitschrift für Sportmedizin*, **35**, 14–24.
246. Zimmermann, J. (1986) Untersuchung zum Nachweis von exogenen Gaben von Testosteron. Institut für Biochemie, Deutsche Sporthochschule Köln. Dissertation.
247. Kuoppasalmi, K., and Karjalainen, U. (1984) *Doping Analysis in Helsinki 1983*, United Laboratories, Helsinki.
248. Todd, J., and Todd, T. (2001) The history, ethics, and social context of doping, in *Doping in Elite Sport* (eds W. Wilson and E. Derse), Human Kinetics, Champaign, pp. 63–128.
249. Todd, T. (1987) Anabolic steroids: The gremlins of sport. *Journal of Sports History*, **14**, 87–107.
250. Neff, C. (1983) Caracas: A scandal and a warning. *Sports Illustrated*, **59**, 18–23
251. De Rose, E.H. (2008) IX Pan–American Games—Caracas: Positive test results. (personal communication).
252. Canadian Broadcasting Corporation (August 24, 1983) *Caught in Caracas*. (http://archives.cbc.ca/sports/drugs_sports/clips/8960/)

253. De Rose, E.H. (2008) Doping in athletes—an update. *Clinical Sports Medicine*, **27**, 107–130, viii–ix.
254. Park, S.J., Kim, Y.J., Pyo, H.S., and Park, J. (1990) Analysis of corticosteroids in urine by HPLC and thermospray LC/MS. *Journal of Analytical Toxicology*, **14**, 102–108.
255. Schänzer, W., Opfermann, G., and Donike, M. (1990) Metabolism of stanozolol: Identification and synthesis of urinary metabolites. *Journal of Steroid Biochemistry*, **36**, 153–174.
256. Ventura, R., Nadal, T., Alcalde, P., and Segura, J. (1993) Determination of mesocarb metabolites by high-performance liquid chromatography with UV detection and with mass spectrometry using a particle-beam interface. *Journal of Chromatography*, **647**, 203–210.
257. Hemmersbach, P., Bjerke, B., Birkeland, K., and Haug, E. (1994) Short report from the doping analysis during the XVII Olympic Winter Games, in *Recent Advances in Doping Analysis* (eds M. Donike, H. Geyer, A., Gotzmann, and U. Mareck-Engelke) Sport&Buch Strauß, Cologne, pp. 393–401.
258. Schänzer, W., Delahaut, P., Geyer, H., *et al.* (1996) Long-term detection and identification of metandienone and stanozolol abuse in athletes by gas chromatography-high-resolution mass spectrometry. *Journal of Chromatography B*, **687**, 93–108.
259. Becchi, M., Aguilera, R., Farizon, Y. (1994) Gas chromatography/combustion/isotope-ratio mass spectrometry analysis of urinary steroids to detect misuse of testosterone in sport. *Rapid Communications in Mass Spectrometry*, **8**, 304–308.
260. Shackleton, C.H., Phillips, A., Chang, T., and Li, Y. (1997) Confirming testosterone administration by isotope ratio mass spectrometric analysis of urinary androstanediols. *Steroids*, **62**, 379–387.
261. Horning, S., Geyer, H., Machnik, M. (1997) Detection of exogenous testosterone by $^{13}C/^{12}C$ analysis, in *Recent Advances in Doping Analysis* (W. Schänzer, H. Geyer, A. Gotzmann, and U. Mareck-Engelke) Sport und Buch Strauß, Cologne, pp. 275–283.
262. Aguilera, R., Becchi, M., Casabianca, H., *et al.* (1996) Improved method of detection of testosterone abuse by gas chromatography/combustion/isotope ratio mass spectrometry analysis of urinary steroids. *Journal of Mass Spectrometry*, **31**, 169–176.
263. Thevis, M., Opfermann, G., Schanzer, W. (2001) Nachweis des Plasmavolumenexpanders Hydroxyethylstärke in Humanurin. *Deutsche Zeitschrift für Sportmedizin*, **52**, 316–320.
264. Thevis, M., Opfermann, G., and Schänzer, W. (2000) Detection of the plasma volume expander hydroxyethyl starch in human urine. *Journal of Chromatography B*, **744**, 345–350.
265. Thevis, M., Opfermann, G., and Schänzer, W. (2000) Mass spectrometry of partially methylated alditol acetates derived from hydroxyethyl starch. *Journal of Mass Spectrometry*, **35**, 77–84.

266. SportScience (2001) Doping Disaster for Finnish Ski Team: A Turning Point for Drug Testing? Available at http://www.sportsci.org/jour/0101/ss.htm. Accessed 11-19-2008.
267. Guddat, S., Thevis, M., Thomas, A., and Schänzer, W. (2008) Rapid screening of polysaccharide-based plasma volume expanders dextran and hydroxyethyl starch in human urine by liquid chromatography-tandem mass spectrometry. *Biomedical Chromatography*, **22**, 695–701.
268. Deventer, K., Van Eenoo, P., and Delbeke, F.T. (2006) Detection of hydroxyethylstarch (HES) in human urine by liquid chromatography-mass spectrometry. *Journal of Chromatography B*, **834**, 217–220.
269. Gutierrez-Gallego, R., and Segura, J. (2004) Rapid screening of plasma volume expanders in urine using matrix-assisted laser desorption/ionisation time-of-flight mass spectrometry. *Rapid Communications in Mass Spectrometry*, **18**, 1324–1330.
270. Bowers, L.D., and Fregien, K. (1993) HPLC/MS confirmation of peptide hormones in urine: An evaluation of limit of detection, in *Recent Advances in Doping Analysis* (eds M. Donike, H. Geyer, A. Gotzmann, U. Mareck-Engelke, and S. Rauth), Sport&Buch Strauß, Cologne, pp. 175–184.
271. Thevis, M., Ogorzalek-Loo, R.R., Loo, J.A., and W. Schänzer. (2003) Doping control analysis of bovine hemoglobin-based oxygen therapeutics in human plasma by LC-electrospray ionization-MS/MS. *Analytical Chemistry*, **75**, 3287–3293.
272. Thevis, M., and Schänzer, W. (2005) Identification and characterization of peptides and proteins in doping control analysis. *Current Proteomics*, **2**, 191–208.
273. Thevis, M., and Schänzer, W. (2007) Mass spectrometry in sports drug testing: Structure characterization and analytical assays. *Mass Spectrometry Reviews*, **26**, 79–107.
274. Thevis, M., and Schänzer, W. (2005) Mass spectrometry in doping control analysis. *Current Organic Chemistry*, **9**, 825–848.
275. Thevis, M., and Schänzer, W. (2007) Current role of LC-MS(/MS) in doping control. *Analytical and Bioanalytical Chemistry*, **388**, 1351–1358.
276. Thevis, M., and Schänzer, W. (2007) Mass spectrometric identification of peptide hormones in doping control analysis. *Analyst*, **132**, 287–291.
277. Gasthuys, M., Alves, S., and Tabet, J. (2005) N-terminal adducts of bovine hemoglobin with glutaraldehyde in a hemoglobin-based oxygen carrier. *Analytical Chemistry*, **77**, 3372–3378.
278. Simitsek, P.D., Giannikopoulou, P., Katsoulas, H., *et al.* Electrophoretic, size-exclusion high-performance liquid chromatography and liquid chromatography-electrospray ionization ion trap mass spectrometric detection of hemoglobin-based oxygen carriers. *Analytica Chimica Acta*, **583**, 223–230.
279. Tsivou, M., Kioukia-Fougia, N., Lyris, E., *et al.* (2006) An overview of the doping control analysis during the Olympic Games of 2004 in Athens, Greece. *Analytica Chimica Acta*, **555**, 1–13.

280. Gam, L.H., Tham, S.Y., and Latiff, A. (2003) Immunoaffinity extraction and tandem mass spectrometric analysis of human chorionic gonadotropin in doping analysis. *Journal of Chromatography B, Analytical Technologies in the Biomedical and Life Sciences*, **792**, 187–196.
281. Thevis, M., and Schänzer, W. (2005) Examples of doping control analysis by liquid chromatography-tandem mass spectrometry: Ephedrines, beta-receptor blocking agents, diuretics, sympathomimetics, and cross-linked hemoglobins. *Journal of Chromatographic Science*, **43**, 22–31.
282. Thevis, M., Thomas, A., Delahaut, P. (2005) Qualitative determination of synthetic analogues of insulin in human plasma by immunoaffinity purification and liquid chromatography-tandem mass spectrometry for doping control purposes. *Analytical Chemistry*, **77**, 3579–3585.
283. Thevis, M., Thomas, A., and Schänzer, W. (2008) Mass spectrometric determination of insulins and their degradation products in sports drug testing. *Mass Spectrometry Reviews*, **27**, 35–50.
284. Thevis, M., Thomas, A., Delahaut, P., Bosseloir, A., and Schänzer, W. (2006) Doping control analysis of intact rapid-acting insulin analogues in human urine by liquid chromatography-tandem mass spectrometry. *Analytical Chemistry*, **78**, 1897–1903.
285. Thomas, A., Thevis, M., Delahaut, P., Bosseloir, A., and Schänzer, W. (2007) Mass spectrometric identification of degradation products of insulin and its long-acting analogues in human urine for doping control purposes. *Analytical Chemistry*, **79**, 2518–2524.
286. Thevis, M., Bredehöft, M., Geyer, H., Kamber, M., Delahaut, P., and Schänzer, W. (2006) Determination of Synacthen in human plasma using immunoaffinity purification and liquid chromatography/tandem mass spectrometry. *Rapid Communications in Mass Spectrometry*, **20**, 3551–3556.
287. Bredehöft, M., Schänzer, W., Thevis, M. (2008) Quantification of human insulin-like growth factor-1 and qualitative detection of its analogues in plasma using liquid chromatography/electrospray ionisation tandem mass spectrometry. *Rapid Communications in Mass Spectrometry*, **22**, 477–485.
288. Thomas, A., Geyer, H., Kamber, M., Schänzer, W., and Thevis, M. (2008) Mass spectrometric determination of gonadotrophin-releasing hormone (GnRH) in human urine for doping control purposes by means of LC-ESI-MS/MS. *Journal of Mass Spectrometry*, **43**, 908–915.
289. Thevis, M., Maurer, J., Kohler, M., Geyer, H., and Schänzer, W. (2007) Proteases in doping control analysis. *International Journal of Sports Medicine*, **28**, 545–549.
290. Starke, K. (2007) *Die Geschichte des Pharmakologischen Instituts der Universität Freiburg*. Springer, Berlin.

2 Mass Spectrometry and the List of Prohibited Substances and Methods of Doping

The detection and identification of prohibited compounds and methods of doping has been regulated for sports drug testing laboratories particularly by guidelines established by the IOC and, since 2001, the WADA.[1] Respective technical documents were created that outlined minimum required performance limits[2] as well as international standards for laboratories (ISL),[3] which accredited doping control laboratories must follow as evaluated in one or more annually conducted proficiency test programs.

2.1 CRITERIA FOR THE MASS SPECTROMETRIC IDENTIFICATION OF PROHIBITED COMPOUNDS

One of the most important aspects in sports drug testing is the obligation to unambiguously determine the presence of a prohibited compound and to prevent, with utmost confidence, false positive test results. Concise recommendations on how to characterize an analyte in doping control specimens by chromatographic/mass spectrometric systems were made by the IOC and WADA.[1] These suggestions concern retention times of target compounds, the number of (fragment/product) ions and allowed deviations of their relative abundance in reference and suspicious sample in MS- and MS/MS-based assays. However, the decision, which criteria shall apply and are subject to method validation, has been with the individual laboratories.

Mass Spectrometry in Sports Drug Testing: Characterization of Prohibited Substances and Doping Control Analytical Assays, By Mario Thevis
Copyright © 2010 John Wiley & Sons, Inc.

TABLE 2.1: Maximum Allowed Deviation of Relative Abundances According to WADA[a]

Relative Abundance (%)	GC-MS (EI)	GC-MS (CI); GC-MS^n; LC- MS; LC-MS^n
>50%	±10% (absolute)	±15% (absolute)
25% to 50%	±20% (relative)	±25% (relative)
<25%	±5% (absolute)	±10% (absolute)

[a]Ref. 1.

2.1.1 Low Molecular Weight Analytes

A general prerequisite of mass spectral analyses recording full or partial scans is the presence of all diagnostic ions with a relative abundance greater than 10% in the spectrum of the suspicious sample and the reference specimen (reference collection urine, spiked urine, or reference material). A "diagnostic ion" is generally defined as the "molecular ion or fragment ions whose presence and abundance are characteristic of the substance and thereby may assist in its identification."[1] Besides full/partial scan analysis, selected ion monitoring (SIM) or tandem mass spectrometry using selected ion transitions or product ion scan experiments are also possible. Commonly, three diagnostic ions are required to identify a target analyte, and the relative abundance of these ions should not differ by more than a defined maximum tolerance window as outlined in Table 2.1. The ranges are divided into categories accounting for the applied instrument including ionization technique and (tandem) mass spectrometer as well as the percentage of the relative abundance. According to these guidelines, ions resulting from EI on a GC-MS system should not deviate by more than ±10% (absolute) when having a relative abundance of more than 50%, between 50 and 25% a relative deviation of ±20% is allowed, and below 25% of relative abundance, the difference between the analysis of the suspicious sample and the comparison reference should not be more than 5% (absolute). When the analyses are conducted using chemical ionization (CI) on a GC-MS instrument, or by means of GC-MS^n, LC-MS, or LC-MS^n, the maximum tolerance windows are expanded to ±15% (absolute), ±25% (relative), and ±10% (absolute), respectively. In all cases, the signal-to-noise ratio of the ions in extracted ion chromatograms must be at least 3. Substantiating evidence in case of the lack of a sufficient number of diagnostic ions is to be provided either by using additional ionization techniques or derivatization of the target analyte.

These recommendations are comparable to criteria that have been established for forensic or toxicological purposes as well as veterinary medicine[4,5] employing data of a seminal work done by Sphon in 1978.[6] Although challenged more recently by demonstrating that every rationally selected additional (product) ion would improve the identification confidence by approximately one order of magnitude,[7] the established guidelines were not modified as the characterization of an analyte does not solely rely on MS data but supporting information is obtained by chromatography. Alternatively, so-called identification points (IPs) have been used in various fields of analytical chemistry for the detection and identification of compounds of interest.[8–10] This system based on IPs appreciates differences in identification power of e.g., low resolution, high resolution/high accuracy, MS and MS^n techniques in contrast to the approach reported above; however, currently IPs are not subject to the identification criteria of low molecular weight compounds in sports drug testing.

2.1.2 High Molecular Weight Analytes

The use of mass spectrometry to determine peptide hormones, proteins,[11–13] and proteases[14,15] was introduced in doping controls recently as demonstrated with various applications, enabling for instance the detection of bovine hemoglobin-based oxygen carriers (HBOCs),[16] synthetic insulins,[17–19] insulin-like growth factors (IGFs),[20] corticotrophins,[21] luteinizing hormone releasing hormone (LHRH),[22] and human chorionic gonadotrophin (hCG).[23] Due to the significantly different nature of high molecular weight analytes compared with low molecular weight compounds, the need for complementary criteria has arisen to demonstrate the "fitness for purpose,"[24] and recommendations on how to approach the peculiar properties of peptides and proteins were made.[25] These suggestions concern the facts that (a) peptides and proteins commonly carry more than one charge, which might require a better MS resolution to accurately determine molecular weights; (b) an analyte is potentially to be characterized using its degradation products, i.e., bottom-up sequencing approaches, which necessitates criteria on how many peptides (and/or how much sequence coverage), and also which "prototypical" peptide is, necessary to unambiguously identify a substance; and (c) sample preparation procedures that commonly employ immunoaffinity purification as well as mass analyzer qualities (i.e., resolution, accuracy, and tandem mass spectrometry) should be taken into consideration and appreciated.[26] Principally, the IP system as suggested also for low molecular weight analytes was included in the first set of recommendations combined with existing guidelines as established by WADA considering the needs for peptide and protein

TABLE 2.2: Suggested IP Values for the Identification of Peptides and Proteins Smaller Than 8 kDa Using Mass Spectrometry and Immunoaffinity Purification[a]

Mass Spectrometric Technique	IP Value
Low resolution MS^n precursor ion/molecular weight determination	1.0
Low resolution MS^n product ion	1.5
High resolution/high accuracy MS^n precursor ion/molecular weight determination	2.0
High resolution/high accuracy MS^n product ion	2.5
Immunoaffinity purification	1.0

[a] Modified from Ref. 25.

identification. The proposed IPs assigned to selected identification tools are listed in Table 2.2.

Generally, the identification of a peptide or protein smaller than 8 kDa should be based on at least 5 IPs. Within the sequence of measurements required for a state-of-the-art confirmation analysis, the chromatographic retention times of target analytes should not deviate by more than 0.4 min or 2% (whichever is smaller), and the determination of the analyte's molecular weight within a tolerance window of ±0.5 Da is mandatory. Subsequently, further identification points are to be obtained either by low- or high-resolution MS^n experiments. The application of immunoaffinity purification, which adds another dimension of selectivity of the assay to the target compound(s), is appreciated with 1 IP. In the case of bottom-up protein identification, a minimum of two peptides including at least one prototypical peptide and a sequence coverage of at least 10% is recommended.

Commonly, proteomics approaches and respective peptide and protein identifications are based on database search algorithms[27–29] that deduce peptide dissociation patterns, amino acid sequences, and a series of modifications from provided genomes. Theoretical spectra are calculated from protein sequences by simulating chemical or enzymatic hydrolysis *in-silico* followed by the prediction of MS/MS spectra resulting from the generated peptides. The applied search algorithms are usually based on a cross-correlation between calculated and observed spectra (e.g., SEQUEST)[30] and/or probabilistic models as used for instance with MASCOT.[31] Databases have been constantly improved and expanded to enable a comprehensive data analysis, and interpretation and validation of proteomics data have been a key point of numerous research projects,[32–34] which is of great interest also for doping control purposes. However, adverse analytical findings resulting from database searches only have not been considered acceptable yet, but a comparison between authentic reference material (or administration study specimens) and doping control samples have been required.

2.2 MODERN MASS SPECTROMETERS IN DOPING CONTROLS: ADVANTAGES AND DISADVANTAGES OF AVAILABLE TECHNIQUES

Modern human sports drug testing laboratories are commonly equipped with a wide variety of mass spectrometry-based analyzers to cope with the demand to determine a constantly increasing number of prohibited compounds and methods of doping. The diverse physicochemical properties of target analytes necessitates the exploration of all possible features of state-of-the-art mass spectrometry to ensure the best possible detection methods and retrospective analyses. Until now, no analytical technique has provided unequivocal results as reliable as those generated by mass spectrometry; however, immunochemical methods have shown superior sensitivities for selected substances and are, thus, also valuable complementary assays in doping controls.

2.2.1 Ionization Techniques in Routine Doping Controls

Numerous ionization techniques have been reported in the last century, which are applicable to modern mass spectrometry. Depending on the method of analyte introduction (e.g., direct inlet, GC, LC, or capillary electrophoresis), different strategies have been employed including EI, CI, thermospray, particle beam, electrospray ionization (ESI), atmospheric pressure chemical ionization (APCI), fast-atom bombardment (FAB), matrix-assisted laser desorption ionization (MALDI), etc.[35] In sports drug testing, only selected approaches have been applied to routine doping control analyses, which are outlined in the following.

2.2.1.1 Electron Ionization (EI) Historically, EI has been the most frequently used ionization technique in doping controls, being present in most gas chromatography-mass spectrometry (GC-MS), gas chromatography-high resolution mass spectrometry (GC-HRMS), and gas chromatography/combustion/isotope ratio mass spectrometry (GC/C/IRMS) instruments. The principle of positively ionizing a solid substance by electron impact was described as early as 1918 by Dempster[36,37] and found perfectly suited also for gaseous and vaporized compounds[38–45] that are to be transferred to and analyzed by mass spectrometry. Experiencing a variety of modifications, the Nier-design as published in 1947 was finally considered the essential model for EI sources.[46] Electrons are released via thermionic emission from a cathode (the filament) and accelerated to an anode to interact by orthogonal orientation within the ion source with compounds eluting from the GC column. Electrons that are commonly released with 70eV (i.e., the potential between filament and anode is 70V), are associated to a wavelength of approximately 1.4Å and efficiently remove an electron from analytes in the ion source. Besides several ionization processes, the most important ones of which are listed below, approximately 10–20eV are transferred to the ionized molecule and trigger extensive fragmentation of the analyte.[47]

$$(1) \quad AB + e^- \rightarrow AB^+ + 2e^-$$

$$(2) \quad AB + e^- \rightarrow AB^{2+} + 3e^-$$

$$(3) \quad AB + e^- \rightarrow AB^-$$

Positive ionization (AB^+) is predominantly accomplished by a mechanism (1) that liberates an electron upon interaction of the accelerated electron released from the filament with the vaporized analyte. In addition, also two electrons can be removed from the substance AB, giving rise to a doubly charged molecule AB^{2+} (2). Moreover, electrons can be incorporated into the molecule AB, resulting in a negatively charged analyte AB^- (3). Most commonly, positive EI prevails in doping control analytical assays, and the generated energetic radical cations incline to dissociate into product ions, which are detected in the mass selective analyzer, giving rise to a characteristic mass spectrum of the respective substance.[48,49] These product ions can be radicals as well as even-electron ions, depending on rearrangement and elimination processes. Comprehensive spectra libraries were generated over decades covering thousands of drugs and metabolites with relevance for sports drug testing, clinical, environmental, and forensic sciences,[50–52] which support, facilitate, and accelerate the identification of target compounds. In addition, EI was shown to be less susceptible to ion suppression effects as observed for instance with other competitive and soft ionization techniques (e.g., electrospray ionization) and, thus, provides an excellent ionization tool for quantitative mass spectrometry. However, the rather "destructive" nature of EI represents a drawback for fragile analytes that dissociate substantially after electron impact, and important information, for example, about the molecular weight, might be lost. Consequently, soft alternatives were elucidated and commonly found in CI, which has frequently been employed in positive and negative modes.

2.2.1.2 Chemical Ionization (CI) The fundamentals of CI were described in 1966 by Munson and Field with regard to methane as CI reaction gas,[53–55] outlining the ion-molecule-reaction character of the applied technique. Primary ions, which are generated by electron ionization of the reactant gas (e.g., methane), undergo reactive collisions with the bulk reactant gas and produce further reactive ion species yielding a so-called ionization plasma. This enables ionization of the target analytes under hydrogen or hydride transfer reactions to yield $[M + 1]^+$ or $[M - 1]^-$, respectively, as well as adduct formation, charge transfer, etc. Moreover, thermal (low-energy) electrons are created that can be captured by molecules and give rise to negatively charged analytes.

A widely used ionization reaction is the proton transfer, which necessitates differences in proton affinity (PA) of reactant gas and target analyte.[56] Most commonly, methane (PA = 5.7 eV), isobutane (PA = 8.5 eV), and ammonia (PA = 9.0 eV) are employed for CI, which need to have lower PA than the analytes to be protonated. Hence, the ionization with isobutane and ammonia is more selective than using methane, and protonation is less exothermic, resulting in considerably reduced dissociation of the analytes.[47] In Figure 2.1, three mass spectra of methamphetamine using EI (a) and CI (b and c) are illustrated, which demonstrate the significantly reduced fragmentation of the target compound when applying chemical ionization, and several applications for the detection of, for example, beta-receptor blocking[57] agents and anabolic steroids,[58,59] were reported.

Alternatively, negative CI has been found useful for a variety of applications due to its selectivity for acidic or electronegative residues of molecules. Thermal electrons are captured under associative or dissociative resonance conditions (which represent energies of the electrons of 0–2 eV or 0–15 eV, respectively), and particularly analytes derivatized with residues bearing halogens are sensitively measured from blood and urine specimens.[60–62] However, neither positive nor negative CI is currently an element of routine doping controls but used only occasionally for specific issues of drug testing. A major reason for that might be the introduction of other soft ionization techniques, especially electrospray ionization (ESI) and atmospheric pressure chemical ionization (APCI) that allow a robust and efficient interfacing of LC and MS systems.

2.2.1.3 Electrospray Ionization (ESI) The principle of generating ions of macromolecules by means of electrospray, a phenomenon that was described by French physicists as early as the 18th century as "charged water spraying" (Fig. 2.2),[63] was reported in 1968, when Dole and co-workers studied "Molecular Beams of Macroions" without mass spectrometry.[64] In 1984, the utility of ESI for the production of ionized

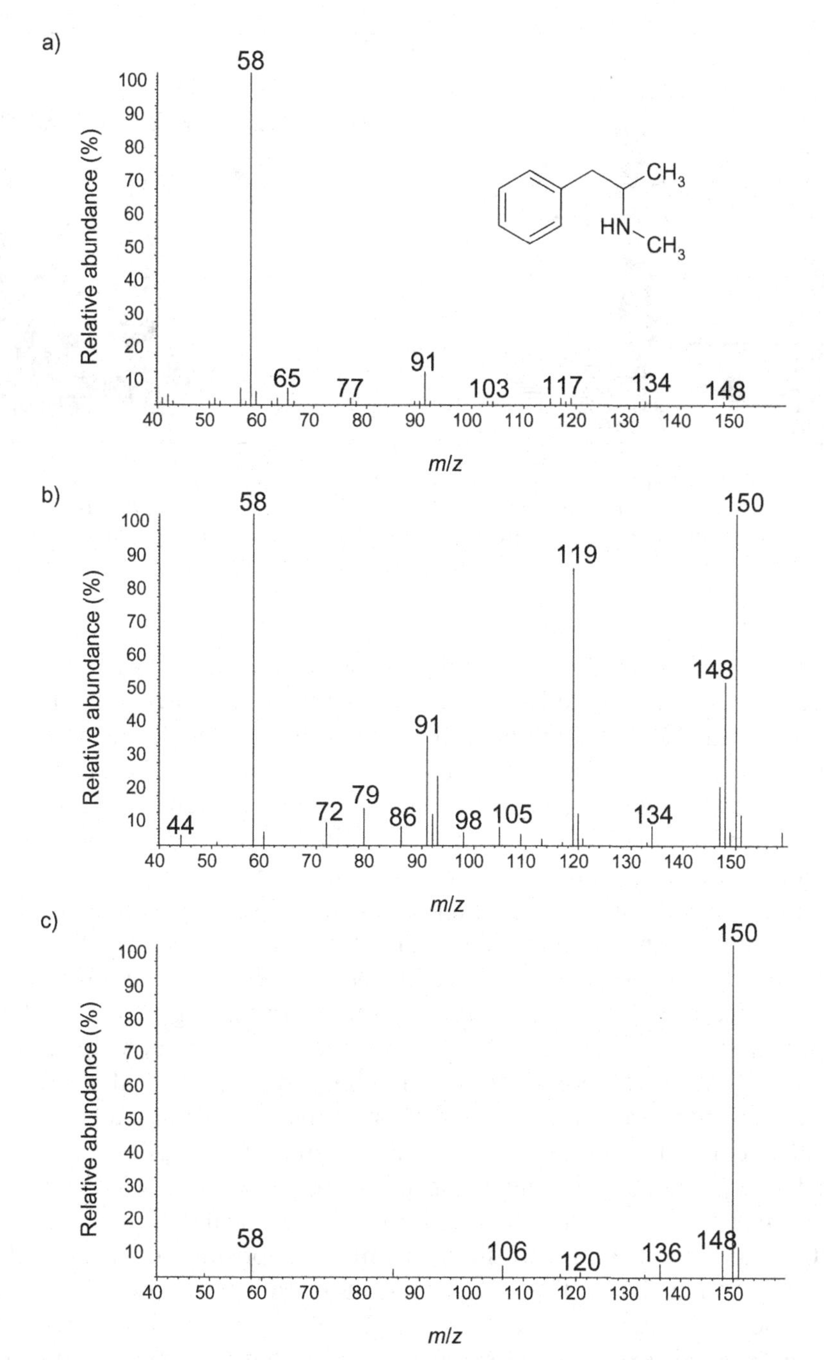

Figure 2.1: Comparison of EI (a) and CI (b and c) mass spectra of methamphetamine (mol wt = 149 Da). The CI gas was methane and ammonia in case of (b) and (c), respectively, resulting in different degrees of dissociation. While the EI mass spectrum does not contain the molecular ion at *m/z* 149, CI gave rise to the $[M + H]^+$ ion at *m/z* 150 using both reagent gases.

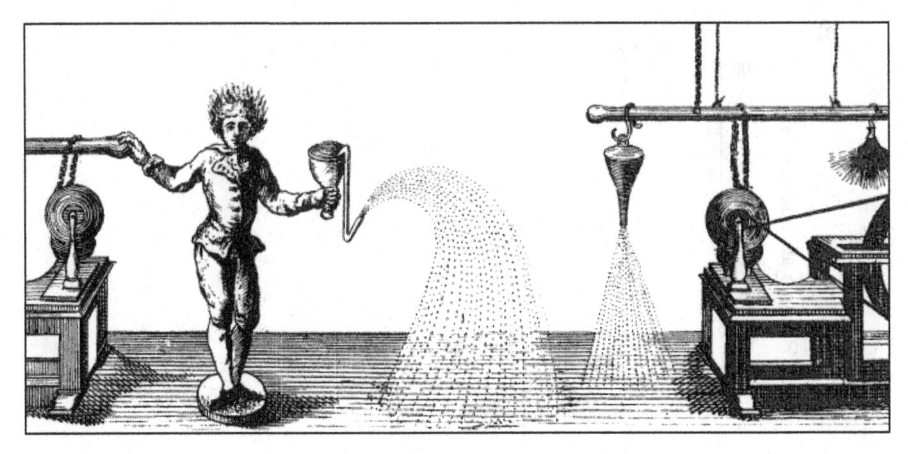

Figure 2.2: First experiments reported in 1775 that demonstrate the influence of voltage on a jet of water. Reproduced with permission from the Conservatoire numérique. Bibliothéque du Conservatoire national des arts and metiers, CNAM 8 SAR 18.

analytes and their detection in a mass spectrometer was demonstrated,[65,66] which initiated a new era of analytical chemistry that allowed the determination of small as well as large (bio)molecules using a soft ionization at atmospheric pressure.[67] First reports on the successful ionization of polymers (polyethylene glycols and proteins)[68–73] outlined the enormous potential of this new technique, which has proven to be one of the most versatile ionization methods in biological, biochemical, and pharmaceutical research and routine analysis. The complex mechanisms contributing to the generation of singly and multiply charged gas-phase ions from solution were investigated in detail, and different theories were considered.[74–76] While liquids containing protonated or deprotonated molecules are sprayed by means of a capillary tip at high voltages (1 kV and higher), they form charged droplets that shrink by solvent evaporation, and repeated droplet disintegration leads to very small and highly charged droplets. Subsequently, gas-phase ions are produced presumably by means of the *charged residue model*[64] and *ion evaporation*[77]. A penetration of an imposed electric field into the liquid of the capillary will lead to the enrichment of positive charges at the surface of the solvent. This causes destabilization of the meniscus, and formation of a Taylor-cone and a so-called jet-emission of droplets, which carry an excess of positive ions. The charged droplets further shrink by solvent evaporation while the charge remains constant. Hence, an increase of the electrostatic repulsion occurs, resulting in offspring of droplets as the Rayleigh stability limit is reached. This phenomenon continues with ongoing evaporation of solvent until very small and highly charged droplets are created. Finally, as proposed by Dole and co-workers,[64] only one ion remains in a singly charged droplet, and the evaporation of solvent gives rise to a gas-phase ion (*charged residue model*). Alternatively, the direct ion emission from droplets with a radius smaller than 10 nm was postulated, also generating gas-phase ions (*ion evaporation*).[78]

The soft ionization of analytes using ESI made this technique perfectly suitable for polar low molecular weight analytes as well as peptides and proteins, and numerous applications have been reported also in sports drug testing ever since commercial instruments equipped with ESI sources became available.[48,49,79–81] However, particular care is required to control possible ion suppressing effects due to the "competitive" ionization occurring in ESI; hence, either stable-isotope labeled internal standards or comprehensive studies on signal suppression or also enhancement are recommended to ensure adequate robustness and reproducibility of analytical results.[82–84]

2.2.1.4 Atmospheric Pressure Chemical Ionization (APCI) Besides ESI, APCI is another important ionization technique applied in modern doping controls. Developed in the early 1970s,[85,86] it demonstrated considerable utility for the ionization of small molecules from liquids at atmospheric pressure by applying a method closely related to conventional CI at reduced pressure. LC systems are commonly connected to mass spectrometers via APCI sources, where a nitrogen beam converts the eluate into a thin fog that is subsequently heated in a quartz tube to vaporize the mobile phase and respective analytes. When passing a corona discharge electrode, reactant ions are generated from the solvent, which ionizes polar and relatively non-polar compounds by typical CI processes such as proton transfer/abstraction and adduct formation.[47] In contrast to ESI, APCI is more commonly applied to low molecular weight compounds, and singly charged analytes are obtained rather than multiply charged species. Its applicability to comparably non-polar substances has made APCI a valuable and complementary method for LC-MS(/MS) in sports drug testing,[48,79] and its reduced sensitivity to ion suppression/enhancement matrix effects has been recognized as an important benefit.

2.2.2 Mass Analyzers in Routine Doping Controls

The need for fast, sensitive, and selective mass analyzers that are commonly combined with chromatographic systems has been a major reason for the use of a variety of mass spectrometers in routine doping controls. Depending on the requirements set by the physicochemical properties of specific classes of target compounds as well as respective minimum required performance limits and urinary or blood concentration levels, different systems have been used in past and present sports drug testing procedures.

2.2.2.1 Single Quadrupole MS The most frequently employed instruments in doping controls include GC-MS systems with single quadrupole mass spectrometers. The principle of the quadrupole mass selective detector operating without a magnetic field was invented by Paul in 1953,[87] and ion separation is based on different trajectory stabilities of charged molecules in an oscillating electric field. Commonly, four circular or hyperbolic rods consisting of various materials (e.g., fused silica coated with a gold layer) are aligned to form the quadrupole unit. Opposing rods are connected while adjacent segments are electrically isolated, and ions are accelerated from the source into the center of the quadrupole. While applying alternating current (ac) to the segments, positive or negative fields are established toward the centerline (z-axis) of the quadrupole. Consequently, positive ions passing through the rods are pushed away with positive and attracted with negative polarization. The extent of ion deflection along x and y directly depends on the applied voltage, its frequency (i.e., the duration of exposure to alternating fields), and the mass of the ions. In addition, to one segment, positive direct current (dc) is applied while the other segment is provided negative dc. As a result, only ions of a distinct mass-to-charge ratio (m/z) can travel through the two-dimensional quadrupole field if the applied dc and ac with a defined frequency are appropriate for a stable oscillating move. The control of these parameters allows for an optimized mass selection, which enables the isolation of a single ion or the recording of a full spectrum by scanning a distinct range of m/z over a given time period.[35,88]

The first commercially available quadrupole mass spectrometer, the Quad 200 RGA (residual gas analyzer), was built by Electronic Associates, Inc. (EAI, Palo Alto, CA) in 1964,[89] and its major advantages over competitive systems were its superior robustness and the comparably low costs. Only 4 years later, the first GC-MS system using a quadrupole mass analyzer, the Model 1015 (Finnigan Instruments Corp., Sunnyvale, CA), was introduced having a mass range from 1 to 750 Da with unit resolution. Scan speeds of 50 ms for a defined mass range were possible, and further improvements of mass limits (approximately 4000 Da) and resolution (full width half maximum at m/z 1000:

ca. 2000)[47] as well as sensitivity have made the GC-quadrupole-MS benchtop machine a successful, robust, and cost-effective analytical instrument in numerous fields, including sports drug testing. However, drawbacks in mass accuracy compared to other mass analyzers such as magnetic sector, time-of-flight, or Orbitrap MS, and the lack of MS/MS capability necessitated alternatives to complement the equipment of modern doping control laboratories.

2.2.2.2 Ion Trap MS The evolution of ion trap mass spectrometry started also in 1953 with the same patent of Paul and Steinwedel that described the quadrupole mass selective detector.[90] The applicability of a three-dimensional quadrupole to "trap" ions was recognized in the late 1950s,[91] followed by several comparable observations using a circular two-dimensional (linear) ion trap.[92]

In principle, a three-dimensional Paul trap includes two end-cap electrodes (one with an ion inlet and one with an ion exit aperture) and a ring electrode, in the center of which the ions are stored by application of an appropriate rf field established between the ring electrode and the two end-cap electrodes.[93] The ion trap is commonly filled with a gas (e.g., helium or argon) at approximately 1 mTorr to damp the trajectories of ions that are generated by an external ion source (e.g., EI or ESI) and transferred into the center of the ion trap.[94] The ion trajectories generally appear in a Lissajous figure (comparable to a three-dimensional 8) as illustrated as early as 1959 by illuminating charged aluminum particles.[95] In addition, the collision of ionized species with inert gases reduces the kinetic energy of the ions, which favors their storage within the trap. The storage of a broad range of m/z ratios is directly dependent on the ac voltage applied to the ring electrode; however, storage of ions alone was not sufficient for mass spectrometric purposes, and the utility of the trap as a mass spectrometer enabling mass selective detection, storage, and ejection evolved over more than three decades.[90] The consecutive ejection of ions that finally enables their mass selective analysis is accomplished by the so-called instability mode, which is based on the successive increase of the ac voltage applied to the ring electrode in combination with an ac voltage applied to the end cap electrodes, causing resonant motion. Ions of defined m/z ratios develop instable trajectories, are ejected through the perforations of the end-cap electrode (ion exit), and detected with an electron multiplier, which yielded the first commercially available ion trap mass spectrometer in 1984.[96]

An alternative to three-dimensional Paul traps, linear (two-dimensional) ion traps have demonstrated great utility as mass spectrometers.[97] One of the first models was derived from bent linear quadrupoles forming a closed circle in the late 1960s,[98] and approximately 20 years later the possibility to store ions in collision cells of tandem mass spectrometers was shown,[99,100] which turned out to be a milestone in the development of linear ion trap (LIT) mass spectrometry. The principle components of a LIT analyzer are a conventional quadrupole and lenses on each end of the rods. Hence, ions are confined in the radial dimension by application of a quadrupolar field and repelled in the axial dimension by an electric field as established at the end of the rods to remain within the LIT.[92] Mass selective ejection of ions is possible either radial or axial, and a great advantage of LITs over Paul traps was found to be the considerably higher trapping capacity and efficiency of ions generated in an external ion source and transferred into the mass analyzer. Thus, space charge effects are significantly reduced and the sensitivity of ion trap-based mass spectrometers greatly enhanced.[47]

In addition to an efficient scan operation mode, ion trap mass spectrometers offer possibilities of MS^n experiments. With the selective removal of ions from the trap, storage of a precursor ion of interest and its resonant activation, collision-induced dissociation (CID) is obtained, giving rise to product ion spectra that provide information comparable to triple quadrupole MS/MS experiments. Moreover, product ions obtained from CID can be further subjected to isolation and dissociation, which enables MS^n analyses and, thus, more insights into structural features of analytes and gas-phase reaction and degradation processes. However, also limitations of ion trap tandem mass spectrometry were reported primarily resulting from the low-mass cut-off in MS^n experiments, which were circumvented for instance by combining in-space dissociation devices with ion trap mass analyzers.[101]

2.2.2.3 Triple Quadrupole MS The selection and isolation of single (product) ions followed by controlled fragmentation using CID provides important information on ionized species and can significantly increase specificity of analytical procedures. For these purposes, the serial combination of quadrupoles has demonstrated distinguished robustness and was first reported in 1978.[102,103] Interfaced by an rf-only quadrupole, which acts as a collision-cell that is filled with a respective collision gas (e.g., nitrogen or argon), two conventional quadrupoles (commonly referred to as Q1 and Q3) are operated in scan or selected ion monitoring (SIM) modes, which offer the options of product ion scans, specific ion transition monitoring, neutral loss as well as precursor ion scans. These features have demonstrated great utility for multi-analyte screening methods, quantitative analytical procedures, and metabolism studies, all of which are of particular importance for sports drug testing assays. Hence, instruments using either a QqQ or QqLIT configuration have frequently been installed in doping control laboratories.

2.2.2.4 Double-Focusing Magnetic Sector MS First established in 1934 by Mattauch and Herzog,[104] double-focusing electric and magnetic sector analyzers of different combinations and geometries have been developed in the past. These utilize the fact that magnetic sectors allow sorting and focusing of ions depending on their momentum-to-charge ratio (which includes the mass of ions), while electric sectors operate independent of the mass of analytes and deflect ions primarily depending on respective kinetic energies. The combination of these features allows high mass resolution (up to 100,000) due to the double-focusing of a beam of ions that share a common mass but are angularly and energetically divergent. Consequently, broadened beams coming from the ion source are brought to focus and pass through extremely narrow collector slits without loss of sensitivity, and are allowed to hit the detector. The Mattauch-Herzog as well as the Nier-Johnson[105] double-focusing MS, developed approximately 20 years later, were equipped with a serial combination of electric (E) and magnetic (B) sectors, and several alternatives followed for instance with reversed geometry (BE)[106,107] or further added sectors. The first double-focusing MS introduced in sports drug testing in 1994 (Finnigan MAT 95) employed a reversed Nier-Johnson geometry (i.e., BE) and was installed for the Olympic Winter Games due to its capability to sensitively and selectively measure a subset of anabolic steroid metabolites (metandienone, methyltestosterone, and stanozolol metabolites) as well as clenbuterol, which were not detected at relevant urinary concentrations using conventional GC-MS quadrupole systems. The high resolution and concurrently accomplished high mass accuracy enabled the specific analysis of these few selected compounds, which were of particular interest for sports drug testing authorities due to an assumed widespread misuse. Indeed, the use of a double-focusing magnetic sector MS became a prerequisite for laboratories that analyze doping control samples at Olympic Games for several years; however, limitations in simultaneously measuring numerous different ions, comparably high maintenance and running costs, and space requirements of these instruments initiated the search for adequate alternatives, which were found in GC-ion trap and GC-triple quadrupole MS systems. Nevertheless, double-focusing magnetic sector MS have been present at all Olympic Games since 1994.[108]

2.2.2.5 Time-of-Flight MS Being one of the early inventions in mass spectrometry,[109–114] time-of-flight (TOF) MS underwent a considerable evolution and experienced a renaissance with the introduction of soft ionization techniques. The correlation between the mass of analytes and their flight time in a drift tube of defined length is exploited and allows the calculation of mass-to-charge ratios over a virtually unlimited mass range. Therefore, ions are accelerated into the flight tube using a pulse, which provides all ions with the same kinetic energy, and as smaller ions travel faster than large one, the precise determination of flight times enables the highly accurate calculation of m/z values. Early TOF-MS designs suffered from low resolution (approximately 140) due to long gating pulses and less adequate electronics for precise flight time determinations, but modern systems have overcome these issues, using, for instance, an orthogonal acceleration of a continuous ion beam,[115–117] which provides a defined packet of ions to be analyzed, and mass resolutions of more than 20,000 and accuracies below 10 ppm are accomplished.[47] Especially in combination with quadrupole mass filters and collision cells, resulting QqTOF-MS instruments have gained attention in sports drug testing but are not as frequently present in doping control laboratories as triple-quadrupole or ion trap mass spectrometers. This is presumably due to higher acquisition costs and less accurate quantitation results compared to triple-quadrupole mass spectrometers. Nevertheless, various applications using (Qq)TOF-MS have recently been reported and demonstrated the utility of this analytical technique.

2.2.2.6 Orbitrap MS More recently, a new mass analyzer termed Orbitrap was introduced,[118] which combines the features of the Kingdon ion storage device,[119] the Paul trap,[87] and the Fourier transform (FT) ion cyclotron resonance instrument.[120,121] With a purely electrostatic field for ion trapping, ions injected perpendicular into the Orbitrap revolve around the central electrode and simultaneously oscillate in the z-direction. The ion motion along the z-axis compares to a harmonic oscillator, and the mass-to-charge ratio of ions is simply related to the frequency of oscillation. Broadband image current detection is followed by a fast FT (FFT) algorithm to yield a mass/charge spectrum with high resolution (up to 150,000) and high accuracy (2 ppm).[122] The combination of the Orbitrap mass analyzer with a LIT and a curved linear ion trap (C-trap) was commercialized in 2005 (LTQ-Orbitrap, Thermo Electron), and its utility for doping control purposes was immediately recognized. Numerous applications in particular with regard to peptide hormone analysis[12,22] and structure characterization of low molecular weight analytes were published,[123,124] and also screening procedures solely using the high resolution/high accuracy capability in full scan modes were reported.[125] Drawbacks were observed in a limited number of simultaneously conducted MS/MS experiments due to comparably long duty cycles and long equilibration times after polarity switching, which have been improved with the successor model of the LTQ-Orbitrap MS.

2.2.2.7 Isotope-Ratio MS (IRMS) The identification and analysis of isotopes of elements using mass spectrometry was first reported in the early 20th century using a "Positive Ray Spectrograph" developed in 1919 by the late Nobel laureate F.W. Aston.[126,127] The determination of exact masses of numerous isotopes followed during the next decades; however, the accurate measurement of isotopic abundances remained complicated due to the use of photographic plates.[128] The introduction of magnetic sector MS equipped with Faraday Cup collectors in the 1940s solved this issue,[46,129] and the variation of natural isotopes depending on material sources was recognized[130–132] and has been studied systematically since.[128,133] Interfacing GC to IRMS instruments via combustion units has allowed sensitive "compound-specific isotope analysis" as described in the late 1970s,[134,135] and the commercialization of such instruments has provided invaluable tools for numerous research areas such as ecology, biogeochemistry, paleontology, and archeology, but also medicine and pharmacology.[133,136]

In modern GC-combustion-IRMS (GC/C/IRMS) systems, the chromatographically separated analytes are subjected to 940°C in an oxidation furnace that converts all carbon of each substance to CO_2. These molecules are subsequently ionized by EI and separated in a magnetic sector before collection in Faraday Cups positioned to accurately measure ions at *m/z* 44, 45, and 46, which represent $^{12}C^{16}O^{16}O^{+\cdot}$, $^{13}C^{16}O^{16}O^{+\cdot}$, and $^{12}C^{16}O^{18}O^{+\cdot}$, respectively. Consequently, carbon isotope ratios of $^{13}C/^{12}C$ can be calculated, which are (per definition) expressed using δ-values. These are defined as

$$\delta^{13}C = (r_{Sample}/r_{Standard} - 1) \times 1000$$

with r = $^{13}C/^{12}C$, and the standard being Vienna Pee Dee Belemnite (PVDB). Hence, δ-values are reported on a "parts-per-thousand" or permille (‰) scale.

The fact that, among others, the carbon isotope ratios (CIRs) vary between compounds derived from different sources such as human biosynthesis, plant biosynthesis, or chemical production, a distinction of compounds with nominally identical structure and composition is possible. This feature in particular has been exploited in sports drug testing, as it allows the detection of synthetic counterparts to naturally occurring anabolic androgenic steroids or their prohormones in human urine. If athletes use, for instance, testosterone to artificially increase their athletic performance, the discrimination of endogenously produced testosterone and respective metabolites from the exogenously provided steroid is accomplished by measuring $\delta^{13}C$ values of the target analytes (e.g., testosterone itself or its major metabolites androsterone, etiocholanolone, etc.). These $\delta^{13}C$ values are further compared to those obtained from so-called endogenous reference compounds (ERCs) that are not influenced by the application of the synthetic testosterone. If a defined difference in $\delta^{13}C$ values between an ERC and the target analyte (the $\Delta\delta^{13}C$ value) is exceeded, an adverse analytical finding is reported. A comprehensive review describing the utility of GC/C/IRMS for doping control purposes was recently published,[137] which outlines the enormous discrimination power of such systems but also the care that is required in sample preparation, analysis, and statistical evaluation.

REFERENCES

1. World Anti-Doping Agency (2004) *Identification Criteria for Qualitative Assays Incorporating Chromatography and Mass Spectrometry*. Available at http://www.wada-ama.org/rtecontent/document/criteria_1_2.pdf. Accessed 07-27-2006.
2. World Anti-Doping Agency (2009) *Minimum Required Performance Limits for Detection of Prohibited Substances*. Available at http://www.wada-ama.org/rtecontent/document/MINIMUM_REQUIRED_PERFORMANCE_LEVELS_TD_v1_0_January_2009.pdf. Accessed 11-24-2008.
3. World Anti-Doping Agency (2009) *The World Anti-Doping Code—International Standard for Laboratories*. Available at http://www.wada-ama.org/rtecontent/document/International_Standard_for_Laboratories_v6_0_January_2009.pdf. Accessed 11-24-2008.
4. U.S. Department of Health and Human Services Food and Drug Administration (2003) *Mass Spectrometry for Confirmation of the Identity of Animal Drug Residues*. Available at http://www.fda.gov/cvm/Guidance/guide118.pdf#search=%22FDA%20mass%20spectrometry%20%22. Accessed 11-9-06.

5. Van Eenoo, P., and Delbeke, F.T. (2004) Criteria in chromatography and mass spectrometry: A comparison between regulations in the field of residue and doping analysis. *Chromatographia*, **59**, S39–S44.
6. Sphon, J.A. (1978) Use of mass spectrometry for confirmation of animal drug residues. *J Assoc Off Anal Chem*, **61**, 1247–1252.
7. Stein, S.E., and Heller, D.N. (2006) On the risk of false positive identification using multiple ion monitoring in qualitative mass spectrometry: Large-scale intercomparisons with a comprehensive mass spectral library. *Journal of the American Society for Mass Spectrometry*, **17**, 823–835.
8. Commission of the European Communities (2002) Commission Decision of 12 August 2002 implementing Council Directive 96/23/EC concerning the performance of analytical methods and the interpretation of results. Available at http://eur-lex.europa.eu/LexUriServ/site/en/oj/2002/l_221/l_22120020817en00080036.pdf. Accessed 9-4-06.
9. Rivier, L. (2003) Criteria for the identification of compounds by liquid chromatography-mass spectrometry and liquid chromatography-multiple mass spectrometry in forensic toxicology and doping analysis. *Analytica Chimica Acta*, **492**, 69–82.
10. Petrovic, M., and Barcelo, D. (2006) Application of liquid chromatography/quadrupole time-of-flight mass spectrometry (LC-QqTOF-MS) in the environmental analysis. *Journal of Mass Spectrometry*, **41**, 1259–1267.
11. Thevis, M., and Schänzer, W. (2005) Identification and characterization of peptides and proteins in doping control analysis. *Current Proteomics*, **2**, 191–208.
12. Thevis, M., Thomas, A., and Schänzer, W. (2008) Mass spectrometric determination of insulins and their degradation products in sports drug testing. *Mass Spectrometry Reviews*, **27**, 35–50.
13. Thevis, M., Kohler, M., and Schänzer, W. (2008) New drugs and methods of doping and manipulation. *Drug Discovery Today*, **13**, 59–66.
14. Thevis, M., Maurer, J., Kohler, M., *et al.* (2007) Proteases in doping control analysis. *International Journal of Sports Medicine*, **28**, 545–549.
15. Thomas, A., Kohler, M., Walpurgis, K., *et al.* (2009) Proteolysis and autolysis of proteases and the detection of degradation products in doping control. *Drug Testing and Analysis*, **1**, 81–86.
16. Thevis, M., Ogorzalek Loo, R.R., Loo, J.A., and Schänzer, W. (2003) Doping control analysis of bovine hemoglobin-based oxygen therapeutics in human plasma by LC-electrospray ionization-MS/MS. *Analytical Chemistry*, **75**, 3287–3293.
17. Thevis, M., Thomas, A., Delahaut, P., *et al.* (2006) Doping control analysis of intact rapid-acting insulin analogues in human urine by liquid chromatography-tandem mass spectrometry. *Analytical Chemistry*, **78**, 1897–1903.
18. Thevis, M., Thomas, A., Delahaut, P., *et al.* (2005) Qualitative determination of synthetic analogues of insulin in human plasma by immunoaffinity

purification and liquid chromatography-tandem mass spectrometry for doping control purposes. *Analytical Chemistry*, **77**, 3579–3585.

19. Thomas, A., Thevis, M., Delahaut, P., *et al.* (2007) Mass spectrometric identification of degradation products of insulin and its long-acting analogues in human urine for doping control purposes. *Analytical Chemistry*, **79**, 2518–2524.
20. Bredehöft, M., Schänzer, W., and Thevis, M. (2008) Quantification of human insulin-like growth factor-1 and qualitative detection of its analogues in plasma using liquid chromatography/electrospray ionisation tandem mass spectrometry. *Rapid Communications in Mass Spectrometry*, **22**, 477–485.
21. Thevis, M., Bredehöft, M., Geyer, H., *et al.* (2006) Determination of Synacthen in human plasma using immunoaffinity purification and liquid chromatography/tandem mass spectrometry. *Rapid Communications in Mass Spectrometry*, **20**, 3551–3556.
22. Thomas, A., Geyer, H., Kamber, M., *et al.* (2008) Mass spectrometric determination of gonadotrophin-releasing hormone (GnRH) in human urine for doping control purposes by means of LC-ESI-MS/MS. *Journal of Mass Spectrometry*, **43**, 908–915.
23. Gam, L.H., Tham, S.Y., and Latiff, A. (2003) Immunoaffinity extraction and tandem mass spectrometric analysis of human chorionic gonadotropin in doping analysis. *Journal of Chromatography. B, Analytical Technologies in the Biomedical and Life Sciences*, **792**, 187–196.
24. Bethem, R., Boison, J., Gale, J., *et al.* Establishing the fitness for purpose of mass spectrometric methods. *Journal of the American Society for Mass Spectrometry*, **14**, 528–541.
25. Thevis, M., Loo, J.A., Loo, R.R., and Schänzer, W. (2007) Recommended criteria for the mass spectrometric identification of target peptides and proteins (<8 kDa) in sports drug testing. *Rapid Communications in Mass Spectrometry*, **21**, 297–304.
26. Norbeck, A., Monroe, M.E., Adkins, J.N., *et al.* (2005) The utility of accurate mass and LC elution time information in the analysis of complex proteomes. *Journal of the American Society for Mass Spectrometry*, **16**, 1239–1249.
27. Yates, J.R., 3rd. (1998) Database searching using mass spectrometry data. *Electrophoresis*, **19**, 893–900.
28. Mann, M., and Wilm, M. (1994) Error-tolerant identification of peptides in sequence databases by peptide sequence tags. *Analytical Chemistry*, **66**, 4390–4399.
29. Nesvizhskii, A.I. (2007) Protein identification by tandem mass spectrometry and sequence database searching. *Methods in Molecular Biology*, **367**, 87–119.
30. Eng, J.K., Mccormack, A.L., and Yates, J.R. (1994) An approach to correlate tandem mass-spectral data of peptides with amino-acid-sequences

in a protein database. *Journal of the American Society for Mass Spectrometry*, **5**, 976–989.

31. Perkins, D.N., Pappin, D.J.C., Creasy, D.M., and Cottrell, J.S. (1999) Probability-based protein identification by searching sequence databases using mass spectrometry data. *Electrophoresis*, **20**, 3551–3567.
32. Nesvizhskii, A.I., Vitek, O., and Aebersold, R. (2007) Analysis and validation of proteomic data generated by tandem mass spectrometry. *Nature Methods*, **4**, 787–797.
33. Zhang, Z., Sun, S., Zhu, X., *et al.* (2006) A novel scoring schema for peptide identification by searching protein sequence databases using tandem mass spectrometry data. *BMC Bioinformatics*, **7**, 222.
34. Salmi, J., Nyman, T.A., Nevalainen, O.S., and Aittokallio, T. (2009) Filtering strategies for improving protein identification in high-throughput MS/MS studies. *Proteomics*, **9**, 848–860; DOI 10.1002/pmic.200800517.
35. Budzikiewicz, H. (1998) *Massenspektrometrie*. Wiley-VCH, Weinheim.
36. Dempster, A.J. (1918) A new method of positive ray analysis. *Physical Review*, **11**, 316–325.
37. Bauer, S.H. (2001) Mass spectrometry in the mid-1930's: Were chemists intrigued? *Journal of the American Society for Mass Spectrometry*, **12**, 975–988.
38. Smyth, H.D. (1922) A new method for studying ionising potentials. *Proceedings of the Royal Society of London. Series A*, **102**, 283–293.
39. Hogness, T.R., Lunn, E.G. (1924) The ionization potentials of hydrogen as interpreted by positive ray analysis. *Proceedings of the National Academy of Sciences of the United States of America*, **10**, 398–405.
40. Hogness, T.R., and Lunn, E.G. The ionization of hydrogen by electron impact as interpreted by positive ray analysis. *Physical Review*, **26**, 44–55.
41. Hogness, T.R., and Lunn, E.G. (1925) The ionization of nitrogen by electron impact as interpreted by positive ray analysis. *Physical Review*, **26**, 786–793.
42. Smyth, H.D., and Mueller, D.W. (1933) The ionization of water vapor by electron impact. *Physical Review*, **43**, 116–120.
43. Smyth, H.D., and Mueller, D.W. (1933) The ionization of sulphur dioxide by electron impact. *Physical Review*, **43**, 121–122.
44. Tate, J.T., Smith, P.T., and Vaughan, A.L. (1935) A mass spectrum analysis of the products of ionization by electron impact in nitrogen, acetylene, nitric oxide, cyanogen and carbon monoxide. *Physical Review*, **48**, 525–531.
45. Nier, A.O., and Hanson, E.E. (1936) A mass-spectrographic analysis of the ions produced in HCl under electron impact. *Physical Review*, **50**, 722–726.
46. Nier, A.O. (1947) A mass spectrometer for isotope and gas analysis. *Review of Scientific Instruments*, **18**, 398–411.

47. de Hoffmann, E., and Stroobant, V. (2007) *Mass Spectrometry: Principles and Applications*. Wiley, Chichester.
48. Thevis, M., and Schänzer, W. (2007) Mass spectrometry in sports drug testing: Structure characterization and analytical assays. *Mass Spectrometry Reviews*, **26**, 79–107.
49. Thevis, M., and Schänzer, W. (2005) Mass spectrometry in doping control analysis. *Current Organic Chemistry*, **9**, 825–848.
50. National Institute of Standards and Technology (NIST) (2008) NIST/EPA/NIH Mass Spectral Library (NIST 08). Available at http://www.nist.gov/srd/nist1.htm. Accessed 11-27-2008.
51. Maurer, H.H., Pfleger, K., and Weber, A.A. (2007) *Mass Spectral Library of Drugs, Poisons, Pesticides, Pollutants and Their Metabolites*. Wiley-VCH, Weinheim.
52. Maurer, H.H. (1988) Massenspektrometrische Datenbanken in der Toxikologie. *Zeitschrift für analytische Chemie*, **330**, 317–318.
53. Munson, M.S.B., and Field, F.H. (1966) Chemical ionization mass spectrometry. I. General introduction. *Journal of the American Chemical Society*, **88**, 2621–2630.
54. Munson, M.S.B., Field, F.H. (1966) Chemical ionization mass spectrometry. II. Esters. *Journal of the American Chemical Society*, **88**, 4337–4345.
55. Field, F.H. (1968) Chemical ionization mass spectrometry. *Accounts of Chemical Research*, **1**, 42–49.
56. Schoengold, D.M., and Munson, M.S.B. (1970) Combination of gas chromatography and chemical ionization mass spectrometry. *Analytical Chemistry*, **42**, 1811–1813.
57. Leloux, M.S., and Maes, R.A. (1990) The use of electron impact and positive chemical ionization mass spectrometry in the screening of beta blockers and their metabolites in human urine. *Biomedical and Environmental Mass Spectrometry*, **19**, 137–142.
58. de Boer, D., Bernal, M.E.G., van Ooyen, R.D., and Maes, R.A.A. (1991) The analysis of trenbolone and the human urinary metabolites of trenbolone acetate by gas chromatography/mass spectrometry and gas chromatography/tandem mass spectrometry. *Biological Mass Spectrometry*, **20**, 459–466.
59. de Boer, D., de Jong, E.G., and Maes, R.A. (1990) Mass spectrometric characterization of different norandrosterone derivatives by low-cost mass spectrometric detectors using electron ionization and chemical ionization. *Rapid Communications in Mass Spectrometry*, **4**, 181–185.
60. Maurer, H.H. (2002) Role of gas chromatography-mass spectrometry with negative ion chemical ionization in clinical and forensic toxicology, doping control, and biomonitoring. *Therapeutic Drug Monitoring*, **24**, 247–254.
61. Choi, M.H., Chung, B.C., Kim, M., *et al.* Determination of four anabolic steroid metabolites by gas chromatography/mass spectrometry with

negative ion chemical ionization and tandem mass spectrometry. *Rapid Communications in Mass Spectrometry*, **12**, 1749–1755.

62. Choi, M.H., Chung, B.C., Lee, W., *et al.* (1999) Determination of anabolic steroids by gas chromatography/negative-ion chemical ionization mass spectrometry and gas chromatography/negative-ion chemical ionization tandem mass spectrometry with heptafluorobutyric anhydride derivatization. *Rapid Communications in Mass Spectrometry*, **13**, 376–380.
63. Jacquet de Malzet, L.S. (1775) *Précis de l'électricité. Extrait expérimental & théorétique des phénomenes électriques*. Jean Thomas Trattner, Vienna.
64. Dole, M., Mack, L.L., Hines, R.L., *et al.* (1968) Molecular beams of macroions. *The Journal of Chemical Physics*, **49**, 2240–2249.
65. Yamashita, M., and Fenn, J.B. (1984) Electrospray ion source. Another variation on the free-jet theme. *Journal of Physical Chemistry*, **88**, 4451–4459.
66. Yamashita, M., and Fenn, J.B. (1984) Negative ion production with the electrospray ion source. *Journal of Physical Chemistry*, **88**, 4671–4675.
67. Fenn, J.B. (2002) Electrospray ionization mass spectrometry: How it all began. *Journal of Biomolecular Techniques*, **13**, 101–118.
68. Wong, S.F., Meng, C.K., and Fenn, J.B. (1988) Multiple charging in electrospray ionization of poly(ethylene glycols). *Journal of Physical Chemistry*, **92**, 546–550.
69. Meng, C.K., Mann, M., and Fenn, J.B. (1988) Of protons or proteins. *Zeitschrift für Physik D*, **10**, 361–368.
70. Fenn, J.B., Mann, M., Meng, C.K., *et al.* (1989) Electrospray ionization for mass spectrometry of large biomolecules. *Science*, **246**, 64–71.
71. Mann, M., Meng, C.K., and Fenn, J.B. (1989) Interpreting mass spectra of multiply charged ions. *Analytical Chemistry*, **61**, 1702–1708.
72. Loo, J.A., Udseth, H.R., and Smith, R.D. (1989) Peptide and protein analysis by electrospray ionization-mass spectrometry and capillary electrophoresis-mass spectrometry. *Analytical Biochemistry*, **179**, 404–412.
73. Smith, R.D., Loo, J.A., Edmonds, C.G., *et al.* New developments in biochemical mass spectrometry: Electrospray ionization. *Analytical Chemistry*, **62**, 882–899.
74. Fenn, J.B., Mann, M., Meng, C.K., *et al.* (1990) Electrospray ionization: Principles and practice. *Mass Spectrometry Reviews*, **9**, 37–70.
75. Wilm, M., and Mann, M. (1994) Electrospray and Taylor-Cone theory, Dole's beam of macromolecules at last? *International Journal of Mass Spectrometry and Ion Processes*, **136**, 167–180.
76. Fenn, J.B. (1993) Ion formation from charged droplets: Roles of geometry, energy, and time. *Journal of the American Society for Mass Spectrometry*, **4**, 524–535.
77. Iribarne, J.V., and Thomson, B.A. (1976) Evaporation of small ions from charged droplets. *Journal of Chemical Physics*, **64**, 2287–2294.

78. Kebarle, P., and Ho, Y. (1997) On the mechanism of electrospray mass spectrometry, in *Electrospray Ionization Mass Spectrometry: Fundamentals, Instrumentation and Applications* (ed R.B. Cole), John Wiley & Sons, New York, pp. 3–63.
79. Thevis, M., and Schänzer, W. (2007) Current role of LC-MS(/MS) in doping control. *Analytical and Bioanalytical Chemistry*, **388**, 1351–1358.
80. Thevis, M., and Schänzer, W. (2005) Examples of doping control analysis by liquid chromatography-tandem mass spectrometry: Ephedrines, beta-receptor blocking agents, diuretics, sympathomimetics, and cross-linked hemoglobins. *Journal of Chromatographic Science*, **43**, 22–31.
81. Thevis, M., and Schänzer, W. (2007) Mass spectrometric identification of peptide hormones in doping control analysis. *Analyst*, **132**, 287–291.
82. Avery, M.J. (2003) Quantitative characterization of differential ion suppression on liquid chromatography/atmospheric pressure ionization mass spectrometric bioanalytical methods. *Rapid Communications in Mass Spectrometry*, **17**, 197–201.
83. Dams, R., Huestis, M.A., Lambert, W.E., and Murphy, C.M. (2003) Matrix effect in bio-analysis of illicit drugs with LC-MS/MS: Influence of ionization type, sample preparation, and biofluid. *Journal of the American Society for Mass Spectrometry*, **14**, 1290–1294.
84. Matuszewski, B.K., Constanzer, M.L., and Chavez-Eng, C.M. (2003) Strategies for the assessment of matrix effect in quantitative bioanalytical methods based on HPLC-MS/MS. *Analytical Chemistry*, **75**, 3019–3030.
85. Carroll, D.I., Dzidic, I., Stillwell, R.N., *et al.* (1975) Atmospheric pressure ionization mass spectrometry: Corona discharge ion source for use in a liquid chromatograph-mass spectrometer-computer analytical system. *Analytical Chemistry*, **47**, 2369–2373.
86. Horning, E.C., Carroll, D.I., Dzidic, I., *et al.* Liquid chromatograph-mass spectrometer-computer analytical systems: A continuous-flow system based on atmospheric pressure ionization mass spectrometry. *Journal of Chromatography*, **99**, 13–21.
87. Paul, W., and Steinwedel, H. (1953) Ein neues Massenspektrometer ohne Magnetfeld. *Zeitschrift für Naturforschung*, **8**, 448–450.
88. Lottspeich, F., and Zorbas, H. (1998) *Bioanalytik*. Spektrum Akademischer Verlag GmbH, Heidelberg-Berlin.
89. Finnigan, R.E. (1994) Quadrupole mass spectrometers: From development to commercialization. *Analytical Chemistry*, **66**, 969A–975A.
90. Todd, J. (1991) Ion trap mass spectrometer—past, present, and future (?). *Mass Spectrometry Reviews*, **10**, 3–52.
91. Fischer, E. (1959) Die dreidimensionale Stabilisierung von Ladungsträgern in einem Vierpolfeld. *Zeitschrift für Physik*, **156**, 1–26.
92. Douglas, D.J., Frank, A.J., and Mao, D. Linear ion traps in mass spectrometry. *Mass Spectrometry Reviews*, **24**, 1–29.

93. March, R.E. (1997) An introduction to quadrupole ion trap mass spectrometry. *Journal of Mass Spectrometry*, **32**, 351–369.
94. Louris, J.N., Amy, J.W., Ridley, T.Y., and Cooks, R.G. (1989) Injection of ions into a quadrupole ion trap mass spectrometer. *International Journal of Mass Spectrometry and Ion Processes*, **88**, 97–111.
95. Wuerker, R.F., Shelton, H., and Langmuir, R.V. (1959) Electrodynamic containment of charged particles. *Journal of Applied Physics*, **30**, 342–349.
96. Stafford, G.C., Jr., Kelley, P.E., Syka, J.E.P., *et al.* (1984) Recent improvements in and analytical applications of advanced ion trap technology. *International Journal of Mass Spectrometry and Ion Processes*, **60**, 85–98.
97. Schwartz, J.C., Senko, M.W., and Syka, J.E. (2002) A two-dimensional quadrupole ion trap mass spectrometer. *Journal of the American Society for Mass Spectrometry*, **13**, 659–669.
98. Church, D.A. (1969) Storage-ring ion trap derived from the linear quadrupole radio-frequency mass filter. *Journal of Applied Physics*, **40**, 3127–3134.
99. Dolnikowski, G.G., Kristo, M.J., Enke, C.G., and Watson, J.T. (1988) Ion-trapping technique for ion/molecule reaction studies in the center quadrupole of a triple quadrupole mass spectrometer. *International Journal of Mass Spectrometry and Ion Processes*, **82**, 1–15.
100. Beaugrand, C., Jaouen, D., Mestdagh, H., and Rolando, C. (1989) Ion confinement in the collision cell of a multiquadrupole mass spectrometer: Access to chemical equilibrium and determination of kinetic and thermodynamic parameters of an ion-molecule reaction. *Analytical Chemistry*, **61**, 1447–1453.
101. Hager, J.W. (2002) A new linear ion trap mass spectrometer. *Rapid Communications in Mass Spectrometry*, **16**, 512–526.
102. Yost, R.A., Enke, C.G. (1979) Selected ion fragmentation with a tandem quadrupole mass spectrometer. *Journal of the American Chemical Society*, **100**, 2274–2275.
103. Kondrat, R.W., and Cooks, R.G. (1978) Direct analysis of mixtures by mass spectrometry. *Analytical Chemistry*, **50**, 81A–92A.
104. Mattauch, J., and Herzog, R. (1934) Über einen neuen Massenspektrographen. *Zeitschrift für Physik*, **89**, 786–795.
105. Johnson, E.G., and Nier, A.O. (1953) Angular abberations in sector shaped electromagnetic lenses for focusing beams of charged particles. (1953) *Physical Review*, **91**, 10–17.
106. Matsuda, H. (1983) High-resolution high-sensitivity mass spectrometers. *Mass Spectrometry Reviews*, **2**, 299–325.
107. Matsuda, H. (1974) Double focusing mass spectrometers of second order. *International Journal of Mass Spectrometry and Ion Physics*, **14**, 219–233.
108. Hemmersbach, P. (2008) History of mass spectrometry at the Olympic Games. *Journal of Mass Spectrometry*, **43**, 839–853.

109. Stevens, W.E. (1946) A pulsed mass spectrometer with time dispersion. *Bulletin of the American Physical Society*, **21**, 22.
110. Cameron, A.E., Eggers, D.F. (1948) An ion "Velocitron." *Review of Scientific Instruments*, **19**, 605–607.
111. Hays, E.E., Richards, P.I., and Goudsmit, S.A. (1951) Mass measurements with a magnetic time-of-flight mass spectrometer. *Physical Review*, **84**, 824–829.
112. Wolff, M.M., and Stephens, W.E. (1953) A pulsed mass spectrometer with time dispersion. *Review of Scientific Instruments*, **24**, 616–617.
113. Katzenstein, H.S., and Friedland, S.S. (1955) New time-of-flight mass spectrometer. *Review of Scientific Instruments*, **26**, 324–327.
114. Wiley, W.C., and McLaren, I.H. (1955) Time-of-flight mass spectrometer with improved resolution. *Review of Scientific Instruments*, **26**, 1150–1157.
115. Guilhaus, M. (1995) Principles and instrumentation in time-of-flight mass spectrometry. *Journal of Mass Spectrometry*, **30**, 1519–1532.
116. Guilhaus, M., Selby, D., and Mlynski, V. (2000) Orthogonal acceleration time-of-flight mass spectrometry. *Mass Spectrometry Reviews*, **19**, 65–107.
117. Dawson, J.H.J., and Guilhaus, M. (1989) Orthogonal-acceleration time-of-flight mass spectrometer. *Rapid Communications in Mass Spectrometry*, **3**, 155–159.
118. Makarov, A. (2000) Electrostatic axially harmonic orbital trapping: A high-performance technique of mass analysis. *Analytical Chemistry*, **72**, 1156–1162.
119. Kingdon, K.H. (1923) A method for the neutralization of electron space charge by positive ionization at very low gas pressures. *Physical Review*, **21**, 408.
120. Hipple, J.A., Sommer, H., Thomas, H.A. (1949) A precise method of determining the Faraday by magnetic resonance. *Physical Review*, **76**, 1877–1878.
121. Gal, J-F. (1996) A historical note on an unrecognized early stage of the development of fast scanning ion cyclotron resonance spectrometers: The resotron. *International Journal of Mass Spectrometry and Ion Processes*, **157/158**, 1–4.
122. Hu, Q., Noll, R.J., Li, H., *et al.* (2005) The Orbitrap: A new mass spectrometer. *Journal of Mass Spectrometry*, **40**, 430–443.
123. Thevis, M., and Schänzer, W. (2008) Mass spectrometry of selective androgen receptor modulators. *Journal of Mass Spectrometry*, **43**, 865–876.
124. Thevis, M., Beuck, S., Thomas, A., *et al.* (2009) Screening for the calstabin-ryanodine-receptor complex stabilizers JTV-519 and S-107 in doping control analysis. *Drug Testing and Analysis*, **1**, 32–42.
125. Virus, E.D., Sobolevsky, T.G., and Rodchenkov, G.M. (2008) Introduction of HPLC/orbitrap mass spectrometry as screening method for doping control. *Journal of Mass Spectrometry*, **43**, 949–957.

126. Aston, F.W. (1919) A positive ray spectrograph. *Philosophical Magazine*, **38**, 707–714.
127. Aston, F.W. (1920) The mass spectra of chemical elements. *Philosophical Magazine*, **39**, 611–625.
128. Budzikiewicz, H., and Grigsby, R.D. (2006) Mass spectrometry and isotopes: A century of research and discussion. *Mass Spectrometry Reviews*, **25**, 146–157.
129. Nier, A.O. (1940) A mass spectrometer for routine isotope abundance measurements. *Review of Scientific Instruments*, **11**, 212–216.
130. Nier, A.O., and Gulbransen, E.A. (1939) Variations in the relative abundance of the carbon isotopes. *Journal of the American Chemical Society*, **61**, 697–698.
131. Nier, A.O. (1950) A redetermination of the relative abundances of the isotopes of carbon, nitrogen, oxygen, argon, and potassium. *Physical Review*, **77**, 789–793.
132. Murphey, B.F., and Nier, A.O. (1941) Variations in the relative abundance of the carbon isotopes. *Physical Review*, **59**, 771–772.
133. Brand, W.A. (1996) High precision isotope ratio monitoring techniques in mass spectrometry. *Journal of Mass Spectrometry*, **31**, 225–235.
134. Sano, M., Yotsui, Y., Abe, H., and Sasaki, S. (1976) A new technique for the detection of metabolites labelled by the isotope ^{13}C using mass fragmentography. *Biomedical Mass Spectrometry*, **3**, 1–3.
135. Matthews, D.E., and Hayes, J.M. (1978) Isotope-ratio-monitoring gas chromatography-mass spectrometry. *Analytical Chemistry*, **50**, 1465–1473.
136. Brenna, J.T., Corso, T.N., Tobias, H.J., and Caimi, R.J. (1997) High-precision continuous-flow isotope ratio mass spectrometry. *Mass Spectrometry Reviews*, **16**, 227–258.
137. Cawley, A.T., and Flenker, U. (2008) The application of carbon isotope ratio mass spectrometry to doping control. *Journal of Mass Spectrometry*, **43**, 854–864.

3 Structure Characterization of Low Molecular Weight Target Analytes—Electron Ionization

The detailed knowledge of mass spectrometric dissociation behaviors of target compounds as well as structurally related substances is of particular importance for the unambiguous identification of analytes, the detection of analogues, and the distinction of relevant compounds from interferences. Traditionally, numerous analytes are subjected to GC-MS analysis in sports drug testing, preferably employing EI; hence, respective spectra and underlying dissociation pathways of analytes were studied in detail with as well as without derivatization of the substances of interest.

3.1 STIMULANTS

One of the first classes of target analytes in sports drug testing programs were stimulating agents.[1] Being the subject of investigations in early horse doping controls, stimulants were also part of the first lists of prohibited substances and methods of doping (see Chapter 1, Table 1.3) due to their evident performance-enhancing but also harmful properties.[2–18] One of the first compounds identified to possess properties of agents now referred to as stimulants was ephedrine,[19] which was isolated in from *Ephedra vulgaris* (Ma Huang) in 1885.[20] In the following years and decades, numerous structurally related agents were prepared, such as amphetamine (1887),[21] methamphetamine (1893),[22] 3,4-methylenedioxymethamphetamine (1912),[23] nikethamide (1922),[24] methcathinone (1928),[25] phentermine (1946),[26] or benzphetamine (1957),[27] which represent only a small percentage of developed stimulants. However, these drugs and related compounds share common

Mass Spectrometry in Sports Drug Testing: Characterization of Prohibited Substances and Doping Control Analytical Assays, By Mario Thevis
Copyright © 2010 John Wiley & Sons, Inc.

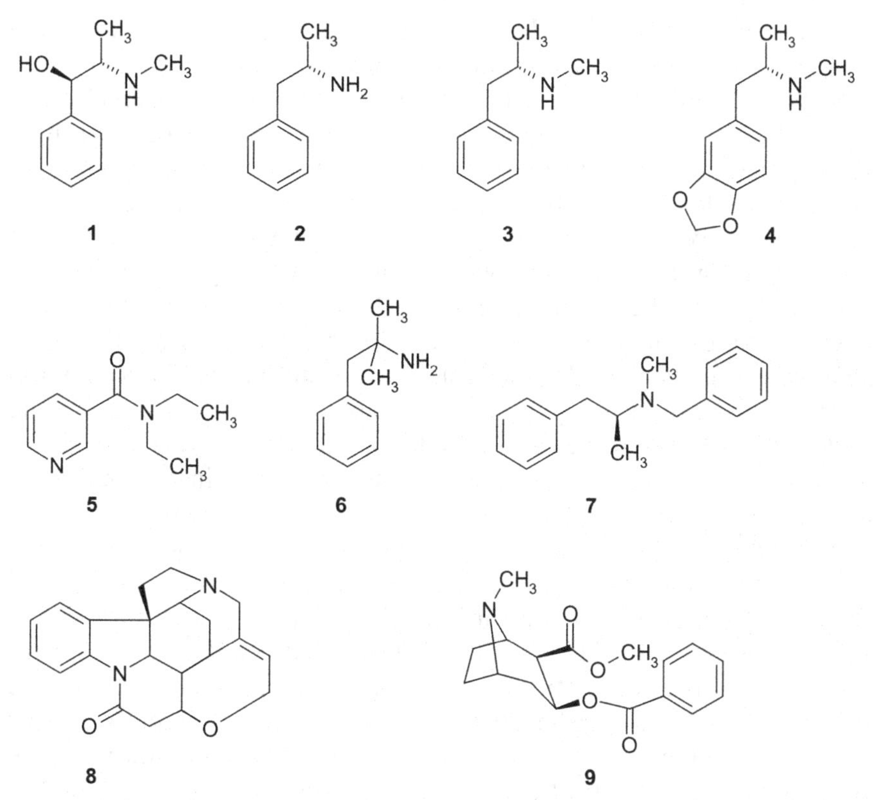

Figure 3.1: Structures of selected stimulants: ephedrine (**1**), amphetamine (**2**), methamphetamine (**3**), 3,4-methylenedioxymethamphetamine (**4**), nikethamide (**5**), phentermine (**6**), benzphetamine (**7**), strychnine (**8**), and cocaine (**9**).

structural features, in particular an aromatic ring system and a (substituted) amino or amide function (Fig. 3.1).

Amines, in general, were the subject of early and comprehensive mass spectrometric investigations, which outlined the likely ionization of amines by EI at the nitrogen atom due to its electron-donating nature.[28–30] Subsequent dissociation processes resulting in prominent peaks that represent fragment ions derived from the so-called α-cleavage (also referred to as β-bond cleavage) were observed, a generic phenomenon that commonly followed a rule of thumb that the largest possible radical is eliminated preferentially from the ionized species.[31–33] A major theoretical argument for this fragmentation behavior has been the great stability of the generated onium ion; however, the discrepancy that the loss of the smaller radical might yield the thermodynamically favored products has not always been considered.[34] Primary alkylamines have been studied in detail with regard to isomerization prior to dissociation and outlined several competing fragmentation processes commonly triggered by an intramolecular hydrogen abstraction and formation of distonic ions.[35] Complex cascades of rearrangements preceding the dissociation of primary amines were demonstrated by means of deuterium labeling experiments and allowed the explanation of a variety of fragment ions in EI mass spectra of analytes besides those resulting from the usually dominating α-cleavage.[36–38]

In addition to these seminal studies, the principle dissociation pathways of various stimulants relevant for doping controls using EI conditions were also elucidated in the past.[30,39–43] As outlined above, a common structural feature of many stimulating agents is the phenylalkylamine nucleus (Fig. 3.1), and the proposed dissociation pathway of ephedrine, which gives rise to an EI mass spectrum as depicted in Figure 3.2a, is shown exemplarily in Scheme 3.1[39,41,231] for a variety of related compounds.

With most stimulants comprising the common core constituents of phenylethylamine, the molecular ion [$M^{+\bullet}$] is hardly or not observed in mass spectra after EI. However, the intermediate existence was postulated to allow the elimination of a hydrogen atom, which yields the ion at *m/z* 164 in case of ephedrine (Fig. 3.2a).[40] The subsequent loss of water (−18 Da) gives rise to *m/z* 146 (Scheme 3.1), which requires the migration of a proton to the hydroxyl function and was determined to originate from the nitrogen atom by means of H/D exchange experiments. The substitution of mobile hydrogens of ephedrine (located at the hydroxyl residue and the secondary amino function) by deuterium atoms and subsequent EI-MS analysis demonstrated the lack of both deuterium atoms in the fragment ion at *m/z* 146 and substantiated the proposed elimination route. The fragment ions at *m/z* 117 and 115 were suggested to result from further dissociation of *m/z* 146 that eliminates HCN (−27 Da) and one or two hydrogen molecules, respectively (Scheme 3.1), which was evidenced by accurate mass measurements. A complementary fragmentation route was postulated to yield the fragment ion at *m/z* 132 via consecutive losses of a methyl group (−15 Da) and water (−18 Da), and possible structures of *m/z* 107, 105, 91, and 77 as hydroxybenzyl, benzoyl, tropylium, and phenyl cations, respectively (Scheme 3.1), were substantiated by stable isotope labeling of the phenyl residue.[39,41] The previously mentioned α-cleavage, i.e., the generation of a double bond between the nitrogen and the α-positioned adjacent carbon atom enabled by the homolytic cleavage of the beta-positioned C–C-linkage, further generated the base peak of the spectrum at *m/z* 58 that presumably consists of an ethylidene

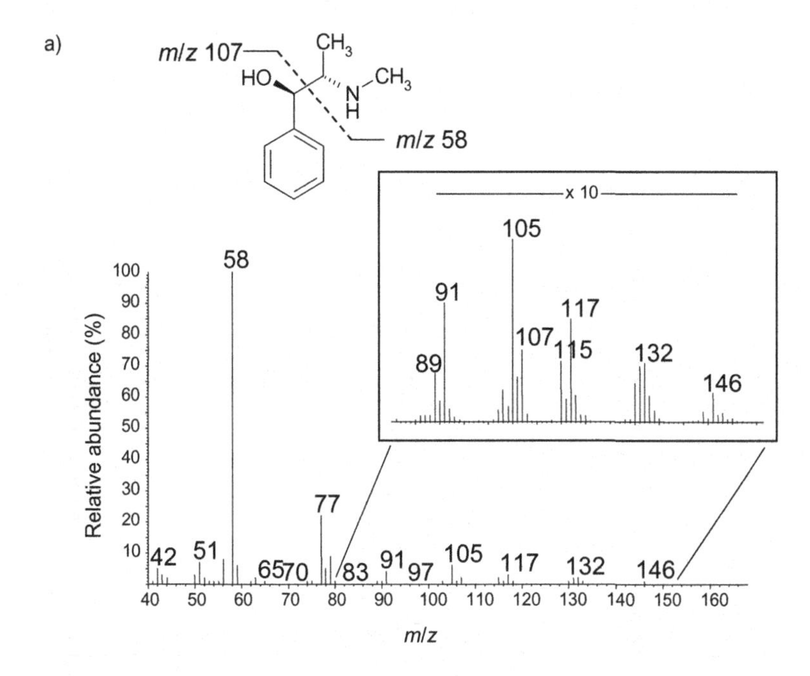

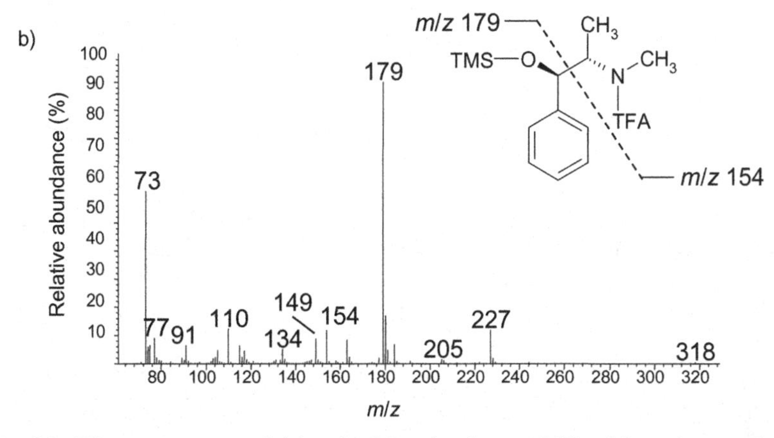

Figure 3.2: EI mass spectra of (a) ephedrine (mol wt = 165) with an inset enlarging the region *m/z* 80–150, and (b) ephedrine-*N*-TFA-*O*-TMS (mol wt = 333). The mixed derivative yields abundant signals obtained from α-cleavages, but preferably initiated at the ether oxygen (*m/z* 179) rather than the acylated nitrogen (*m/z* 154).

methyl ammonium ion (Scheme 3.1). The modification of particular sites of ephedrine or related stimulants causes characteristic mass shifts of diagnostic fragment ions, hence providing important information on possible alterations of known and prohibited compounds and allowing for their detection and identification. Selected stimulants and

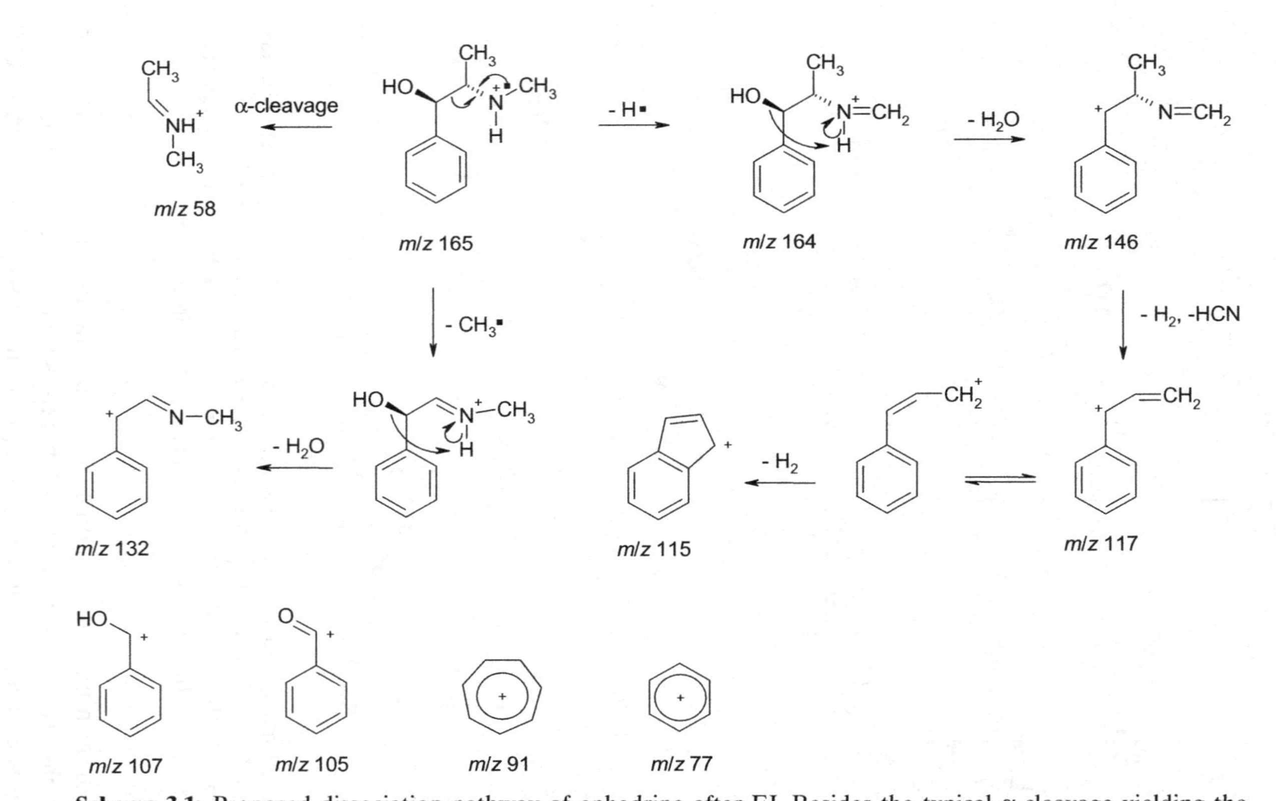

Scheme 3.1: Proposed dissociation pathway of ephedrine after EI. Besides the typical α-cleavage yielding the fragment ion at *m/z* 58, several informative ions are generated that indicate particular structural features of the analyte.[39,41,231]

corresponding fragment ions derived from EI are summarized in Table 3.1. The addition of methylenedioxy and methyl groups to amphetamine, for instance, which leads to 3,4-methylenedioxymetamphetamine (Fig. 3.1), gives rise to a mass spectrum containing an abundant fragment at *m/z* 58 accounting for the above identified ethylidene methyl ammonium ion. Moreover, the proposed tropylium ion at *m/z* 91, which comprises the phenyl residue, is incremented by 46 mass units due to the methyenedioxy functionality to *m/z* 135 (Table 3.1).[44] The addition of an ethyl group to amphetamine plus trifluoromethyl substitution of the phenyl residue yields fenfluramine (Table 3.1), which generates a base peak at *m/z* 72 and a minor fragment at *m/z* 159 that correspond to *m/z* 44 and 91 of amphetamine, respectively.[45]

A general problem of stimulant analysis using GC-EI-MS has been the low abundance of molecular ions of analytes as well as the limited number of diagnostic fragments. In addition, issues in terms of stereoisomer differentiation using gas chromatography arose. In order to approach these problems, various derivatization strategies have been developed during the last decades. Based on findings of Hofmann in 1881 that described the acylating properties of bisacyl amides,[46] preferably fluorinated acyl derivatives have been extensively used to modify amphetamine and related drugs for chromatographic and mass spectrometric analyses. In particular, trifluoroacetylation (TFA) and heptafluorobutyration (HFB) have commonly been utilized,[42,46–49] and, for improved separation of enantiomers, hydroxylated amphetamine- or ephedrine-like substances, also mixed derivatives obtained from the reaction with Mosher's acid (α-methoxy-α-(trifluoromethyl)-phenylacetyl chloride) and *O*-trimethylsilylation (TMS) and/or *N*-acylation have been employed.[50–53] The derivatization of stimulants entails modified mass spectrometric behaviors compared to respective underivatized compounds (Table 3.1); however, the principle dissociation pathways via α-cleavage remain in most cases. In Figure 3.2b, the EI mass spectrum of ephedrine-*N*-TFA-*O*-TMS is illustrated demonstrating the influence of chemically modified functional groups on the general fragmentation route. The mixed derivative of ephedrine also yields a base peak generated by an α-cleavage at *m/z* 179, but this is initiated by the ionization of the ether oxygen.[32] Nevertheless, an ion at *m/z* 154 is also found that corresponds to *m/z* 58 in Figure 3.2a representing the classical nitrogen-induced α-cleavage of underivatized ephedrine analogues.[54] Accordingly, fragmentation routes have been described for HFB derivatives,[42,47–49] and heptafluorobutyration of amine functions causes mass shifts of 196 u (in contrast to 96 u using trifluoroacetylation) in case of fragment ions derived from typical

TABLE 3.1: Characteristic Fragment Ions of Selected Stimulants with and without Derivatization Using EI[f]

Compound	Underivatized				N-TFA, O-TMS or Mixed N-TFA / O-TMS Derivative			
	Mol wt (Da)	Fragment Ions (m/z)			Mol wt (Da)	Fragment Ions (m/z)		
Amfepramone	205	190	100[a]	105	—	—	—	—
Amphetamine	135	91	65	44[a]	231[b]	140[a]	118	91
Benzphetamine	239	224	148[a]	91	—	—	—	—
Cathine	151	107	105	44[a]	319[c]	179	163	140[a]
Cocaine	303	198	182	105	—	—	—	—
Chlorphentermine	183	125	107	58[a]	279[b]	166	154[a]	114
Dimethylamphetamine	163	91	72[a]	65	—	—	—	—
Ephedrine	165	105	77	58[a]	333[c]	179	154[a]	110
Ethylamphetamine	163	148	91	72[a]	259[b]	168[a]	140	91
Etilefrine	181	121	77	58[a]	421[d]	406	267	179
Fencamphamine	215	186	115	98	311[b]	242	170	142
Fenfluramine	231	216	159	72[a]	327[b]	308	168[a]	140
Fenproporex	188	97[a]	92	57	284[b]	193[a]	140	118

Furfenorex	229	138[a]	91	81	—	—	—	—
P-hydroxyephedrine	181	107	71	58[a]	421[d]	267	193	154[a]
Methylenedioxyamphetamine	179	136	77	44[a]	275[b]	162	140[a]	135
Methylenedioxymethamphetamine	193	193	135	58[a]	289[b]	162	154[a]	135
Methylamphetamine	149	134	91	58[a]	245[b]	154[a]	118	110
Methylephedrine	179	105	77	72[a]	251[e]	236	149	72[a]
Methylphenidate	233	91	84[a]	56	329[b]	180[a]	126	67
Nikethamide	178	177	106	78	—	—	—	—
Phendimetrazine	191	176	85	57	—	—	—	—
Phenmetrazine	177	177	77	71	273	167	98	70
Phentermine	149	134	91	58[a]	245[b]	230	154[a]	91
Strychnine	334	334	319	306	—	—	—	—
Tuaminoheptane	115	100	55	44[a]	211[b]	196	140[a]	69

[a] α-cleavage,
[b] *N*-TFA derivative,
[c] *N*-TFA/*O*-TMS derivative,
[d] *N*-TFA/bis-*O*-TMS derivative,
[e] *O*-TMS derivative.
[f] Ref. 231.

α-cleavages. Consequently, ions at *m/z* 44 and 58 of amphetamine and metamphetamine are incremented to *m/z* 240 or 254, respectively, while principal dissociation routes correspond to those described for TFA derivatives (Table 3.1).

In contrast to those stimulants related to ephedrine and amphetamine, strychnine was among the first drugs "officially" (mis)used in sports (see Chapter 1) bearing an entirely different chemical structure (Fig. 3.1, **8**). Strychnine was reportedly used in a medical context as early as 1540 and represented the active ingredient of various over-the-counter tonics and laxatives until the early 1960s.[55,56] In its pure form it was first isolated from plants (such as the bean *Strychnos ignatii*) in 1818[57] and employed since for various purposes (e.g., in poison baits for rodents)[58] but also to stimulate athletic performance.[59] Its considerably different chemical composition compared to ephedrine, amphetamine, etc. results in a distinct mass spectrometric behavior that yields an abundant molecular ion and only few fragment ions upon EI (Fig. 3.3a),[30,60] the composition and origin of which has hardly been studied in detail yet. However, GC-tandem mass spectrometry was conducted in a case of fatal strychnine poisoning, and MS^n experiments allowed the determination of a dissociation pathway that outlined the initial eliminations of 15, 28, and 57 mass units from the molecular ion at *m/z* 334.[55] Due to the lack of high resolution/high accuracy MS data and stable isotope labeling, only rough assumptions on the fragmentation route and composition of ions were possible, which attributed these losses to a methyl radical, ethylene or carbon monoxide, and ethylmethylamine, respectively. In addition, numerous studies on the alkaloids of *Strychnos* in general[61–63] as well as indole derivatives[64] were conducted, and few fragment ions of lower *m/z* ratio but higher abundance in the EI mass spectrum of strychnine such as *m/z* 162, 144, 143, and 130 were attributed to particular sites of the analyte. While *m/z* 162 was suggested to originate from the right part of strychnine, the ions at *m/z* 144, 143, and 130 were proposed to comprise the indole nucleus as illustrated in Scheme 3.2.[61–64]

Further to strychnine, cocaine (Fig. 3.1, **9**) has been one of the early stimulating drugs of abuse in sports. Produced naturally in the leaves of the coca bush (*Erythroxylon coca Lam*), cocaine was consumed by indigenous people in the Peruvian Andes for over a thousand years by chewing coca leaves in the course of religious rituals but also to increase strength and energy. In the mid-16th century, cocaine was introduced into the "Old World" by Spanish conquerors, and it required more than three centuries for chemists and physicians to purify the active principle from the plant.[16] In 1855, cocaine was isolated for the first time

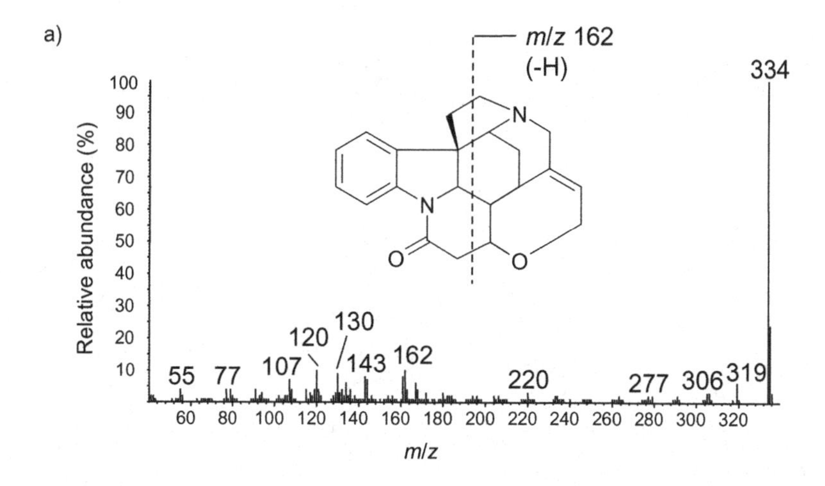

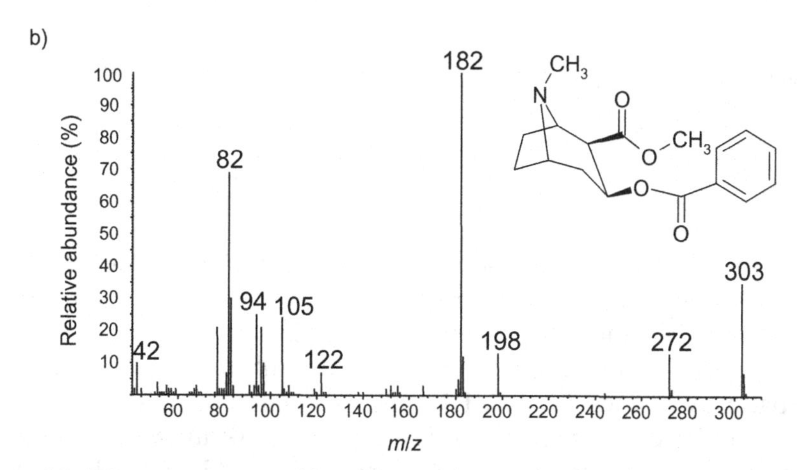

Figure 3.3: EI mass spectra of (a) strychnine (mol wt = 334) and (b) cocaine (mol wt = 303).

and termed "Erythroxyline,"[65] before its chemical characterization was accomplished in 1860 followed by its renaming to "Cocaine."[66] Being used to delay the onset of fatigue or for local anesthesia as well as a variety of other health and disease conditions in the late 19th century, the drug found its way also into sports (see Chapter 1) due to its assumed ergogenic effects similar to amphetamine. However, the only study that is close to a systematic evaluation of the effect of cocaine on athletic performance was conducted in 1884 by Sigmund Freud, who concluded that muscle strength and reaction times were improved for several hours.[16]

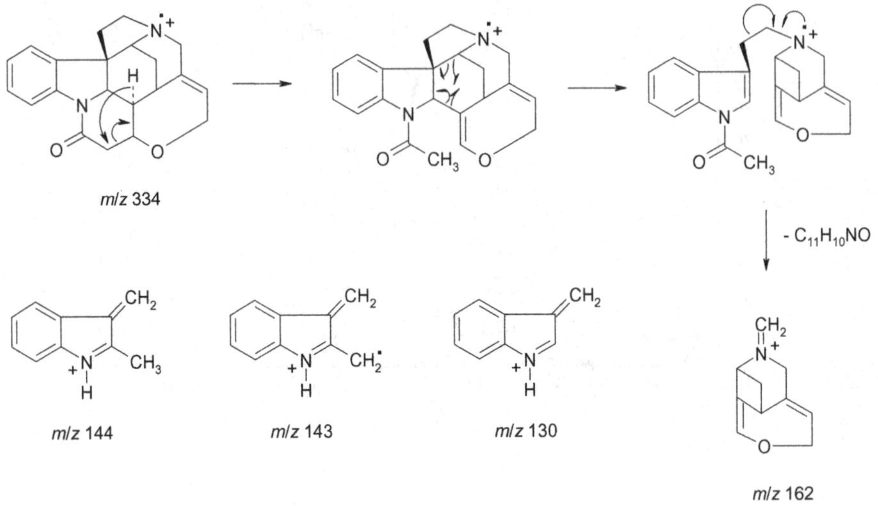

m/z 334

- $C_{11}H_{10}NO$

m/z 144 *m/z* 143 *m/z* 130 *m/z* 162

Scheme 3.2: Suggested structures of selected fragment ions of strychnine after EI.[61–64]

Chemically, cocaine represents a derivative of the tropane alkaloids and was the subject of several EI-MS studies.[67,68] Its EI mass spectrum is illustrated in Figure 3.3b containing a molecular ion at *m/z* 303 and characteristic fragment ions at *m/z* 198, 182, 105, and 82. The dissociation pathway of cocaine after EI was investigated using stable isotope labeling and accurate mass measurements, which resulted in a proposed fragmentation route as depicted in Scheme 3.3.[67–69] In accordance with alkaloids mentioned above, electron ionization is likely at the nitrogen atom and triggers predominantly an α-cleavage that causes the loss of a benzoate radical yielding the fragment ion at *m/z* 182. Alternatively, the tropine nucleus was shown to produce the 1-methyl-2,3-dihydro-1H-pyrrole odd- and even-electron cation (*m/z* 83 and 82, respectively), which evidently included reciprocal hydrogen transfer reactions (e.g., McLafferty rearrangement).[69] Locating the initial charge at the carbonyl residues, either the elimination of a methoxy radical (–31 Da) yielding *m/z* 272 or the generation of the benzoyl cation at *m/z* 105 were suggested. Moreover, the retention of the charge at the ecgonine core of cocaine combined with the neutral loss of the benzoyl radical resulted in the ion at *m/z* 198.

Due to the fact that cocaine is considerably metabolized after administration, its major degradation products (ecgonine, benzoylecgonine,

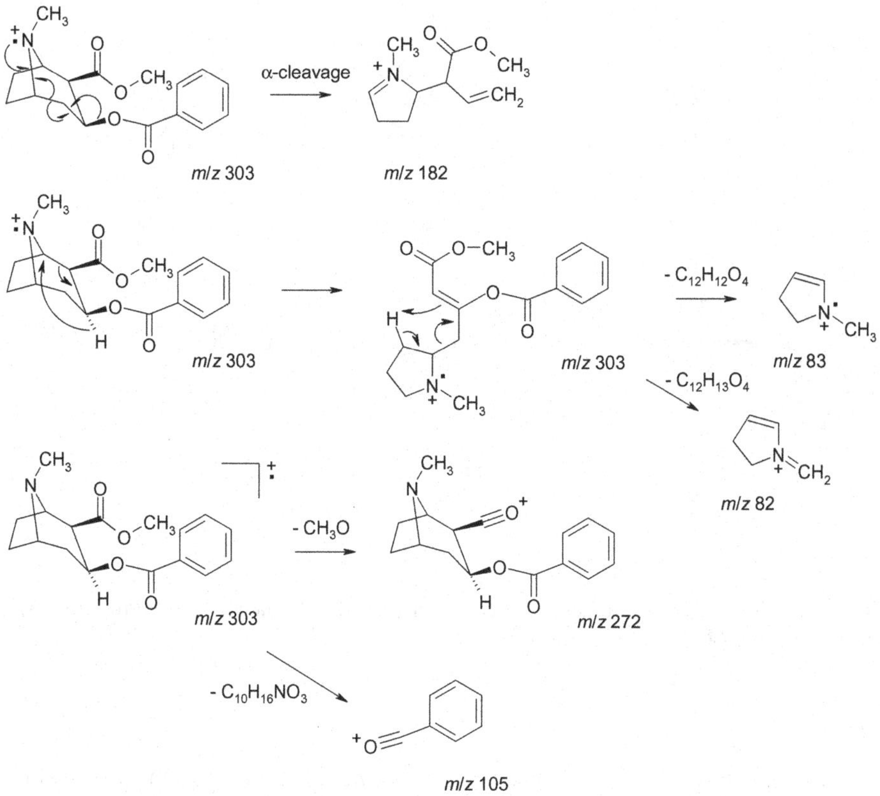

Scheme 3.3: Proposed dissociation pathway of cocaine after EI. Different fragmentation routes are possible depending on the site of ionization.[67–69]

and ecgonine methyl ester) are common targets in sports drug testing and follow the same mass spectrometric fragmentation patterns.

3.2 NARCOTICS

Narcotic agents such as morphine and heroin (Fig. 3.4, **1** and **2**, respectively) have also been prohibited since regulations regarding the use and misuse of drugs in human sports were established. Opium, the dried juice of poppy seeds derived from *Papaver somniferum* was supposedly known for its medicinal utility since 1552 B.C. by Egyptians, and Theophrastus evidently reported on its effects and value in the 3rd century B.C. Introduced to Western Europe in the 11th and 12th century A.D., it became the major component of remedies such as the infamous *laudanum*, a mixture of opium, wine, and spices, as prepared

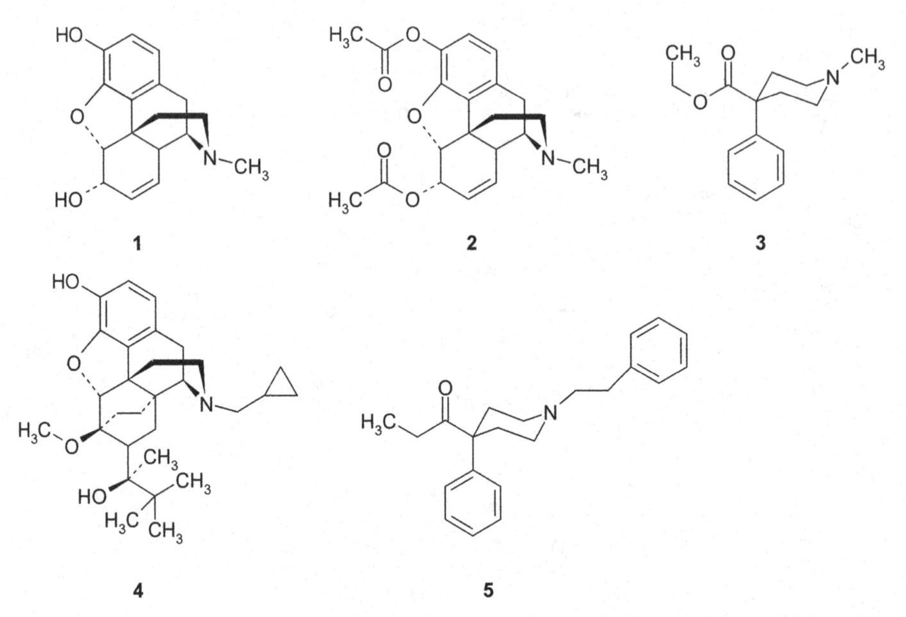

Figure 3.4: Structures of selected narcotics: morphine (**1**), heroin (**2**), pethidine (**3**), buprenorphine (**4**), and fentanyl (**5**).

by Paracelsus in 1520[16] or the *Tinctura opii crocata*, which contained opium, saffron, cloves, cinnamon, and diluted wine.[70] The most important active ingredient, morphine, was isolated from opium in 1803,[71,72] structurally characterized in 1925, and synthetically prepared in 1956.[73] Although not fully elucidated by then, derivatives of morphine were prepared as early as 1874 yielding the well-known diacetylmorphine (heroin),[74] which was commercialized by Bayer in 1898. One of the first entirely synthetic analgesics was pethidine (meperidine, Fig. 3.4, **3**), which was prepared in 1939[75] and considered the ideal candidate to substitute morphine in numerous clinical regards. However, its limited potency and duration of action as well as considerable toxicity[76] have initiated the ongoing search for alternatives,[77] in particular for the treatment of severe acute as well as chronic pain. Over the last decades, numerous additional semi-synthetic opiates such as buprenorphine (Fig. 3.4, **4**)[78] and fully synthetic opioids (e.g., fentanyl and its derivatives, Fig. 3.4, **5**)[79] were developed and also included in the list of prohibited substances and methods of doping (Table 3.2).[80] Although narcotics are not perceived as ergogenic drugs, their misuse in sports has been suspected due to the common occurrence of musculoskeletal injuries and the athlete's pressure to perform and compete.[16]

TABLE 3.2: Characteristic Fragment Ions of Selected Narcotics Using EI

Compound	Mol wt (Da)	Fragment Ions (*m/z*)				
Buprenorphine	467	449	435	410	378	55
Dextromoramide	392	265	165	128	100	55
Fentanyl	336	245	202	189	146	105
Hydromorphone	285	228	214	171	114	96
Heroin	369	327	310	268	204	162
Methadone	309	294	223	165	72	—
Morphine	285	268	215	162	124	115
Oxycodone	315	258	230	201	140	70
Pentazocine	285	217	202	110	70	—
Pethidine	247	246	218	174	172	71

The mass spectrometric behavior of morphine under EI conditions was studied in great detail by different groups in the mid-1960s, and comprehensive dissociation routes to major fragment ions were presented based on deuterium labeling and high resolution MS data.[81,82] In two studies, initial ionization of morphine was suggested at the nitrogen atom, which induces typical α-cleavages as reported for other alkaloids also (see above), and concurrent rearrangement and fragmentation pathways were described yielding the most abundant ions shown in the EI spectrum of morphine (Fig. 3.5a). The fissions of the C-8–C-14 and C-9–C-10 bonds were suggested to be favored due to the formation of an allylic or benzylic radical cation. Subsequently, cascades of dissociation occur yielding the ions at *m/z* 268, 215, 174, 162, 124, 115, and 70, which are illustrated in Scheme 3.4.[81,82]

Locating the radical in benzylic position at C-10, a retro Diels-Alder rearrangement (route **a**) was suggested to result in a benzofuran structure with a 5-(ethyl-methyl-amino)-penta-1,3-dien-1-ol substituent. The side chain undergoes cyclization with subsequent elimination of the 3,4-dimethyl-benzofuran-7-ol radical (–161 Da) to yield the ion at *m/z* 124. Alternatively, the cleavage of the linkage between C-13 and C-15 was proposed (route **b**), which allowed the loss of the 4-methyl-benzene-1,2-diol radical (–123 Da) and the generation of *m/z* 162,[81] the most abundant fragment ion found in the EI mass spectrum of morphine. Here, the origin of the hydrogen being shifted to the leaving group was proved to originate from C-6 as well as the hydroxyl function linked to C-6 by means of deuterium labeling.[82] The α-cleavage yielding the cation with a radical positioned at C-14 was postulated to initiate a variety of dissociations as outlined in routes **c-f** (Scheme 3.4). The loss

a)

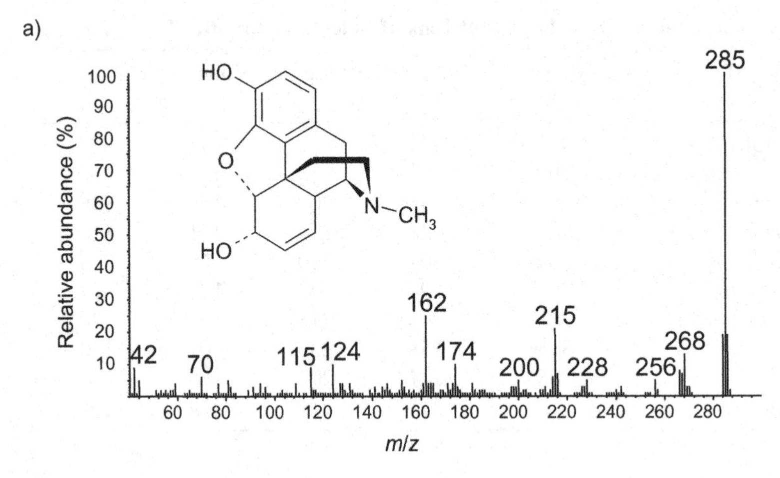

b)

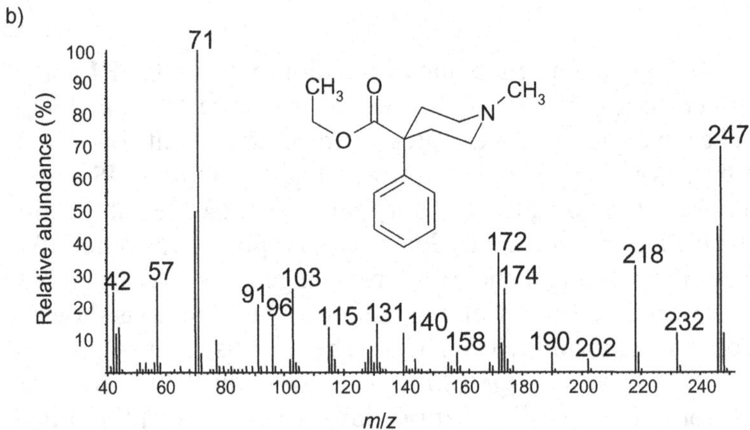

Figure 3.5: EI mass spectra of (a) morphine (mol wt = 285) and (b) pethidine (mol wt = 247).

of a hydroxyl radical (–17 Da) gave rise to *m/z* 268 (route **c**), and the above mentioned retro Diels-Alder rearrangement (route **a**) following the migration of a hydrogen from C-10 to C-14 enabled the release of buta-1,3-dien-1-ol (–70 Da) to generate the ion at *m/z* 215 (route **d**). The additional loss of a methyl radical from *m/z* 215 was suggested to yield *m/z* 200; however, the proposed residue (at the nitrogen atom) was not substantiated using $-CD_3$-labeled morphine,[83] and the actual origin was not determined. The loss of isocyanomethane (–41 Da) from *m/z* 215 gave rise to *m/z* 174, which was postulated to comprise a structure related to the radical cation of 3-ethyl-4-methyl-benzofuran-7-ol. The generation of *m/z* 115 was not elucidated in-depth (route **e**), and its

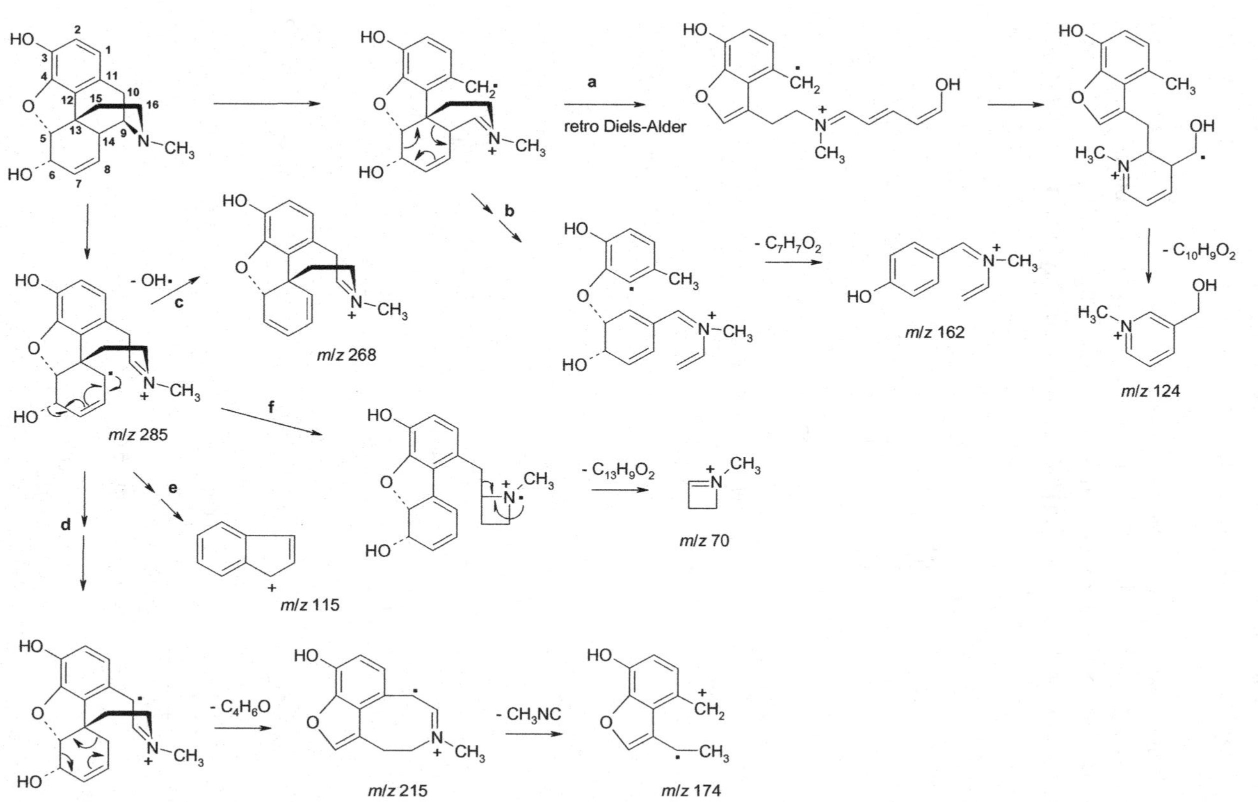

Scheme 3.4: Proposed dissociation routes of morphine after EI as obtained from deuterium-labeling experiments and HRMS.[81,82]

structure (the indene nucleus) was suggested based on its elemental composition determined as C_9H_8. In contrast, the dissociation pathway to the fragment ion at *m/z* 70 was suggested to include the formation of a 1-methyl-azetidine residue and the loss of $C_{13}H_9O_2$ (–215 Da, route **f**).[81]

The EI mass spectrum of pethidine (Fig. 3.5b) contains a variety of characteristic fragment ions,[84–87] the generation of which is predominantly but not exclusively initiated by α-cleavage after ionization of the nitrogen atom (Scheme 3.5).[84–88] A comparably abundant signal at

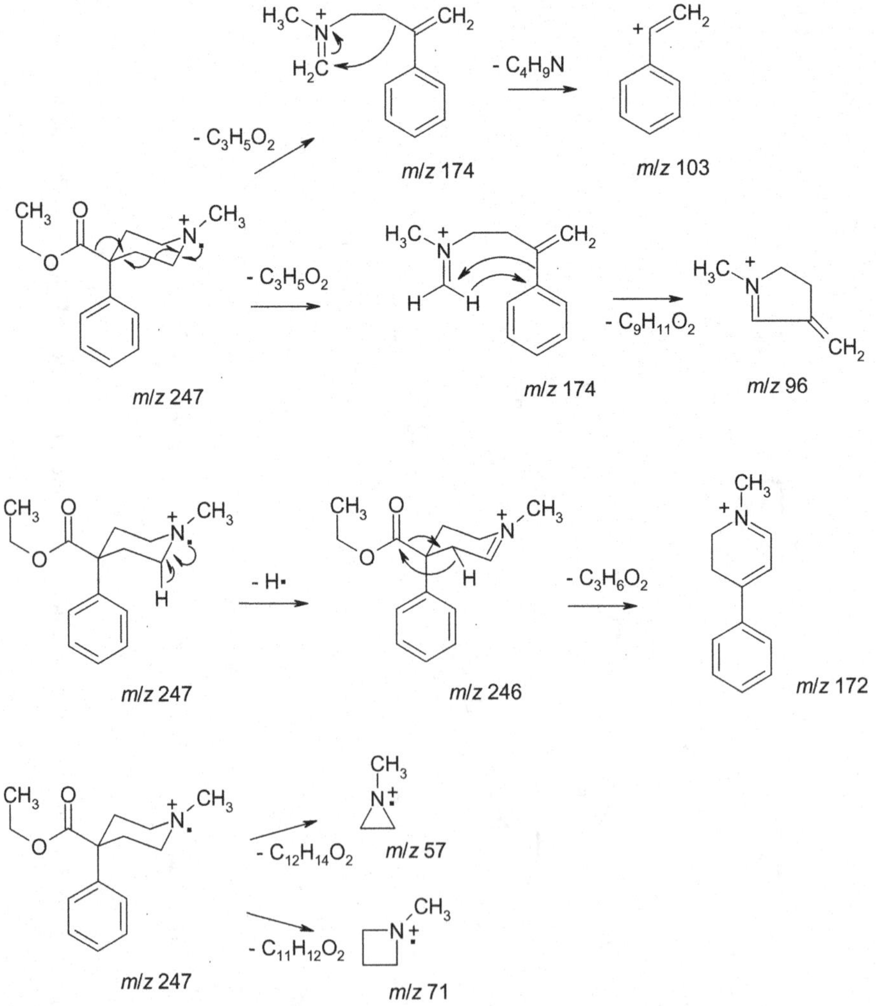

Scheme 3.5: Proposed dissociation routes of pethidine after EI as substantiated by deuterium-labeling experiments.[84–88]

m/z 246 is observed, which results from the loss of a hydrogen radical (−1 Da). A subsequent elimination of ethylene from the ester function yields *m/z* 218, which could, however, also be generated directly from the intact molecule by EI. The ion at *m/z* 172 was suggested to originate from *m/z* 246 by a neutral loss of formic acid ethyl ester (−74 Da) giving rise to the cation of 1-methyl-4-phenyl-2,3-dihydro-pyridine. Here, the neighboring hydrogen is shifted to the leaving group as demonstrated with deuterium labeling at the carbons 3 and 5 of the piperidine residue.[88] The immediate release of the radical of formic acid ethyl ester from the molecular ion at *m/z* 247 yields the fragment at *m/z* 174, which subsequently dissociates into *m/z* 103 and 96. Also in this case, the retention of 2 and 4 deuterium atoms, respectively, supported the proposed dissociation route forming the cations of vinyl-benzene (*m/z* 103) and 1-methyl-4-methylene-3,4-dihydro-2H-pyrrole (*m/z* 96). The most abundant fragment ion is observed at *m/z* 71, which was attributed to the radical cation of 1-methyl-azetidine, which accounts as a neutral radical for the leaving group eliminated from *m/z* 174 to *m/z* 103. Accordingly, the fragment at *m/z* 57 was assigned to 1-methyl-aziridine (Scheme 3.5).

3.3 ANABOLIC ANDROGENIC STEROIDS

Research in steroid biochemistry has been of particular interest for more than 140 years with regard to the desire of rejuvenation and reversion of age-related decline in men. Reports from 1869 about suggestions to inject semen into the blood of elderly men to improve mental and physical powers and subsequent experiments with saline extracts of dog testicles proved the overwhelming demand and search for the chemical fountain of youth.[89] These attempts were not successful for a variety of endocrinological and chemical reasons; however, the hormone quest had started[90] and received a considerable impulse with the identification of the anabolic androgenic principle testosterone in 1935.[91,92] Ever since, a major goal of early[93,94] as well as recent[95,96] developments in steroid biochemistry was the separation of anabolic and androgenic effects concerted with tissue selectivity of potential therapeutics that shall enable the treatment or prevention of debilitating diseases, muscular dystrophy, benign prostate hyperplasia,[97] or osteoporosis.[98–100] In the course of steroid manufacturing, thousands of steroid candidates were prepared and tested for benefits and undesirable effects, and the advantage of steroid administration in sports was suspected and later demonstrated numerous times since the 1950s.[101,102]

Anabolic androgenic steroids have been prohibited in sports since 1975, and mass spectrometry in particular has enabled their unambiguous determination. Although the analysis of most steroids relevant for doping controls is possible without derivatization by means of GC-MS, various chemical modifications were tested to improve chromatographic and mass spectrometric properties of these analytes, which was accomplished in particular by trimethylsilylation of hydroxyl and carbonyl residues yielding respective ether and enol-ether functions.[103–114] Seminal and comprehensive work on the elucidation of dissociation pathways of derivatized and underivatized steroids upon EI was conducted by Budzikiewicz and associates[115] as well as others.[116,117] Detailed studies investigating fragmentation pathways were performed and fundamental information on steroid decomposition behaviors in the gas phase were obtained. The influence of derivatization on the dissociation routes was observed in an early stage of steroid mass spectrometry,[107,118,119] and due to the enormous complexity and amount of data, only a selection is compiled in the following.

3.3.1 Unsaturated 3-Keto-Steroids

The core structure of most steroidal agents relevant for doping controls is the testosterone nucleus, which comprises the androst-4-ene scaffold with 3-oxo and 17-hydroxyl functions (Fig. 3.6, **1**), yielding the EI mass spectrum depicted in Figure 3.7a. Numerous characteristic ions are observed at *m/z* 288 ($M^{+\cdot}$), 273 (M^{+}-15), 270 ($M^{+\cdot}$-18), 246 ($M^{+\cdot}$-42), 203 ($M^{+\cdot}$-85), and 124, with the latter representing the base peak. Its generation was suggested to start with an ionization at the carbonyl oxygen followed by homologous fission of the bond between C-9 and C-10.[115] By means of deuterium labeling at C-8, the migration of the respective hydrogen to carbon C-10 was substantiated, and a subsequent McLafferty rearrangement including the hydrogen located at C-11 was proposed, enabling the formation of the fragment ion at *m/z* 124 as illustrated in Scheme 3.6.[115,120,121] The fragmentation of ionized testosterone to *m/z* 203 was postulated to be initiated by the removal of a π-electron from the α,β-unsaturated system followed by the same migration of the C-8-located hydrogen to C-10. The combined release of C-1, C-2, C-3, C-10, and C-19 plus an additional hydrogen presumably originating from C-14 yields the ion $[M\text{-}85]^{+\cdot}$ at *m/z* 203. Starting from the same precursor ion, the elimination of ketene (–42 Da) forming a 4-member ring structure was proposed to generate the fragment at *m/z* 246 (Scheme 3.6),[120,121] and the common loss of a water molecule (–18 Da) or methyl radical (–15 Da) gives rise to *m/z* 270 and 273, respectively. The release of water was shown to predominantly result from 1,3- and 1,4-elimination processes,[122,123] and deuterium labeling experiments proved both angular methyl residues to contribute equally to the generation of $[M\text{-}15]^{+\cdot}$.[124] In contrast to the more common eliminations of 15, 18, or 42, which were observed also with other α,β-unsaturated steroid nuclei (e.g., 3-keto-1-ene steroids), the

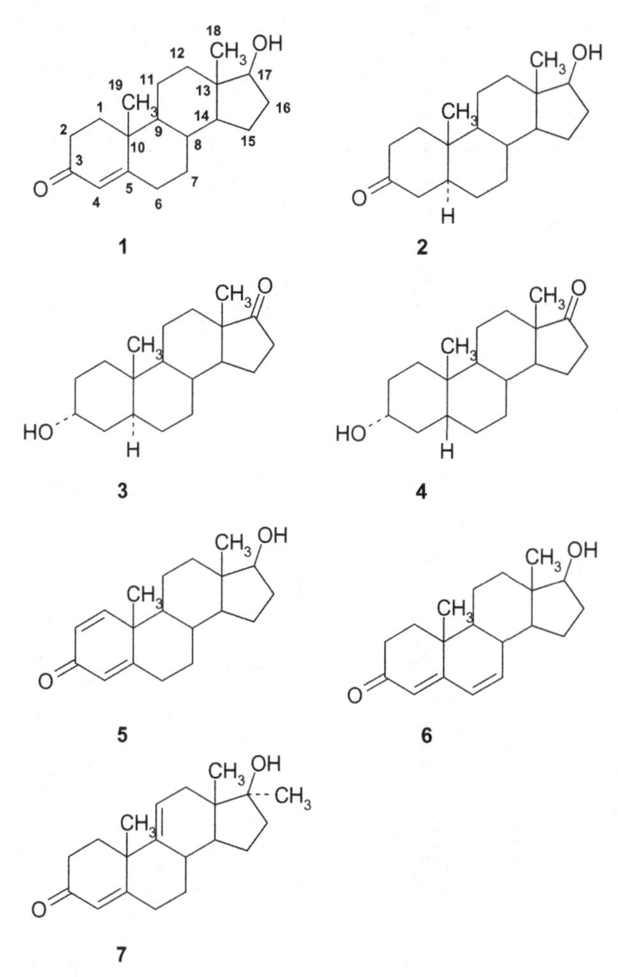

Figure 3.6: Structures of selected steroidal compounds: testosterone (**1**), 5α-dihydrotestosterone (**2**), androsterone (**3**), etiocholanolone (**4**), 1-dehydrotestosterone (**5**), androsta-4,6-dien-17β-ol-3-one (**6**), and 17α-methylandrosta-4,9(11)-dien-17β-ol-3-one (**7**).

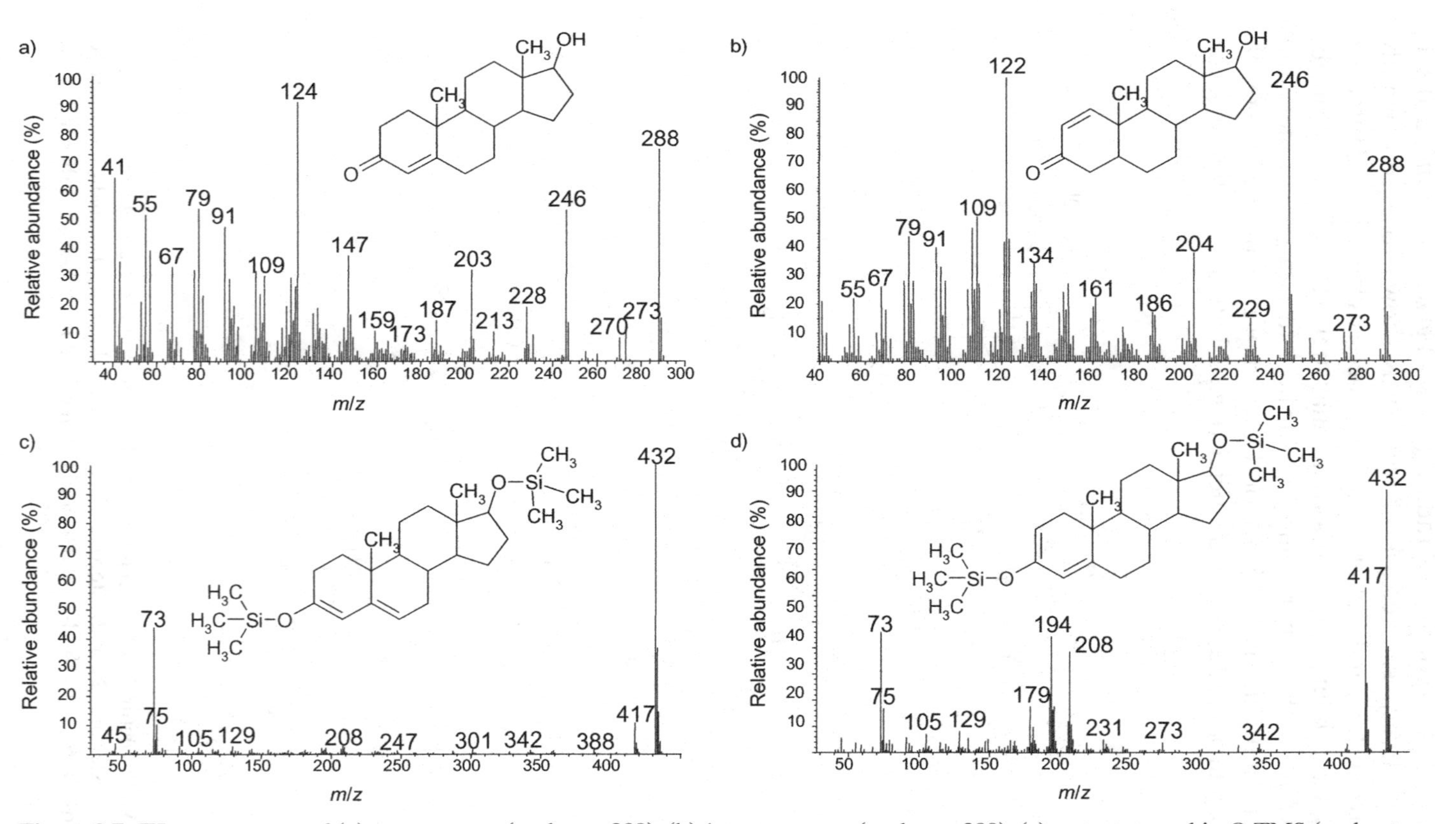

Figure 3.7: EI mass spectra of (a) testosterone (mol wt = 288), (b) 1-testosterone (mol wt = 288), (c) testosterone-bis-*O*-TMS (androsta-3,5-diene-3,17β-diol-bis-*O*-TMS isomer, mol wt = 432), and (d) testosterone-bis-*O*-TMS (androsta-2,4-diene-3,17β-diol-bis-O-TMS isomer, mol wt = 432).

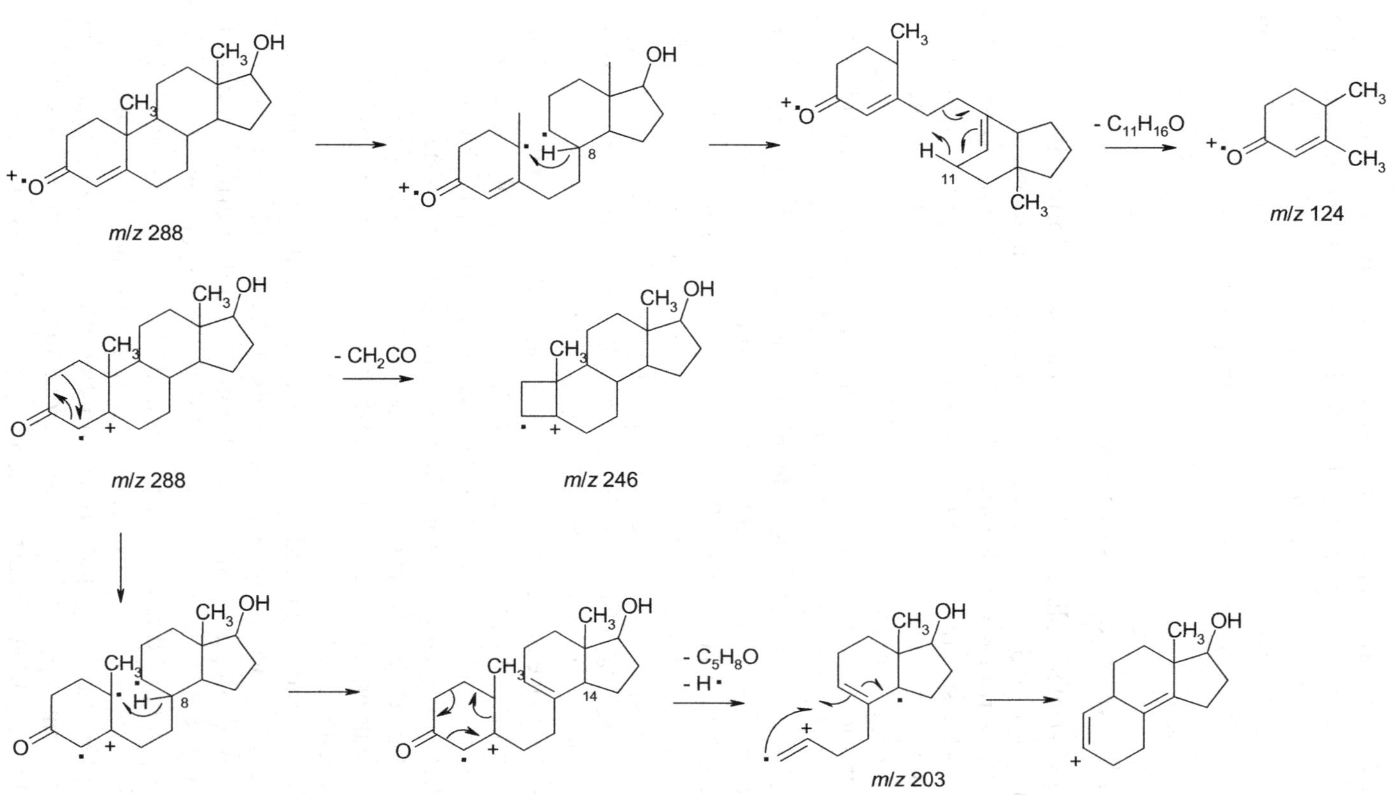

m/z 288

$- C_{11}H_{16}O$

m/z 124

m/z 288

$- CH_2CO$

m/z 246

$- C_5H_8O$

$- H^{\bullet}$

m/z 203

Scheme 3.6: Suggested dissociation routes of testosterone under EI conditions yielding common as well as structure-specific fragment ions.[115,120,121]

fragment ions at *m/z* 124 and 203 demonstrated considerable specificity for a testosterone-related steroid structure.[125] Locating the double bond between C-1 and C-2 instead of C-4 and C-5 entails the formation of a fragment ion at *m/z* 122 under EI conditions (Fig. 3.7b), which is in accordance to the mechanism described for 3-keto-4-ene structures but necessitates the migration of the hydrogen positioned at C-5 instead of C-11.[115,126]

3.3.2 α,β-Saturated Keto-Steroids

The phase-I-metabolism of testosterone leads primarily to 5α-androstan-17β-ol-3-one (dihydrotestosterone), 5α-androstan-3α-ol-17-one (androsterone), and 5β-androstan-3α-ol-17-one (etiocholanolone) (Fig. 3.6, **2-4**, respectively), which represent important parameters of the so-called urinary steroid profile in sports drug testing (see Chapter 6).

Dihydrotestosterone is a saturated 3-keto steroid that generates an abundant molecular ion at *m/z* 290 and fragments of considerable intensity at *m/z* 275 (M^+-15), 273 (M^+-18), and 231 (M^+-59, elimination of the steroidal D-ring).[127] Although stereoisomers are known to yield highly comparable mass spectra, the 5α-configuration can unequivocally be differentiated from the 5β-analogue by an intense signal at *m/z* 220, the generic M^+-70 fragment ion of 3-keto steroids with 5β-configuration that was studied in detail by Budzikiewicz and Djerassi using deuterium labeling experiments.[128] In contrast to testosterone and dihydrotestosterone, the carbonyl function of the testosterone metabolites androsterone and etiocholanolone is located at C-17 (Fig. 3.6, **3-4**). Consequently, a different fragmentation behavior was observed yielding one fragment ion at *m/z* 246 (M^+-44) that particularly characterizes the 17-keto function.[128,129] The elimination of C-16 and C-17 including the migration of two hydrogens to the leaving group allowed the formation of the indicative fragment as described by Egger and Spiteller in 1966 (Scheme 3.7).[127]

Further to these selected testosterone metabolites, numerous additional naturally occurring as well as chemically modified steroids have been isolated, prepared, and tested for human and veterinary medicine as well as research. The knowledge of their mass spectrometric behavior, their principal dissociation routes and, thus, structure-specific fragmentation, is essential for efficient and comprehensive sports drug testing, in particular in light of possibly prepared designer analogues. Fragment ions typical for selected steroid nuclei, which indicate but do not exclusively identify certain steroid core structures are summarized in Table 3.3.

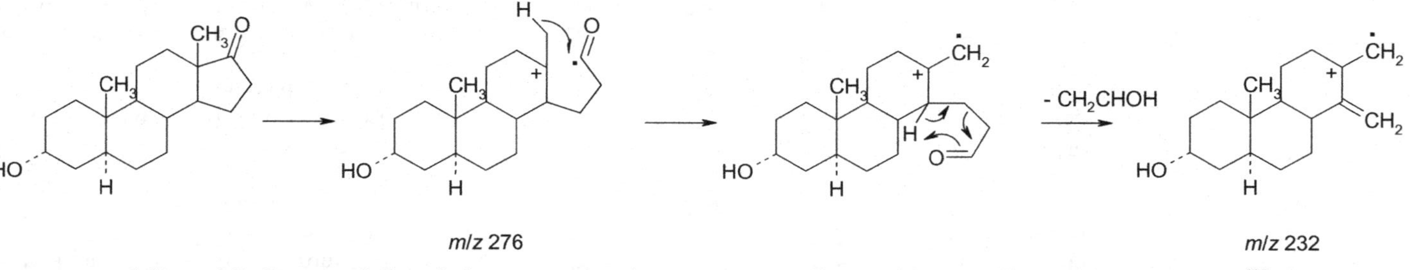

Scheme 3.7: Suggested dissociation route of androsterone under EI conditions yielding the ion at *m/z* 246.[127]

TABLE 3.3: Characteristic Fragment Ions of Selected Underivatized Steroids Using Electron Ionization [a]

Steroid Nucleus	Representative Compound	Mol wt (Da)	Fragment Ions (m/z)						
			M^+	M^+-15	M^+-18				
3-keto	5α-androstane-3,17-dione	288	288[b]	273	270	255	244	224	217
	5α-dihydrotestosterone	290	290[b]	275	272	231	199	163	123
	5β-dihydrotestosterone	290	290[b]	275	272	229	220	201	161
3-keto-4-ene	Testosterone	288	288[b]	273	270	246	203	124	109
	Methyltestosterone	302	302[b]	287	284	269	245	229	124
	Nandrolone	274	274[b]	—	256	231	215	160	110
3-keto-1-ene	1-testosterone	288	288	273	270	246[b]	204	122	109
3-keto-1,4-diene	1-dehydrotestosterone	286	286	—	—	253	227	147	122[b]
	Metandienone	300	300	—	282	267	242	161	122[b]
3-keto-4,6-diene	6-dehydrotestosterone	286	286[b]	271	268	253	227	151	136
3-keto-4,9-diene	17α-methyl-androsta-4,9(11)-dien-17β-ol-3-one	300	300	285	282	267	242[b]	227	215
17-keto	Androsterone	290	290[b]	275	272	257	246	139	215
	Etiocholanolone	290	290[b]	—	272	257	246	244	215

[a] Ref. 130.

[b] Base peak.

— Relative abundance less than 2%.

3.3.3 3-Keto-1,4-Diene, 3-Keto-4,6-Diene, and 3-Keto-4,9(11)-Diene Steroids

Steroids comprising a 3-keto-1,4-diene structure such as present for instance in case of 1-dehydrotestosterone (boldenone, androsta-1,4-dien-17β-ol-3-one, Fig. 3.6, **5**) or metandienone (17α-methyl-androsta-1,4-dien-17β-ol-3-one, Table 3.3) have frequently caused adverse analytical findings in sports. One of the major indicative fragment ions of agents bearing this structural feature is found at *m/z* 122, derived from a fission between C-9 and C-10 as well as C-6 and C-7 and commonly generating the base peak of respective EI mass spectra (Table 3.3).[115,121] In addition, the loss of a water molecule (–18 Da) or a methyl radical (–15 Da) was hardly observed but the combination of both resulting in $[M\text{-}33]^+$ ions. Moreover, a diagnostic ion at $[M\text{-}41]^+$ was reported for 1,4-androstadien-3-one and 17-alkylated analogues, the generation of which was demonstrated to include the release of C-9, C-11, and C-12 as the propene radical;[121] however, in EI spectra of 17-hydroxylated steroids such as boldenone, the ion was not detected.

In contrast to 3-keto-1,4-diene-based steroids, analogues such as androsta-4,6-dien-17β-ol-3-one (6-dehydrotestosterone, Fig. 3.6, **6**) generate abundant molecular ions upon EI and a characteristic fragment ion at *m/z* 136 (Table 3.3), which is proposed to originate from cleavages between C-9 and C-10 and C-7 and C-8 accompanied by two hydrogen migrations from the leaving group to the remaining cation.

The introduction of a double bond between C-9 and C-11 into the testosterone nucleus as in case of 17α-methylandrosta-4,9(11)-dien-17β-ol-3-one (Fig. 3.6, **7**) causes a considerable change of the fragmentation pathway compared for instance to androsta-4,6-dien-17β-ol-3-one. While the molecular ion of androsta-4,6-dien-17β-ol-3-one is base peak in the respective EI mass spectrum, the dissociation of 17α-methyl-androsta-4,9(11)-dien-17β-ol-3-one yields predominantly ions at *m/z* 242 and 227 on expense of the relative abundance of its molecular ion. The elimination of the entire steroidal D-ring or major parts including C-16 and C-17 were proposed to establish suitable leaving groups generating the intense peaks (M^+-58) and (M^+-73) in the spectrum of 17α-methyl-androsta-4,9(11)-dien-17β-ol-3-one (Table 3.3).[130]

3.3.4 Steroid Derivatization

The derivatization of steroids has significantly improved their trace analysis in doping controls by ameliorating chromatographic as well as mass spectrometric properties of target compounds. Trimethylsilylation (TMS) of steroids by means of *N*-methyl-*N*-trimethylsilyltrifluoroacetamide (MSTFA)[104,131] or its mixture with ammonium iodide and ethanethiol, which promotes the *in-situ* formation of the highly reactive trimethyliodosilane (TMIS), has demonstrated great utility and convenience in sample preparation and GC-MS analysis. Consequently, almost all currently employed assays are based on the detection of trimethylsilylated steroid derivatives,[132–135] the fragmentation behavior of which was has been subject of numerous early and recent studies.[104–111,113,114,136,137]

3.3.4.1 Unsaturated 3-Keto-Steroids The modification of steroids to trimethylsilylated counterparts causes a considerable alteration of dissociation pathways, and a summary of fragment ions typically generated by distinct steroid nuclei is presented in Table 3.4. In order to illustrate the impact of derivatization and its influence on the gas-phase decomposition, the EI mass spectrum of testosterone-bis-TMS (androsta-3,5-diene isomer) is depicted in Figure 3.7c. In contrast to the underivatized testosterone, the spectrum is dominated by an abundant molecular ion at *m/z* 432 and the fragment ion at *m/z* 73 representing the trimethylsilyl radical (which is commonly observed in spectra of trimethylsilylated analytes). Diagnostic fragment ions representing particular features of steroids are thus of minor intensity; however, ions such as *m/z* 129 are valuable indicators for instance for a derivatized 3- or 17-hydroxyl function. The origin of steroidal fragments at *m/z* 129 was investigated in detail using stably deuterated or structurally closely related compounds[106,107,111,125,138], and the proposed mechanisms of ion formations are shown in Scheme 3.8a.

As reported for underivatized steroids, the loss of a methyl radical from the molecular ion was due to the release of either of the angular methyl groups of the androstane nucleus. In case of TMS-derivatized substances, the loss of 15 u is frequently attributed to the homolytic cleavage of an Si-C-bond; however, detailed studies concerning the dissociation of the bis-*O*-TMS derivative of 19-norandrosterone revealed the almost exclusive release of the C-18 methyl group upon EI as proved by the introduction of $^{2}H_{9}$-TMS moieties.[111]

3.3.4.2 17-Alkylated Steroids and Enol-TMS Derivatives The presence of fragment ions at *m/z* 130 and 143 in EI mass spectra of TMS-derivatized analytes indicate a 17-methyl residue in trimethylsilylated 17-hydroxy steroids (Table 3.4) resulting from D-ring cleavage.[113,139,140] The proposed generation of *m/z* 143 is in accordance with

TABLE 3.4: Characteristic Fragment Ions of Selected Trimethylsilylated Steroids Using Electron Ionization[a]

Steroid Nucleus	Representative Compound	Mol wt (Da)	Fragment Ions (m/z)						
			M^+	M^+-15	M^+-15 -90				
3-keto	5α-dihydrotestosterone-bis-TMS	434	434[b]	419	329	239	202	143	142
3-keto-4-ene	Testosterone-bis-TMS (androsta-3,5-dien-3,17β-diol-bis-O-TMS)	432	432[b]	417	—	342	208	129	105
	Testosterone-bis-TMS (androsta-2,4-dien-3,17β-diol-bis-O-TMS)	432	432[b]	417	327	231	194	179	129
	Nandrolone-bis-TMS (estra-3,5-dien-3,17β-diol-bis-O-TMS)	418	418[b]	403	313	287	194	182	129
3-keto-1-ene	1-testosterone (androsta-1,3-dien-3,17β-diol-bis-O-TMS)	432	432	417	327	206	194[a]	181	129
3-keto-1,4-diene	1-dehydrotestosterone-bis-TMS(androsta-1,3,5-trien-3,17β-diol-bis-O-TMS)	430	430	415	325	299	229	206[b]	191
3-hydroxy-1-ene	androst-1-ene-3α,17β-diol-bis-O-TMS	434	434	419	329	195	143[b]	142	127
17-methyl	17α-methyl-5α-androstane-3α,17β-diol-bis-O-TMS	450	435	345	270	228	143[b]	130	—
17-keto	Androsterone-bis-O-TMS	434	434	419[b]	329	239	182	169	105
	Etiocholanolone-bis-O-TMS	434	434[b]	419	329	239	182	169	105
	19-norandrosterone-bis-O-TMS	420	420	405	315	225	182	169	129

[a] Ref. 130.
[b] Base peak.
— Relative abundance less than 2%.

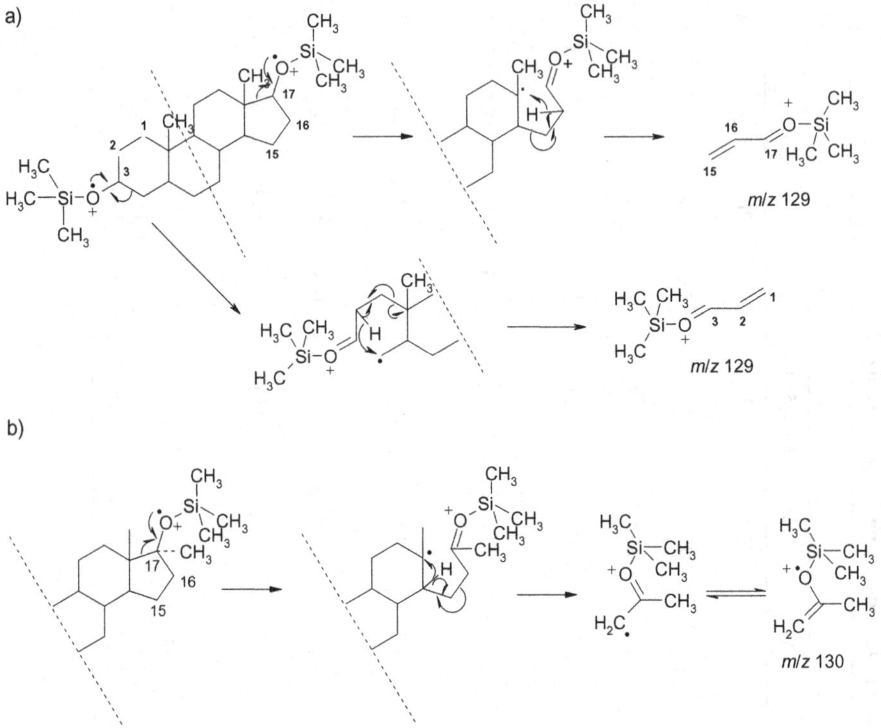

Scheme 3.8: Proposed dissociation route of androstane-3α,17β-diol bis-*O*-TMS under EI conditions yielding the ions at *m/z* 129 (a) and 130 (b).

the formation of the above mentioned D-ring fragment ion at *m/z* 129 supposedly composed by C-15, C-16, and C-17 (including the substituents) as demonstrated by $^2H_9$-TMS-labeling.[113] The suggested dissociation pathway resulting in the radical cation at *m/z* 130 is depicted in Scheme 3.8b.

In addition to the formation of *m/z* 143 by D-ring decomposition, fragments at *m/z* 142 and 143 are obtained from various steroid-TMS and steroid-enol-TMS by A-ring cleavage, e.g., steroids bearing a 3-hydroxy-androst-4-ene[106] or 3-hydroxy-androst-1-ene nucleus (Table 3.4).[141] For the latter in particular, deuterium exchange experiments demonstrated the presence of carbons 2, 3, and 4 in the respective fragment.

The ion at *m/z* 169 has been found characteristic for steroids comprising a 17-keto function (such as androsterone and etiocholanolone) that was derivatized to the corresponding enol-TMS ether (Table 3.4).[142,143] It supposedly originates from C-ring dissociation between C-8 and C-14 as well as C-12 and C-13 containing the entire D-ring including the angular C-18 function.

TMS-derivatization of steroids with a 3-keto-androst-1-ene or 3-keto-androst-4-ene yields enol-TMS ethers with 3-hydroxy-1,3-diene or 3-hydroxy-2,4-diene structure that give rise to mass spectra containing a characteristic fragment ion at *m/z* 194 (Table 3.4) as shown for testosterone in Figure 3.7d.[141,144] A fragmentation route including the fissions of the bonds between C-6 and C-7 as well as C-9 and C-10 was suggested as substantiated by H/D-exchange experiments, which resulted in dissociation pathways that were in close accordance to those postulated for comparable underivatized steroids.[145,146] Moreover, steroidal A-rings comprising a 1,4-dien-3-one scaffold are also enolized by means of trimethylsilylation and generate an intense fragment ion at *m/z* 206 as observed for instance with 1-dehydrotestosterone (Table 3.4) or metandienone.[141,144,147] Also here, deuterium labeling at C-1, 2, 4, and 6 provided evidence for the suggested cleavages of linkages between the carbons C-7 and C-8 as well as C-9 and C-10 that yield the diagnostic fragment ion at *m/z* 206.

3.4 SELECTIVE ANDROGEN RECEPTOR MODULATORS (SARMs)

Besides anabolic androgenic steroids, compounds that are structurally not related to the steroid nucleus were identified to enable the tissue-selective stimulation of androgen receptors with enhanced anabolic but significantly reduced androgenic properties. These agents were categorized as selective androgen receptor modulators (SARMs) and include various chemical core structures such as arylpropionamides, quinolines, and bicyclic hydantoins.[148] None of these substances have yet completed clinical trials and entered the pharmaceutical market, but their potential for misuse in sports has been considered and the entire class of SARMs has been prohibited in sports since 2008. The most advanced subgroup of SARMs, which was introduced in 1998,[149] includes arylpropionamides, e.g., S-4 and S-22 (see Chapter 4);[150] however, numerous alternative drug candidates were described such as LGD-2226[151,152] or BMS-564929,[153,154] which are based on 2-quinolinones or hydroxybicyclohydantoins (Fig. 3.8). In contrast to arylpropionamide-derived SARMs, LGD-2226, and BMS-564929 possess suitable GC-MS properties, and their fragmentation patterns as TMS derivatives under EI conditions were studied in detail to enable the development of GC-MS-based detection methods.[155]

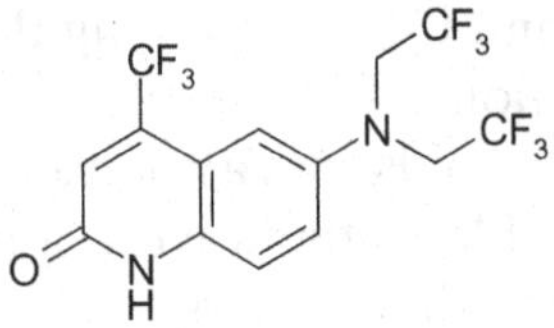

Figure 3.8: Structures of selective androgen receptor modulators: LGD-2226 (**1**), and BMS-564929 (**2**).

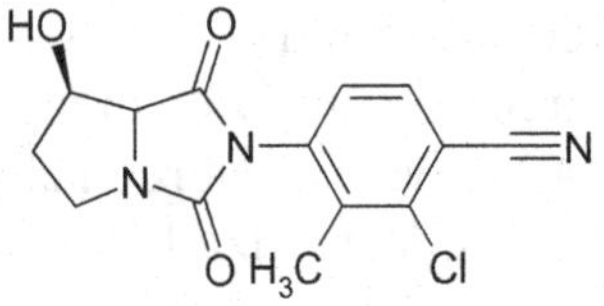

3.4.1 2-Quinolinone-Based SARMs

The trimethylsilylated LGD-2226 (Fig. 3.8, **1**), which is obtained by enolization of the 2-keto function, yielded a molecular ion at *m/z* 464, and further characteristic losses of 69 ($\cdot CF_3$), 84 (F_3C-CF_3), and 152 (F_3C-CH_2-CF_3) with or without the elimination of a methyl radical were observed (Fig. 3.9a). Moreover, informative fragment ions were generated at *m/z* 449 (M^+-15), 395 (M^+-69), 365 (M^+-15 -84), 311 (M^+-69 -84), 297 (M^+-15 -152), and 269 (M^+-15 -152 -28), which were proposed to originate from dissociation processes as depicted in Scheme 3.9a.[155] The initial loss of a methyl radical from the TMS residue triggered a dissociation pathway that first yielded the fragment ion at *m/z* 449. Subsequently, 1,1,1,3,3,3-hexafluoropropane (–152 Da) or 1,1,1-trifluoroethane (–84 Da) were eliminated giving rise to fragment ions at *m/z* 297 and 365, respectively. Consecutively, both product ions formed the ion at *m/z* 269 by losses of a methyleneamine radical (–28 Da) and 2,2,2-trifluoroethylideneamine radical (–96 Da), respectively.

3.4.2 Hydroxybicyclic Hydantoin-Derived SARMs

The mass spectrometric behavior of hydantoins upon EI was described in various studies due to their considerable medical relevance, for instance, in the treatment of epilepsy.[156,157] However, dissociation pathways of bicyclic hydantoins such as BMS-564929 (Fig. 3.8, **2**) initiated

a)

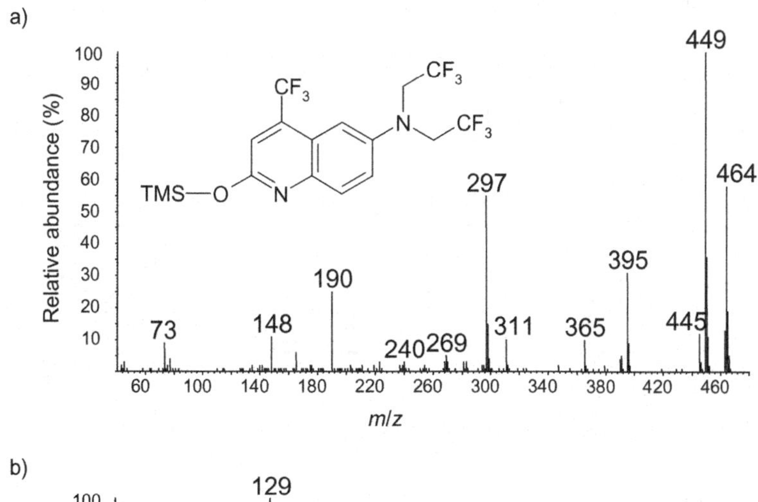

b)

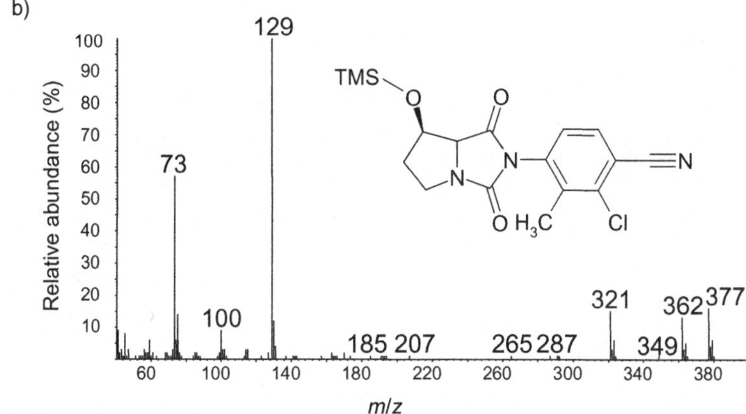

Figure 3.9: EI mass spectra of (a) LGD-2226 mono-TMS (mol wt = 464), and (b) BMS-564929 mono-TMS (mol wt = 377).

by electron impact deviate significantly from common fragmentation routes of regular hydantoin-derived structures (Fig. 3.9b). To a very minor extent, the molecular ion found at *m/z* 377 eliminated carbon monoxide (–28 Da), which gave rise to the product ion at *m/z* 349, representing a dissociation route that is frequently observed in mass spectra of hydantoins and related structures.[157] In contrast, the loss of 56 Da was found to be unique to the TMS-derivatized bicyclic nucleus of BMS-564929 and suggested to originate from the release of propenal. A possible pathway is depicted in Scheme 3.9b,[155] which necessitates the removal of the condensed alkyl ring structure and the migration of the trimethylsilyl residue from the leaving group (propenal) to the hydantoin core. The base peak of the EI-mass spectrum is formed

a)

m/z 464

- 15 u

m/z 449

- $CF_3CH_2CF_3$

m/z 297

- CH_3CF_3

- NCH_2

m/z 365

- 96 u

m/z 269

m/z 464

- CF_3

m/z 395

- CH_3CF_3

m/z 311

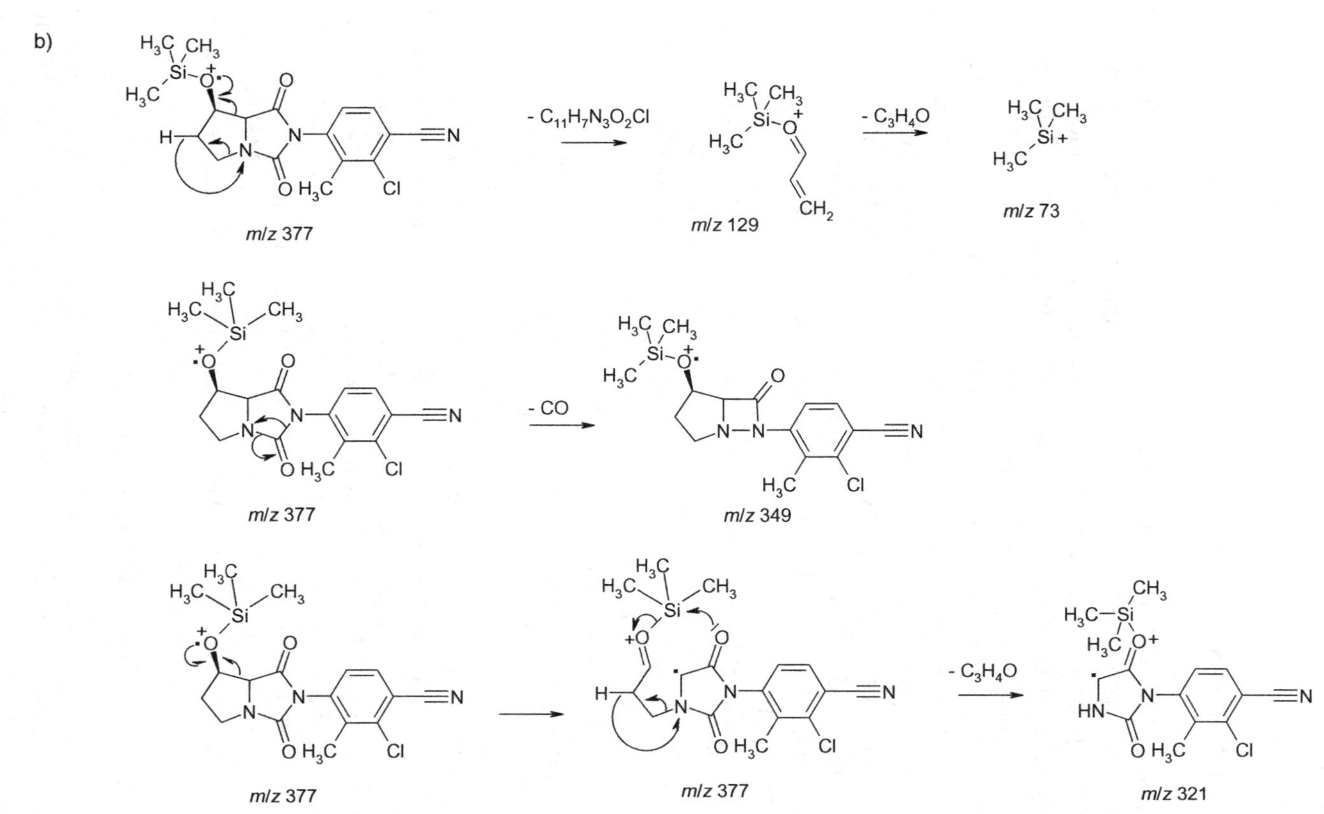

Scheme 3.9: Suggested dissociation pathways of (a) LGD-2226 mono-TMS, and (b) BMS-564929 mono-TMS after EI.[155]

by the ion at *m/z* 129, which is suggested to result from the elimination of the radical of 2-chloro-4-(2,5-dioxo-imidazolidin-1-yl)-3-methyl-benzonitrile (–248 Da).

3.5 DIURETICS

The dating of the discovery of diuretic agents in general has been rather difficult; nevertheless, first reports on the medicinal use of mercury-based compounds were authored in pre-Christian times, and comprehensive clinical utility was observed in the 16th century when agents including mercury were employed to treat syphilis and dropsy.[158,159] The incidental recognition of the diuretic effect of calomel, mercury(I) chloride, in 1885 was followed by systematic studies on the effect of the antisyphilitic agent merbaphen, the diethylbarbiturate derivative of 2-chloro-4-chloromercuriphenoxyacetic acid (Fig. 3.10, **1**)[160] in the early 20th century and the preparation of further mercury-based diuretic compounds such as mersalyl (*o*-[N-(3-hydroxymercuri-2-methoxypropyl)-carbamoyl]phenoxyacetic acid).

In the late 19th century, the diuretic activity of caffeine was observed,[161] leading to the development of several xanthine-derived diuretic agents such as theophylline, and numerous subsequent studies regarding fully synthetic substances stimulating the elimination of water yielded a variety of drugs such as acetazolamide[162] (1950, Fig. 3.10, **2**), hydrochlorothiazide[163] (1959, Fig. 3.10, **3**), furosemide[164] (1959, Fig. 3.10, **4**), triamterene[165] (1961, Fig. 3.10, **5**), or amiloride[166] (1966, Fig. 3.10, **6**).

The elevated diuresis, as induced by different mechanisms by diuretic agents, has been employed to counteract hypertension and selected kidney diseases for several decades. Due to the capability of diuretics to mask other banned compounds by urine dilution rather than any performance-enhancing property, these drugs were added to the list of prohibited substances and methods of doping in 1988. Early attempts to detect the misuse of diuretics in sports were based on GC-MS, and particularly methylated analytes were subjected to EI dissociation studies.

3.5.1 Thiazide-Derived Drugs

Benzothiadiazines such as hydrochlorothiazide (Fig. 3.10, **3**), which are also referred to as thiazides, represent one of the most commonly prescribed and observed classes of diuretic agents. They consist of a common nucleus of 1,2,4-benzothiadiazine-1,1-dioxide with a sulfon-

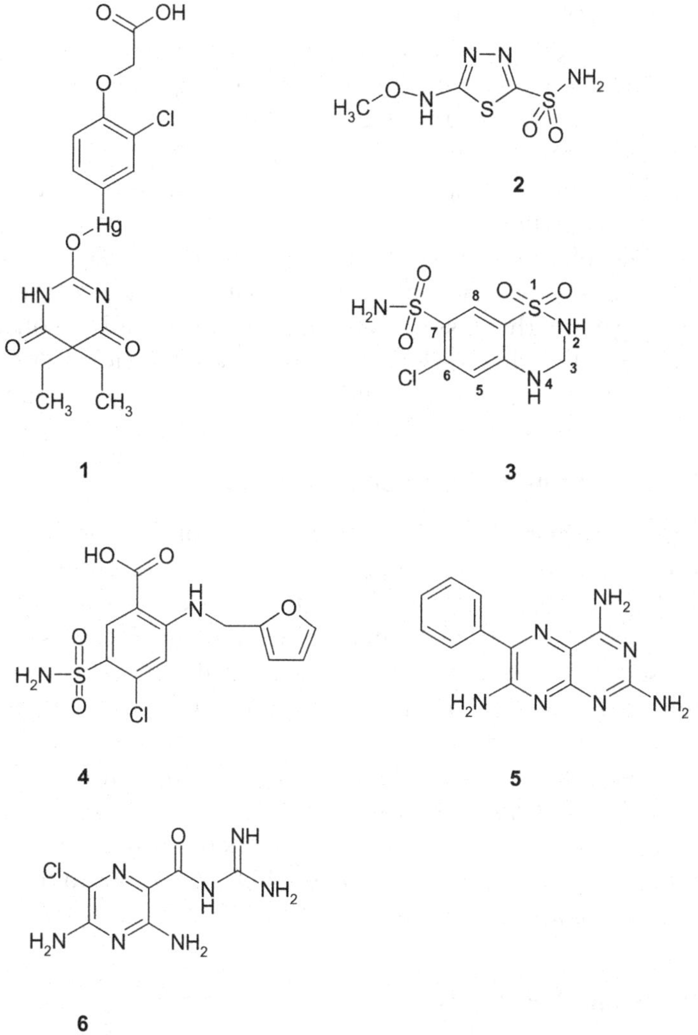

Figure 3.10: Structures of selected diuretic agents: merbaphen (**1**), acetazolamide (**2**), hydrochlorothiazide (**3**), furosemide (**4**), triamterene (**5**), and amiloride (**6**).

amide residue at position 7 and either a chlorine atom or a trifluoromethyl function at position 6. Because these analytes and most of the structurally different diuretic agents are not suitable for GC without adequate derivatization, studies regarding EI mass spectra of underivatized diuretics are rare; however, a comprehensive study on dissociation routes of selected diuretic agents using high resolution MS was published in 1987,[167] which supported the interpretation of EI mass spectra of methylated derivatives of a series of diuretics, and, in combination with trideuteromethylation, substantiated proposed fragmentation pathways.[168] Permethylated hydrochlorothiazide generates an intense $M^{+\cdot}$ at *m/z* 353 (monoisotopic mass) upon EI (Fig. 3.11a) that was suggested to eliminate the methyl methyleneamine (–43 Da) to yield the abundant fragment ion at *m/z* 310. Here, the loss of N-2 is proposed to produce a four-member ring structure that subsequently releases sulfur dioxide (–64 Da), giving rise to *m/z* 246 as illustrated in Scheme 3.10a.[167,168] Moreover, the elimination of a dimethylamine radical is suggested, generating the fragment ion at *m/z* 202. Alternatively, a dissociation route starting with the loss of a hydrogen atom and a consecutive release of sulfur dioxide (–64 Da) was postulated that yields a benzimidazole nucleus, resulting in the abundant ion at *m/z* 288 (Scheme 3.10a).

3.5.2 Benzoic Acid-Derived Loop Diuretics

Loop diuretics, which act at the loop of Henle of the nephron, belong to the most potent diuresis-stimulating agents and commonly comprise a substituted benzoic acid core structure. Typical representatives are furosemide (Fig. 3.10, **4**) and bumetanide, and the EI mass spectrum and proposed dissociation routes of the latter after permethylation are illustrated in Figure 3.11b and Scheme 3.10b,[167,168] respectively. An abundant molecular ion is detected at *m/z* 406, which is suggested to eliminate a proply radical (–43 Da) from the nitrogen-linked n-butyl residue by means of an α-cleavage, and the release of dimethylamine (–45 Da) from the sulfonamide residue yields the fragment at *m/z* 318. The subsequent loss of sulfur dioxide (–64 Da) is proposed to generate the characteristic ion at *m/z* 254.

3.5.3 Potassium-Sparing Diuretics

In contrast to hydrochlorothiazide and bumetanide, triamterene represents a potassium-sparing diuretic agent with a pteridine nucleus and no sulfonamide function. Its six-fold methylated derivative preferably eliminates a hydrogen radical from its molecular ion at *m/z* 337 to form the base peak at *m/z* 336 (Fig. 3.11c). Additionally, the loss of a methyl radical (–15 Da), which results in *m/z* 322, and the consecutive release of methyl methyleneamine (–43 Da) and a methyl residue, were suggested to form the fragments at *m/z* 294 and 279, respectively (Scheme 3.10c).[167,168]

Due to the enormous physicochemical heterogeneity of diuretic agents, no common characteristic fragmentation pathway has been established using EI-MS (Table 3.5). Only the sub-class of thiazides was

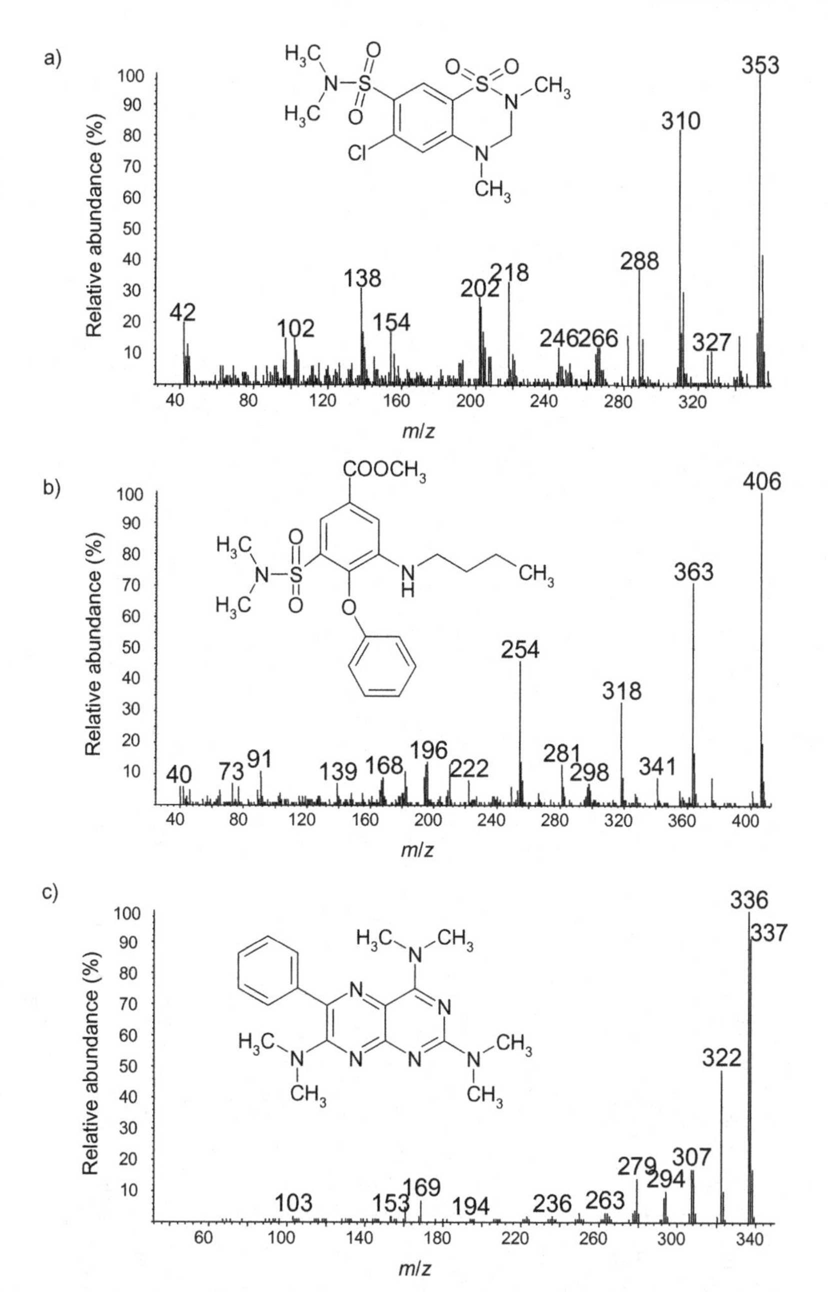

Figure 3.11: EI mass spectra of permethylated diuretic agents: (a) hydrochlorothiazide (mol wt = 353), (b) bumetanide (mol wt = 406), and (c) triamterene (mol wt = 337).

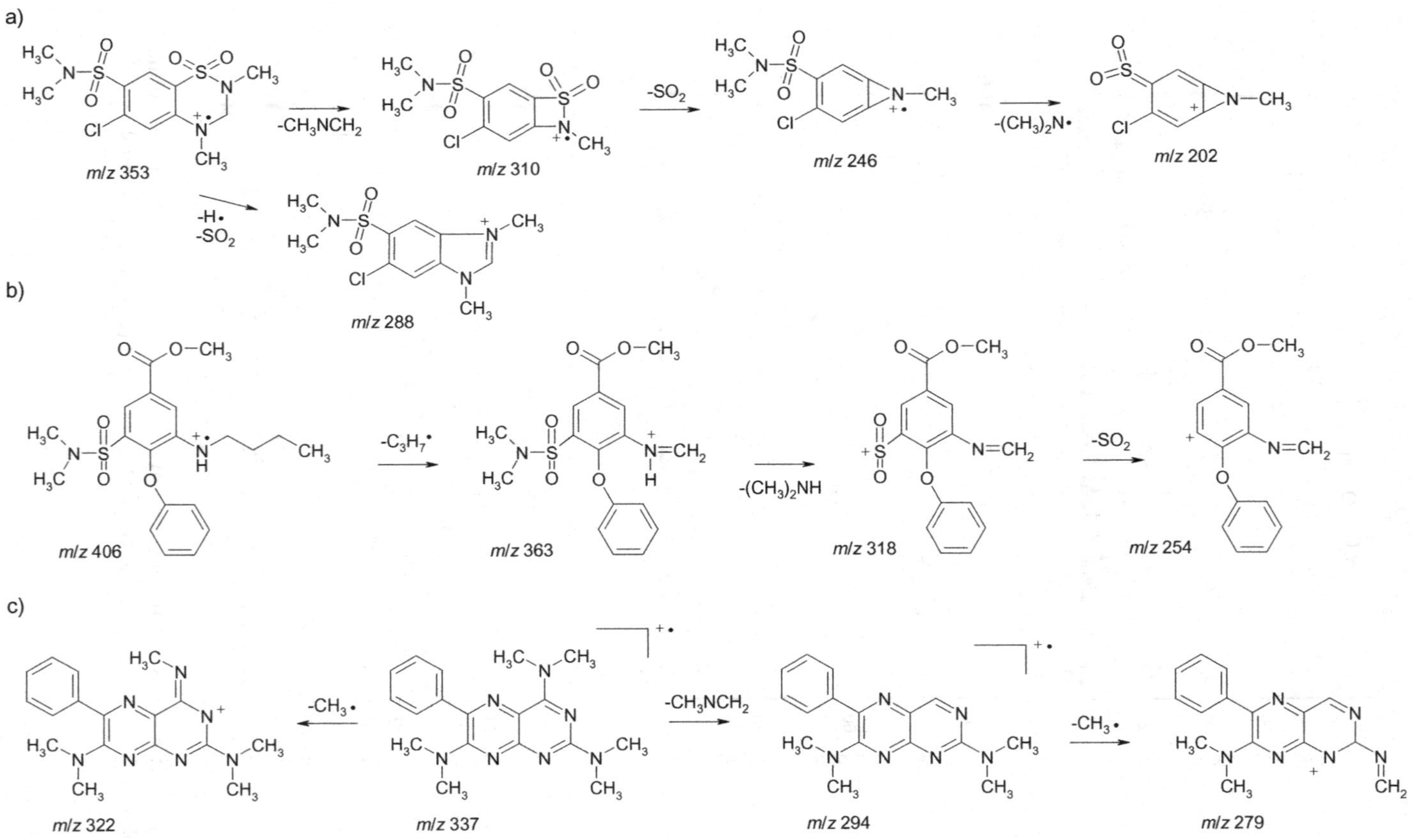

a)

m/z 353 → $-CH_3NCH_2$ → *m/z* 310 → $-SO_2$ → *m/z* 246 → $-(CH_3)_2N\bullet$ → *m/z* 202

m/z 353 → $-H\bullet$, $-SO_2$ → *m/z* 288

b)

m/z 406 → $-C_3H_7^\bullet$ → *m/z* 363 → $-(CH_3)_2NH$ → *m/z* 318 → $-SO_2$ → *m/z* 254

c)

m/z 322 ← $-CH_3\bullet$ ← *m/z* 337 → $-CH_3NCH_2$ → *m/z* 294 → $-CH_3\bullet$ → *m/z* 279

Scheme 3.10: Suggested dissociation routes of three diuretic agents: (a) hydrochlorothiazide, (b) bumetanide, and (c) triamterene.[167,168]

TABLE 3.5: Characteristic Fragment Ions of Selected Permethylated Diuretics Using Electron Ionization

Compound	Mol wt (Da)	Chemical Class	Fragment Ions (*m/z*)			
Acetazolamide-tris-CH_3	264	arylsulfonamide	249[a]	108	83	43
Chlorthalidone-tetrakis-CH_3	394	benzene-sulfonamide	363[a]	287	255	176
Ethacrynic acid-methyl ester	316	phenoxyacetic acid	316	281	261[a]	243
Triamterene-hexakis-CH_3	337	pteridine	336[a]	322	307	294
Clopamide-bis-CH_3	373	sulfamoyl-benzamide	139	127	111[a]	55
Indapamide-tris-CH_3	407	sulfamoyl-benzamide	246	161[a]	132	91
Bumetanide-tris-CH_3	406	sulfamoyl benzoic acid	406[a]	363	318	254
Furosemide-tris-CH_3	372	sulfamoyl benzoic acid	372	96	81[a]	53
Piretanide-tris-CH_3	404	sulfamoyl benzoic acid	404	295[a]	266	219
Xipamide-tetrakis-CH_3	410	sulfamoylbenzoyl-aniline	410[a]	379	290	121
Althiazide-tetrakis-CH_3	440	thiazide	352[a]	244	145	42
Bendroflumethiazide-tetrakis-CH_3	477	thiazide	386[a]	278	91	42
Buthiazide-tetrakis-CH_3	409	thiazide	352[a]	244	42	—
Chlorothiazide-tris-CH_3	337	thiazide	337[a]	245	339	230
Hydrochlorothiazide-tetrakis-CH_3	353	thiazide	353	310[a]	288	218
Hydroflumethiazide-tetrakis-CH_3	387	thiazide	387[a]	344	252	172
Methychlothiazide-tris-CH_3	402	thiazide	352[a]	244	—	—
Polythiazide-tris-CH_3	481	thiazide	352[a]	244	42	—
Trichlormethiazide-tetrakis-CH_3	435	thiazide	352[a]	244	42	—

[a] Base peak.

shown to produce a fragment ion at *m/z* 352 that is considered characteristic for C-3-substituted benzothiadiazine derivatives, although not observed in EI spectra generated from permethylated bendroflumethiazide or chlorothiazide (Table 3.5). Its structure is suggested to originate from an ionization of N-4 (in accordance to the ionization of hydrochlorothiazide, Scheme 3.10a) and a subsequent elimination of the C-3-linked side chain as a radical leaving group. Hence, thiazides with a 6-chloro-7-sulfamoyl-1,2,4-benzothiadiazine 1,1-dioxide nucleus (e.g., althiazide or buthiazide) produce the common fragment ion at *m/z* 352, as summarized in Table 3.5. Still, the influence of alterations on ionization sites, and thus on charge-driven as well as charge-remote dissociation routes, is high and does not allow the determination of more structure-specific fragment ions as observed, for instance, by stimulants or steroidal drugs.

3.6 β_2-AGONISTS

The search for potent beta-adrenoceptor agonists (β_2-agonists) for the treatment of asthma or chronic bronchitis has a long history and presumably started with the first reports on dyspnea in Chinese textbooks from 2600 B.C.[169] The isolation and characterization of adrenaline in the late 19th century followed by the observation that catecholamines such as adrenaline (Fig. 3.12, **1**) alleviate symptoms of asthma initiated a comprehensive quest for analogues that were particularly useful for the correction of conditions thought to cause asthmatic seizures. Numerous synthetic drugs were developed such as isoprenaline (isoproterenol, 1943, Fig. 3.12, **2**), fenoterol (1962, Fig. 3.12, **3**),[170] albuterol (salbutamol, 1966, Fig. 3.12, **4**),[171] and clenbuterol (1966, Fig. 3.12, **5**),[172,173] to specifically target the beta-2-adrenergic receptor (β_2-receptor). In 1992, the class of β_2-agonists was added to the list of prohibited compounds due to assumed performance-enhancing effects as concluded from animal experiments in particular for salbutamol and clenbuterol,[174–177] and numerous studies were conducted to detect these agents by means of MS-based approaches. Due to their rather poor GC properties, various sophisticated derivatization techniques were developed, yielding a series of chemically modified analytes. The employed strategies included trimethylsilylation, different *N*- and/or *O*-acylation,[178–180] intramolecular cyclization by means of formaldehyde,[181] chloromethyldimethylchlorosilane,[182] or methylboronic acid,[183] which yielded tetrahydroisoquinolines, 2-(dimethyl)-silamorpholines, and 5-phenyl-[1,3,2]oxazaborolidines, respectively, and mixed derivatives. Due to the comprehensiveness and heterogeneity of derivatized β_2-agonists, no summary is given, but more details can be found in articles of Polettini[178] as well as Damasceno and colleagues.[179]

The fragmentation of clenbuterol by EI is shown as an example for β_2-agonists derivatized by trimethylsilylation (Fig. 3.13a) and cyclization by means of methylboronic acid (Fig. 3.13b). Both derivatives yield informative and characteristic ions, and routes of formation were suggested as illustrated in Scheme 3.11.[183–185] Electron ionization can occur at different sites of clenbuterol and, thus, entail different fragmentation pathways. The most abundant fragment ion the EI mass spectrum of the bis-TMS derivative of clenbuterol is found at *m/z* 86, which is assumed to represent a typical α-cleavage yielding the *tert*.-butyl methylenamine cation (Scheme 3.11a). Further to this intense ion, characteristic fragments are observed at *m/z* 210, 227, and 300, which were all demonstrated to result from *m/z* 335 in MS^n and deuterium labeling experiments.[184,185] The precursor ion at *m/z* 420 is proposed to eliminate

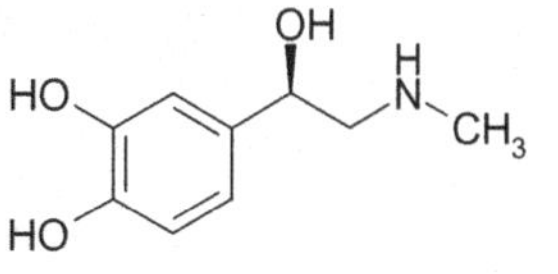

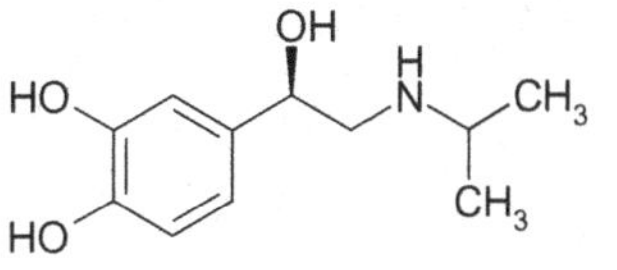

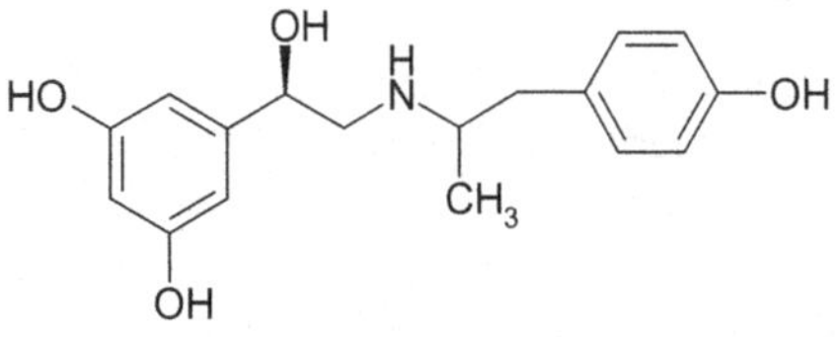

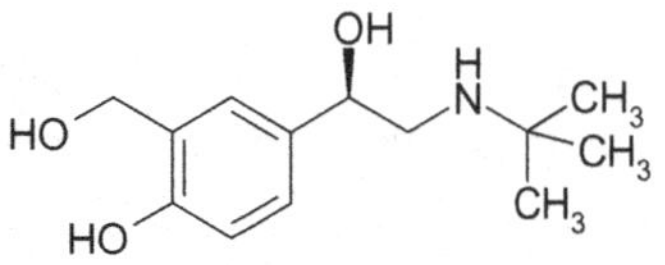

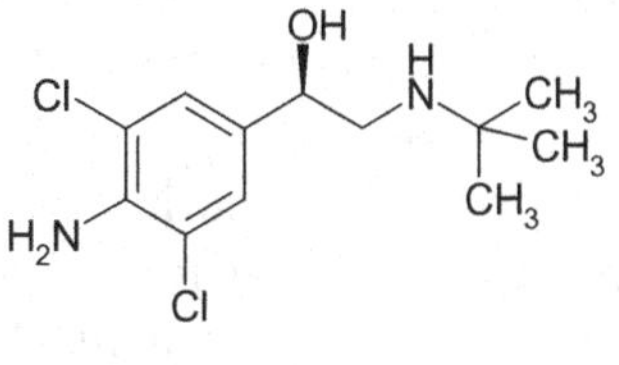

Figure 3.12: Structures of selected adrenoceptor agonists: adrenaline (**1**), isoprenaline (**2**), fenoterol (**3**), salbutamol (**4**), and clenbuterol (**5**).

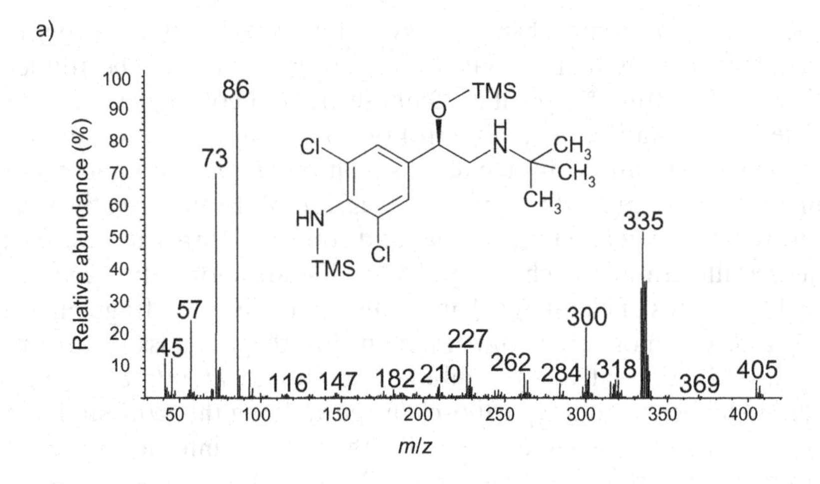

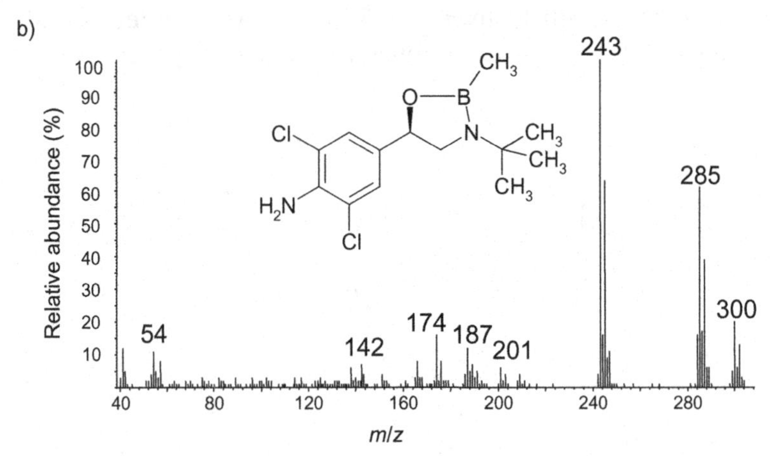

Figure 3.13: EI mass spectra of derivatized clenbuterol: (a) clenbuterol-bis-TMS (mol wt = 420), and (b) clenbuterol methylboronate-derivative (mol wt = 300).

tert.-butyl methylenamine (–85 Da) to produce *m/z* 335 that subsequently releases a chlorine radical (–35 Da) or trimethylchlorosilane (–108 Da) to give rise to *m/z* 300 and 227, respectively. The former of both further dissociates by the loss of methane (–16 Da) or trimethylsilanol (–90 Da) forming the fragments at *m/z* 284 and 210.

In contrast to the bis-TMS derivative of clenbuterol, its methyl boronate analogue generates a comparably abundant molecular ion at *m/z* 300 and few intense fragments[182] derived from the loss of a methyl radical (–15 Da) and methyl oxoboron (–42 Da) at *m/z* 285 and 243, respectively. Evidence for their composition was obtained by HRMS,[183] and a suggested dissociation route is illustrated in Scheme 3.11b.

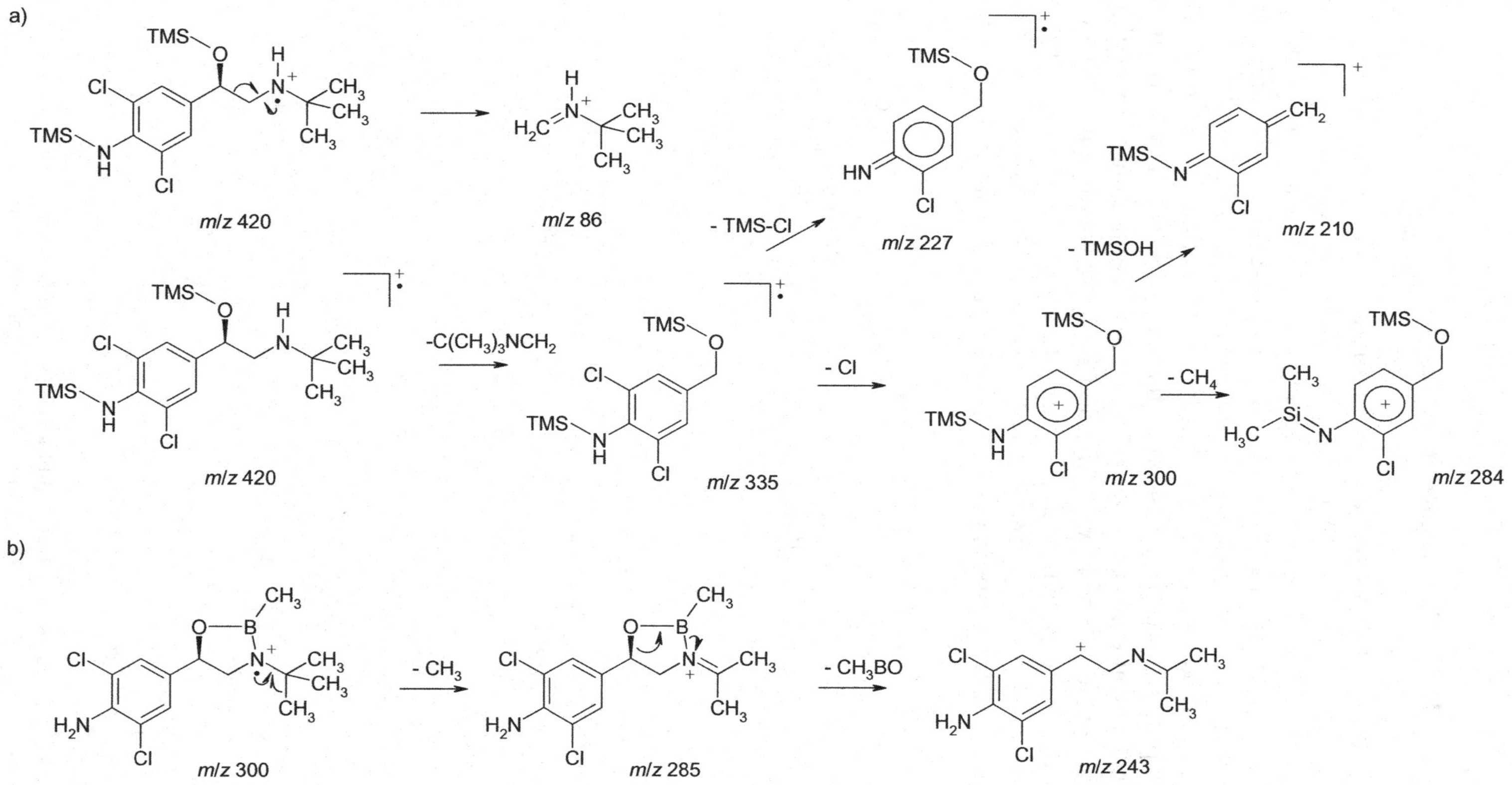

a)

TMS–O, H, N+, CH3, CH3, CH3, Cl, TMS–N–H, Cl

m/z 420

→ $H_2C{=}N^+(H)C(CH_3)_3$

m/z 86

- TMS-Cl →

HN, TMS–O, Cl

m/z 227

- TMSOH →

TMS–N, CH2, Cl

m/z 210

TMS–O, H, N, CH3, CH3, CH3, Cl, TMS–N–H, Cl

m/z 420

-$C(CH_3)_3NCH_2$ →

TMS–O, Cl, TMS–N–H, Cl

m/z 335

- Cl →

TMS–O, TMS–N–H, +, Cl

m/z 300

- CH_4 →

TMS–O, CH3, H_3C–Si=N, +, Cl

m/z 284

b)

CH3, O–B, N+, CH3, CH3, CH3, Cl, H_2N, Cl

m/z 300

- CH_3 →

CH3, O–B, N+, CH3, CH3, Cl, H_2N, Cl

m/z 285

- CH_3BO →

Cl, H_2N, Cl, +, N, CH3, CH3

m/z 243

Scheme 3.11: Proposed fragmentation routes of different derivatives of clenbuterol: (a) clenbuterol-bis-TMS, and (b) clenbuterol methylboronate.[183–185]

3.7 β-RECEPTOR BLOCKING AGENTS

Structurally related to the above mentioned β_2-agonists, β-receptor blocking agents have been developed for the treatment of various disease conditions such as *angina pectoris*, hypertension, and cardiac arrhythmia.[186] The first agent recognized to possess β-receptor antagonistic activity was dichloroisoprenaline (Fig. 3.14, **1**) in 1958,[187] which was never marketed but substituted by the first clinically approved β-blocker propranolol (Fig. 3.14, **2**) in the early 1960s.[188] It was derived from the typical core structure of β_2-agonists containing an arylethanolamine nucleus, which was extended by an oxymethylene unit that caused the β-receptor blocking nature of various compounds synthesized accordingly. While propranolol proved to be a non-selective β-blocker, i.e., targeting both β_1- and β_2-adrenergic receptors, other drugs

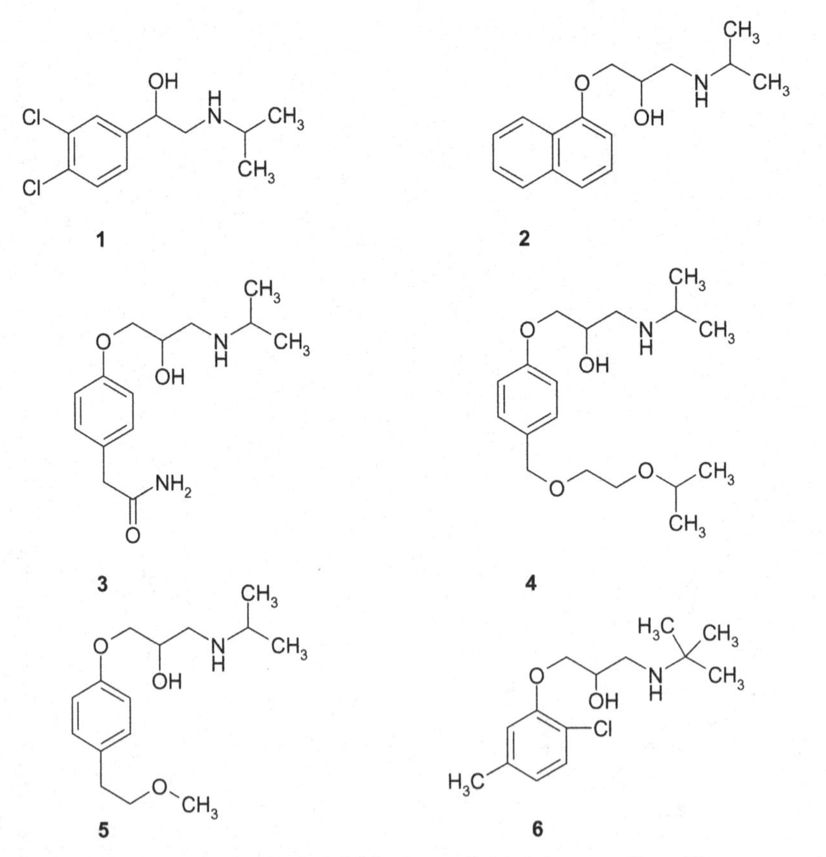

Figure 3.14: Structures of selected β-blockers: dichloroisoprenaline (**1**), propranolol (**2**), atenolol (**3**), bisoprolol (**4**), metoprolol (**5**), and bupranolol (**6**).

such as atenolol (Fig. 3.14, **3**),[189] bisoprolol (Fig. 3.14, **4**),[190] or metoprolol (Fig. 3.14, **5**)[191] were found to be β_1-receptors specific. Further to these, other β-receptor blocking agents comprising the phenylethanolamine core of β_2-agonists were prepared (e.g., nifenalol and sotalol, Table 3.6) but proved to be the less potent agents.

Although not beneficial for numerous sports, β-blockers have been recognized to be performance-enhancing in archery and shooting through their antagonistic properties that inhibit the normally adrenaline-mediated sympathetic actions. Consequently, symptoms of anxiety

TABLE 3.6: Characteristic Fragment Ions of Selected β-blockers after Derivatization Using EI[a]

Compound	Derivative	Mol wt (Da)	Diagnostic (Fragment) Ions (*m/z*)		
Acebutolol	4	576	561	284	129
Alprenolol	3	417	402	284	129
Befunolol	4	559	544	284	129
Betaxolol	3	475	460	284	129
Bisoprolol	3	493	332	284	129
Bucumolol	1	377	362	233	86
Bufetolol	1	395	380	279	86
Bufuralol	1	333	318	247	86
Bunitrolol	1	320	305	176	86
Bupranolol	1	343	328	227	86
Butofilolol	2	455	435	368	86
Carazolol	3	466	368	284	129
Carteolol	2	436	421	235	86
Cloranolol	1	363	348	292	86
Esmolol	3	463	448	284	129
Indenolol	3	415	325	284	129
Mepindolol	3	430	430	284	129
Metipranolol	3	507	507	284	129
Metoprolol	3	435	420	284	129
Moprolol	3	407	407	284	129
Nifenalol	3	392	377	224	194
Oxprenolol	3	433	433	284	129
Penbutolol	1	363	348	101	86
Pindolol	3	416	254	284	129
Propranolol	3	427	427	284	129
Sotalol	4	512	497	344	272
Timolol	1	388	373	186	86
Toliprolol	3	391	376	284	133

1 = *O*-TMS; 2 = bis-*O*-TMS; 3 = *N*-TFA—*O*-TMS; 4 = *N*-TFA—bis-*O*-TMS.

[a] Refs. 54 and 232.

are reduced, and the decreased heart rate as well as tremor might allow a better targeting. Since 1988, β-blockers have been prohibited in sports, and detection methods commonly based on selective derivatization strategies were applied to uncover the misuse of these drugs.[46,51,53,192] While hydroxyl functions were usually trimethylsilylated, the amino residue was acylated using several different moieties such as acetyl,[193] trifluoroacetyl (TFA),[194] pentafluoropropionyl (PFP),[195] or heptafluorobutyl (HFB)[196] groups. In addition, the formation of cyclic boronates as reported for β_2-agonists (*vide supra*) was described.[197] The EI mass spectra of propranolol- and metoprolol-*N*-TFA-*O*-TMS are depicted in Fig. 3.15 a and b, both of which contain abundant signals at *m/z* 284 and 129 that are highly diagnostic for the derivatized side chain of β-blockers comprising an oxypropanolamine core with terminal isopropyl function as shown in Table 3.6 with a variety of related substances. Accordingly, analogues bearing a terminal *tert.*-butyl residue such as bupranolol (Fig. 3.14, **6**) are not *N*-trifluoroacetylated but still *O*-trimethylsilylated and yield a characteristic fragment ion at *m/z* 86 (Fig. 3.15c, Table 3.6).

The fragmentation routes to the most common and abundant product ions with diagnostic properties are illustrated in Scheme 3.12[194] using propranolol-*N*-TFA-*O*-TMS as an example. In contrast to several of the above reported compounds, the most prominent fragment ion observed with various derivatives of β-blockers is not derived from an α-cleavage but from a fission of the alkyl side chain at the ether linkage. An initial ionization at the nitrogen atom was suggested to cause the formation of a four-member ring structure[194] that eliminates the substituted aryl nucleus and yields the common fragment ion at *m/z* 284 in case of *N*-TFA-*O*-TMS derivatized β-blockers. Supporting information for this proposal was obtained from *N*,*O*-bis-TFA analogues as well as deuterium-labeling experiments in various studies,[198,199] which gave rise to corresponding fragment ions of identical origin but different elemental composition and further outlined the subsequent release of propene (–42 Da) to yield the common ion at *m/z* 242 (Fig. 3.15 a and b). Moreover, *m/z* 284 was found to produce the fragment at *m/z* 129 in MS/MS experiments, which necessitates the neutral loss of 2,2,2-trifluoro-*N*-isopropyl-acetamide (–155 Da) giving rise to the trimethylsilylated propen-2-ol cation. The low abundant ion at *m/z* 168 was proposed to originate from the molecular ion as a result of a typical α-cleavage, the corresponding counterpart of which was found at *m/z* 72 in case of underivatized analytes[200–202] as well as *m/z* 86 in spectra of those β-blockers comprising a *N*-linked *tert.*-butyl residue such as bupranolol (Fig. 3.15c, Table 3.6).[193] The fact that numerous β-receptor

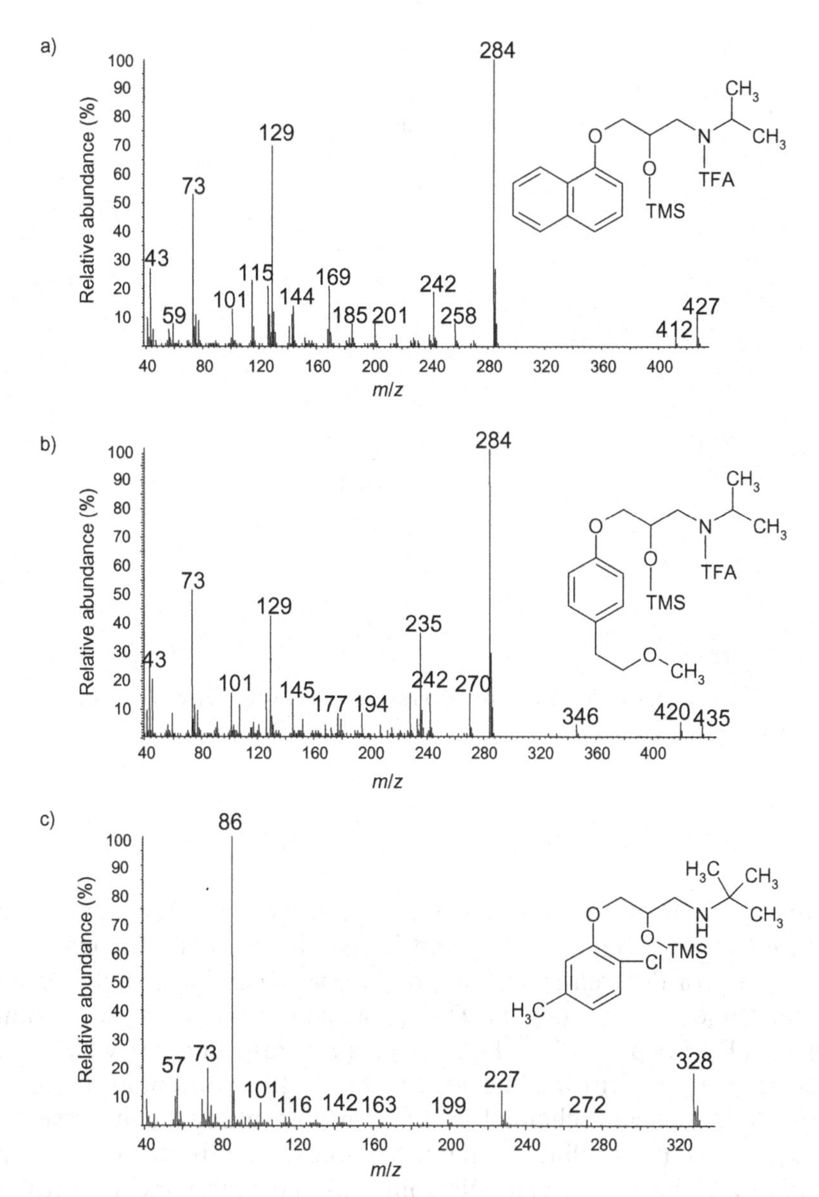

Figure 3.15: EI mass spectra of selected β-blockers: (a) propranolol-*N*-TFA-*O*-TMS (mol wt = 427), (b) metoprolol-*N*-TFA-*O*-TMS (mol wt = 435), and (c) bupranolol-*O*-TMS (mol wt = 343).

blocking agents bear these structural features that lead to highly similar EI mass spectra with characteristic fragment ions resulted in various screening approaches focused on these target ions, which allowed a comprehensive coverage of β-blockers (see Chapter 6).

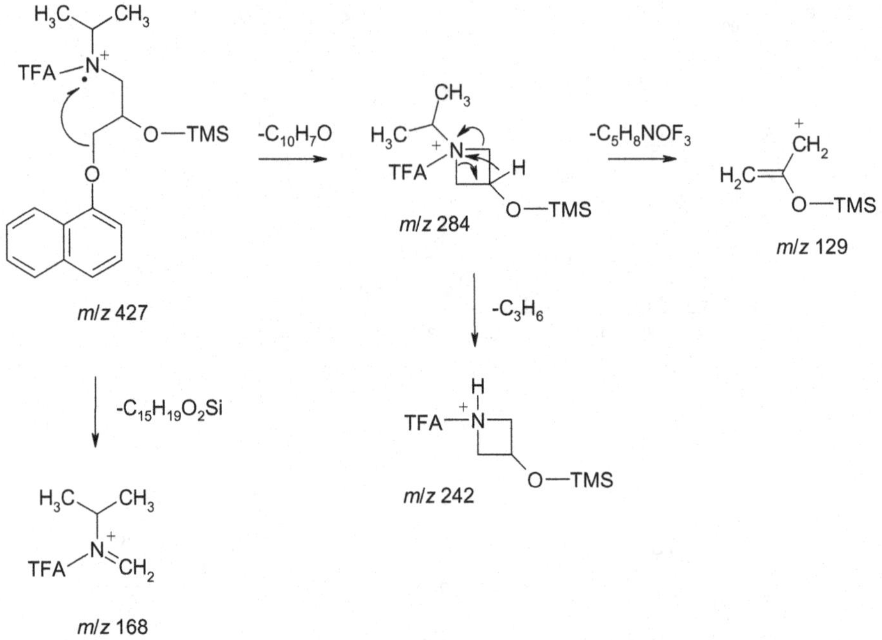

Scheme 3.12: Proposed fragmentation routes of selectively derivatized propranolol-*N*-TFA-*O*-TMS.[194]

3.8 CALCIUM-CHANNEL MODULATORS (RYCALS)

In the early 1990s, the cardioprotective effects of 1,4-benzothiazepine derivatives were reported[203,204] and initiated the search for lead drug candidates for the treatment and prevention of cardiac arrhythmia and related sudden heart deaths. Two promising substances were found termed JTV-519 and S-107 (Fig. 3.16, **1** and **2**, respectively), which have been studied in various clinical trials due to their potential to correct intracellular calcium-channel leaking and resulting health issues.[205] These rycals (ryanodine-calstabin complex stabilizers) restore the decreased affinity between calstabin and the ryanodine receptor in calstabin-depleted ryanodine-based calcium channels, which play an essential role in the modulation of cytosolic calcium concentrations in muscle cells and, thus, in corresponding muscle contractions. The recognition of the influence of rycals also on skeletal muscle and their potential to counteract exercise-induced fatigue[206] has initiated consideration of these agents also with regard to doping controls, although they are currently not banned by WADA. Initial studies on the mass spectrometric behavior of S-107 and its putative metabolite (Figure 3.16, **2** and **3**) were conducted, which is particularly due to its GC-

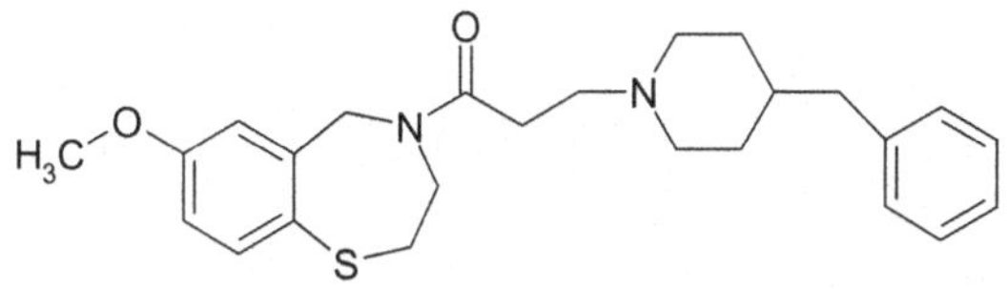

Figure 3.16: Structures of the rycals JTV-519 (**1**), S-107 (**2**), and its desmethylated analogue (**3**).

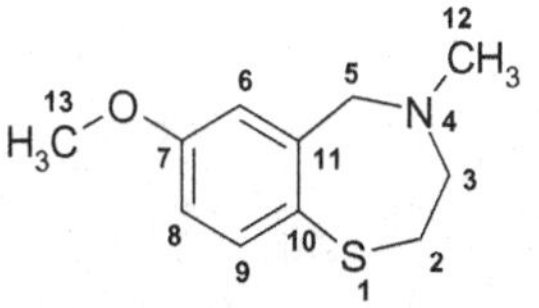

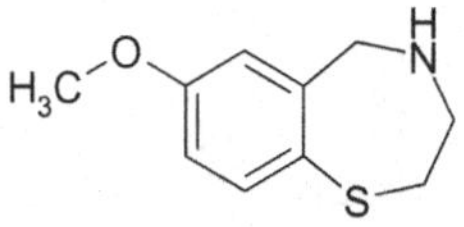

compatible physicochemical properties in contrast to JTV-519. The EI mass spectrum of S-107 contains an abundant molecular ion at *m/z* 209, and further characteristic fragment ions for instance at *m/z* 208 (–1 Da), 194 (–15 Da), 181 (–28 Da), and 151 (–58 Da) as shown in Figure 3.17a. Since EI is likely to occur at the sulfur atom yielding the molecular ion at *m/z* 209, several different dissociation pathways are feasible, triggered either by the loss of a hydrogen atom (–1 Da), a methyl radical (–15 Da) or ethylene (–28 Da) giving rise to the above mentioned fragment ions at *m/z* 208, 194 and 181, respectively.[207] The elimination of a hydrogen atom can occur at different positions in the molecule; however, C-5 and C-12 were shown to be unlikely in deuterium labeling experiments, but the release of a hydrogen atom from C-2 or any of the carbons comprising the aromatic ring would allow delocalizing the introduced charge. A major product ion derived from *m/z* 208 was found at *m/z* 151, which was suggested to result from the loss of 1-methyl-aziridine (–57 Da) yielding the cation of 3-methoxy-7-thia-bicyclo[4.2.0]octa-1,3,5-triene (Scheme 3.13a). In addition to the fragment ion at *m/z* 151, another abundant product ion was observed at *m/z* 180, which was proposed to resemble 5-methoxy-2-methyl-2,3-dihydro-1,2-benzisothiazole as a result of the loss of ethylene (–28 Da, Scheme 3.13a).[207]

The elimination of a methyl residue (–15 Da) was studied in great detail and required several deuterium labeling experiments to reveal the fact that the initial loss of CH_3▪ did not include C-12 or C-13 but must result from the 7-member ring structure, which presumably forms a new 6-member ring structure representing 6-methoxy-3-methyl-3,4-

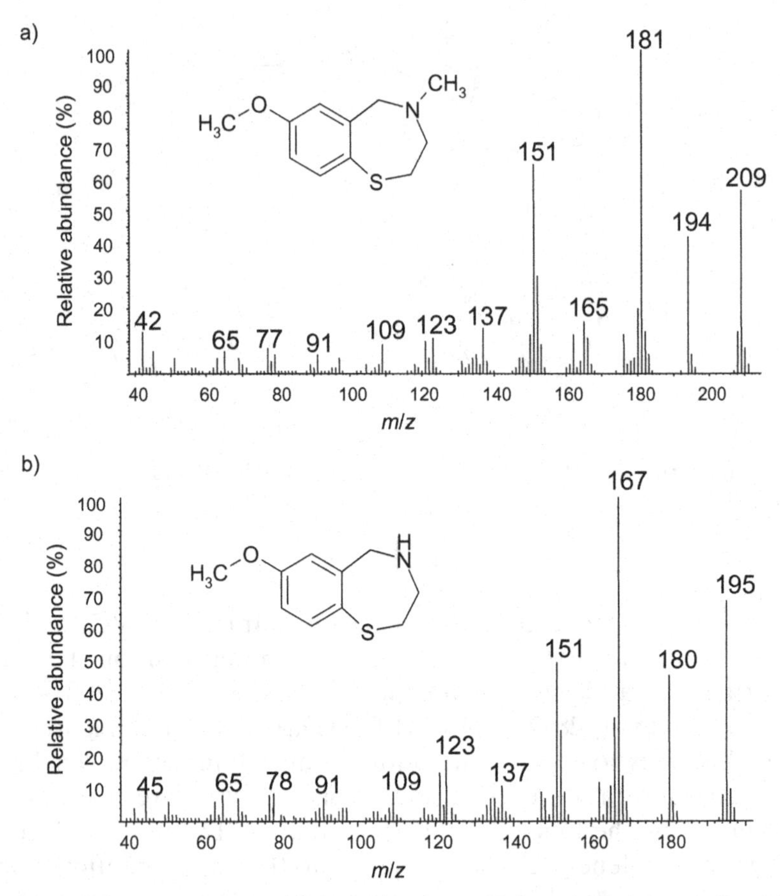

Figure 3.17: EI mass spectra of (a) S-107 (mol wt = 209), and (b) desmethylated S-107 (mol wt = 195).

dihydro-2H-1,3-benzothiazine as suggested in Scheme 3.13b. Here, the release of the methyl function is suggested to include an intermediate ring opening, which rearranges to the proposed 1,3-benzothiazine nucleus due to the aligned orientation of the adjacently located substituents. In addition, the loss of 41 Da from *m/z* 194 was observed to yield the fragment ion at *m/z* 153, which required the elimination of methyl isocyanide (CH_3NC), which was found possible from both suggested structures of *m/z* 194, accompanied by the migration of the hydrogen atom located at C-5.[207]

The initial loss of ethylene (–28 Da) from the molecular ion was proposed to include the carbons C-2 and C-3 (Scheme 3.13c) to generate a new bicyclic structure of the radical cation of 5-methoxy-2-

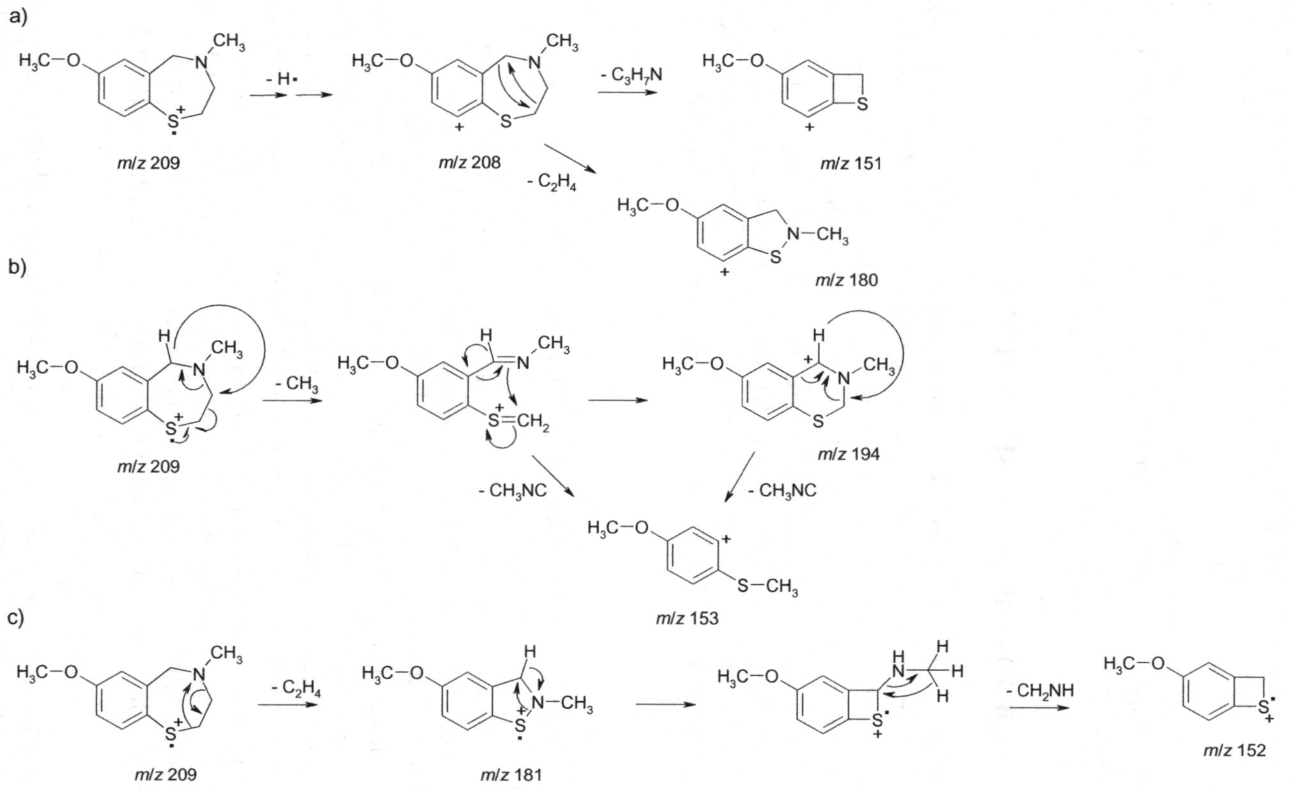

Scheme 3.13: Suggested dissociation routes of S-107 after EI.

methyl-2,3-dihydro-1,2-benzisothiazole (*m/z* 181), which subsequently eliminated methyleneamine (–29 Da) to form a fragment ion at *m/z* 152 (Fig. 3.17a). Here, the migration of a hydrogen atom from C-5 to the adjacent nitrogen accompanied by the intermediate formation of 3-methoxy-7-thia-bicyclo[4.2.0]octa-1,3,5-trien-8-yl)-methylamine was suggested, which was followed by the release of methyleneamine that requires a hydrogen shift from the *N*-linked methyl function to C-5 (Scheme 3.13c).[207]

3.9 CARBOHYDRATE-BASED AGENTS

Only a few drugs based on carbohydrate structures have been considered relevant for doping controls, including mannitol (Fig. 3.18, **1**) as diuretic agent, polysaccharides such as dextran (Fig. 3.18, **3**) and hydroxyethyl starch (Fig. 3.18, **4**) as plasma volume expanders, and glycerol (Fig. 3.18, **2**) as a hyperhydration drug. Although not of low molecular weight, the polysaccharides will be discussed in this section as the analytical approaches and respective mass spectrometric studies are very similar to those employed for mannitol and glycerol.

3.9.1 Mannitol

Mannitol ([2*R*,3*R*,4*R*,5*R*]-hexane-1,2,3,4,5,6-hexol) has been employed primarily for its diuretic properties in cases of acute renal failure, and elevated intracranial or intra-ocular pressure. It is commonly administered parenterally at dosages of 50–200 g, which increase the osmotic pressure of the plasma and urine flow rates with negligible added energy value.[208] If applied intravenously, mannitol has been prohibited as diuretic and masking agent according to anti-doping regulations. Its detection and quantification based on mass spectrometric approaches has been necessary due to its natural occurrence, and different derivatization strategies including acetylation,[209,210] trimethylsilylation,[211] and n-butyl-diboronate formation[212] was used. The EI mass spectrum of the per-acetylated derivative of mannitol did not include the molecular ion (Fig. 3.19a) but abundant fragments resulting from cleavages of the carbon backbone and eliminations of acetic acid and ketene (Scheme 3.14).[209,210,219]

The ions at *m/z* 361 and 73 were considered as complementary products, resulting from fissions between C-1–C-2 or C-5–C-6 and retaining the charge at either of the breakdown products. The fragments at *m/z* 289 and 145 were suggested to be generated in a similar way, which necessitates the cleavage of C-2–C-3 or C-4–C-5 bonds, while the ion

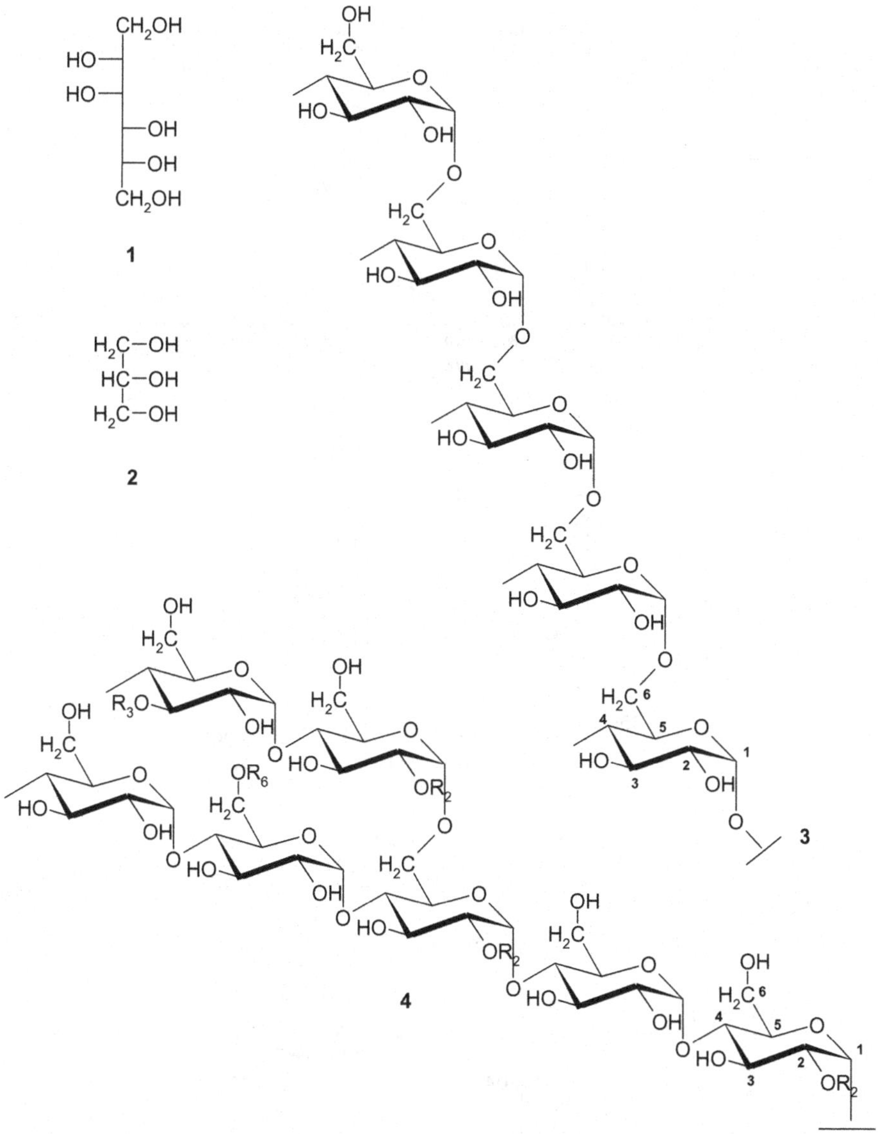

Figure 3.18: Structures of the carbohydrate-based drugs mannitol (**1**), glycerol (**2**), dextran (**3**), and hydroxyethyl starch (**4**), with R = CH_2CH_2OH.

at *m/z* 217 is most likely the product of the homolytic dissociation of the C-3–C-4 linkage as substantiated by accurate mass measurements (Scheme 3.14a).[209] Consecutive eliminations of acetic acid (–60 Da) and/or ketene (–42 Da) were proposed to yield additional abundant fragment ions, e.g., at *m/z* 259 (*m/z* 361-60-42), 187 (*m/z* 289-60-42), and 115 (*m/z* 217-60-42), which characterize peracetylated hexitols but not

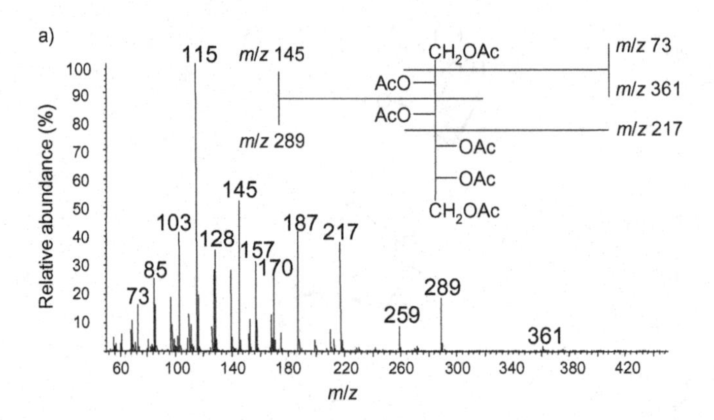

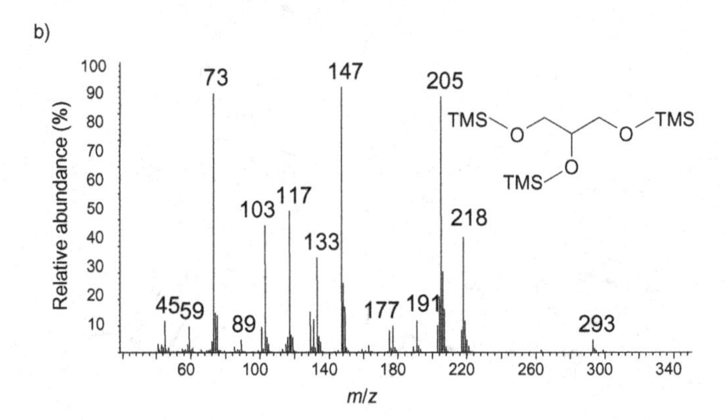

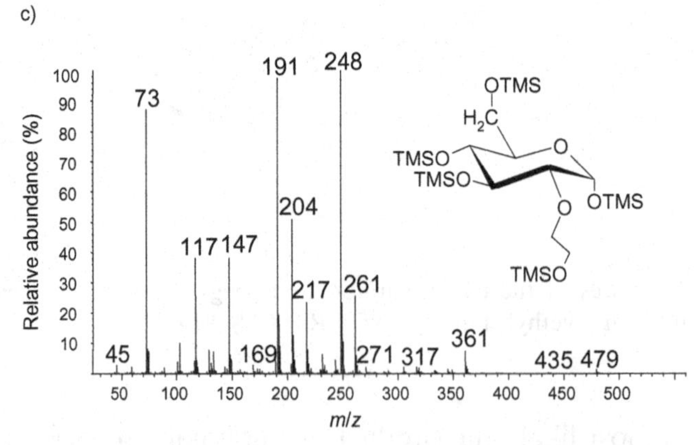

Figure 3.19: EI mass spectra of (a) peracetylated mannitol (mol wt = 434), (b) glycerol-tris-TMS (mol wt = 308), (c) 2-hydroxyethyl glucose-pentakis-TMS (mol wt = 584), and (d) partially methylated alditol acetate of 1,4-linked 2-hydroxyethylated glucose (mol wt = 394). (e) EI mass spectrum of partially methylated alditol acetate of 1,6-linked glucose (mol wt = 350) derived from dextran.

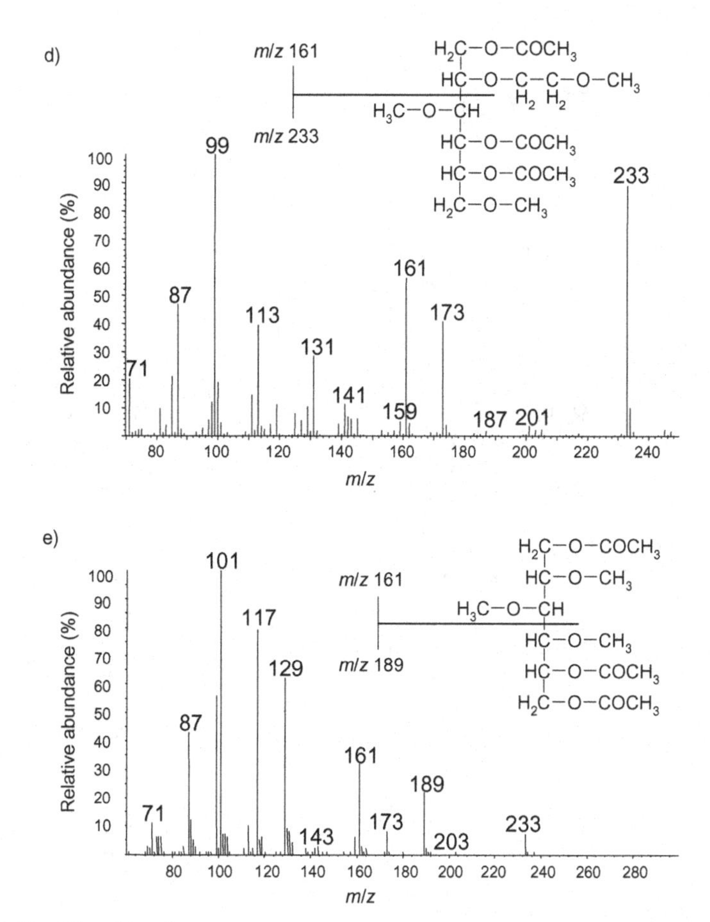

Figure 3.19: (*Continued*)

specifically the stereoisomer mannitol.[210] Chromatographic separation was still required to unambiguously identify and isolate the target analyte from analogues such as allitol, altritol, sorbitol, galactitol, and iditol (see Chapter 6).

3.9.2 Glycerol

Glycerol (1,2,3-propanetriol) represents the structural backbone of a variety of lipids, and occurs naturally in the human organism at serum levels of 4.6–27.6 μg/mL.[208] Its urinary concentrations have been considered negligible under healthy physiological conditions, but a significant correlation between urinary glycerol levels and oral or intravenous administrations were reported. Clinically, glycerol has

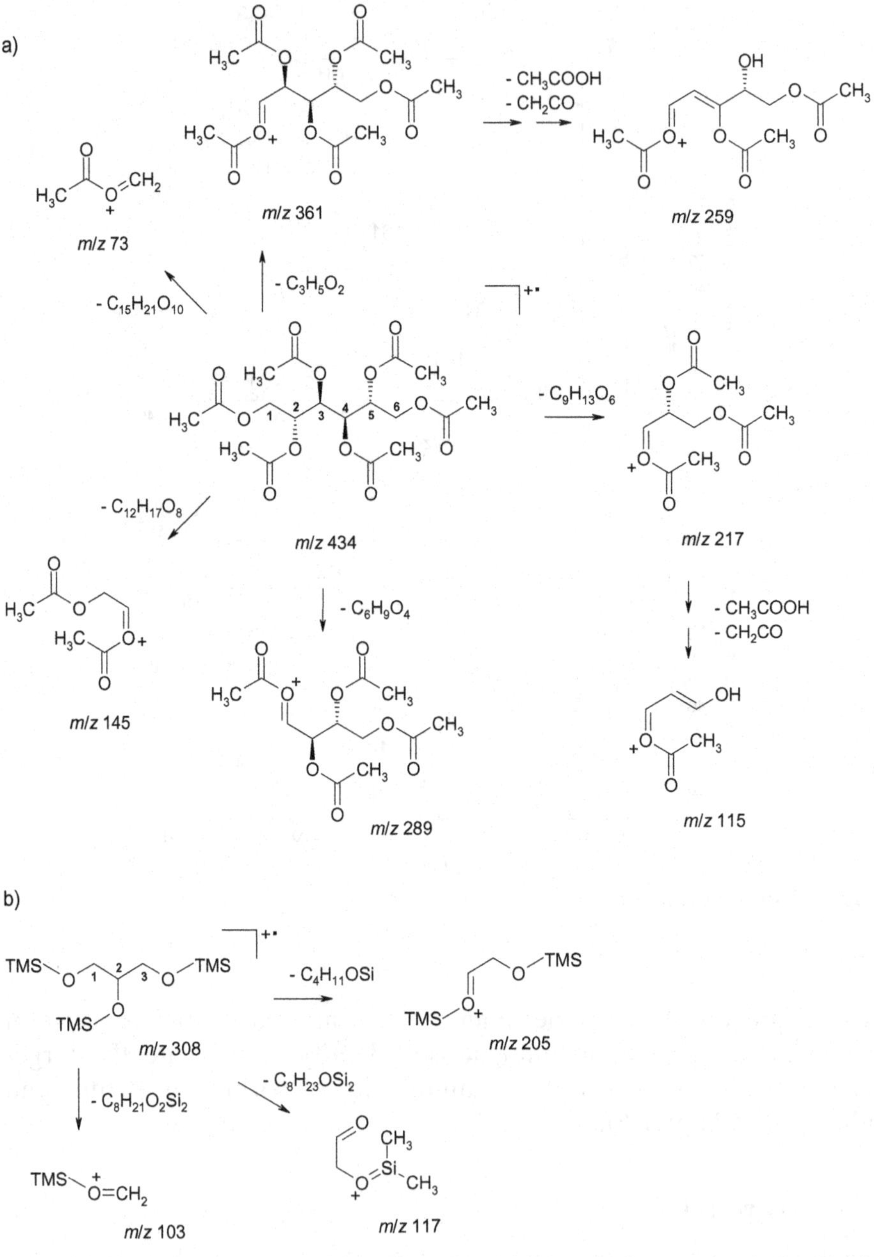

Scheme 3.14: Suggested dissociation pathways of (a) per-acetlyated mannitol, and (b) tris-TMS-glycerol after EI.[209,210,219]

been used to treat cerebral edema, glaucoma, or intracranial hypertension,[213–215] and hyperhydration that supposedly influences athletes' thermoregulation or cardiovascular strains was accomplished by means of oral co-administration of fluids and glycerol. Although yielding controversial results, several studies described the ergogenic effect of glycerol, e.g., in Olympic distance triathlon[216] or cycling,[217] and the use of glycerol has been considered relevant for sports drug testing. Various approaches to measure glycerol from different specimens were reported, and GC-MS using acetylation,[218] trimethylsilylation,[219] *tert.*-butyldimethylsilyl (tBDMS)[220,221] and heptafluorobutyryl (HFB)[222] derivatives was commonly employed. For doping control purposes, TMS-derivatization has been the method of choice due to the widespread use of trimethylsilylation, and a sufficient amount of characteristic fragment ions is obtained using EI (Fig. 3.19b). While the molecular ion (*m/z* 308) is not observed, the common fragments M^+-15 and M^+-90 are found at *m/z* 293 and 218, representing the elimination of a methyl radical and trimethylsilanol, respectively. The EI mass spectrum further contains typical ions at *m/z* 205, 117, and 103, which were identified to originate from dissociations comparable to those observed in case of acetylated mannitol, mainly resulting from C–C-bond fissions and neutral loss of TMSOH (Scheme 3.14b) as demonstrated using deuterium-labeled glycerol.[219]

3.9.3 Hydroxyethyl Starch and Dextran

Plasma volume expanders were studied for more than a century and included various compounds such as gum Arabic (also referred to as gum acacia), colloidal gelatin, polyvinylpyrrolidone, dextran, hydroxyethyl starch (HES), and albumin.[223,224] Initiated in the late 19th century,[225] first medicinal applications of plasma volume expanders/plasma volume substitutes were reported in 1915, when colloidal gelatin was employed for the treatment of hypovolemic shock.[226] In the following decades, plasma volume expander administration was further recommended for hemodilution, treatment of burns, management of disturbed capillary blood circulation, and for the cryoprotection of frozen stored erythrocytes. The most suitable synthetic alternatives to plasma or plasma fractions were found to be dextrans of various molecular weights (40,000–70,000 Da) and HES since 1945 and 1957, respectively. Due to their plasma volume expanding effect, these agents have been categorized as masking agents by WADA as artificially increased hematocrit and hemoglobin values can be corrected to levels below established thresholds.

Although the patent for the synthesis of HES from starch was filed as early as 1920, its first application as plasma volume expander was reported in 1957.[223] HES is based on α-1,4-linked glucose residues bearing hydroxyethyl functions at C-2, C-3, and/or C-6 (Fig. 3.18, **4**) with different degrees of substitution. As hydroxyethylated starch does not occur naturally, hydroxyethylated glucose as obtained from acidic hydrolysis of the polysaccharide was chosen as a target for screening purposes in sports drug testing. The per-TMS derivative of 2-hydroxyethylated glucose provided distinct and diagnostic fragment ions at *m/z* 248 and 261 in addition to fragments commonly observed in analyses of trimethylsilylated saccharides such as *m/z* 191, 204, and 217 (Fig. 3.19c). In accordance to acetylated or partially methylated glucose, the latter ions were proposed to comprise propene-1,3-diol (or, alternatively, cyclopropane-1,2-diol) and ethene-1,2-diol nuclei (Scheme 3.15a).[227,228] In the presence of hydroxyethyl residues, those fragments are incremented by 44 Da to yield the earlier mentioned characteristic ions at *m/z* 248 and 261.[227]

In addition to the detection of hydroxyethylated monosaccharides as per-TMS derivatives, partially methylated alditol acetates (PMAAs) were prepared from urinary HES to prove the polymeric structure and substantiate the application of the plasma volume expander. The intact polysaccharide is permethylated prior to acidic hydrolysis, and the obtained partially methylated monosaccharides are reduced to corresponding alditols, which are finally acetylated to allow the differentiation of monomers from polysaccharides using GC-MS. Distinct fragment ions that characterize particular features of PMAAs allow the localization of former linkages between monosaccharides as well as modifications of the monosaccharide units. In the case of PMAAs derived from 1,4-linked glucose and its 2-hydroxyethylated analogue, abundant fragment ions are obtained after EI resulting from C–C-bond cleavages and subsequent eliminations of ketene (–42 Da) and acetic acid (–60 Da). Ions at *m/z* 233, 173, and 99 were commonly found, while the fragment at *m/z* 117 of 1,4-linked glucose (data not shown), which represented the counterpart to *m/z* 233, was incremented by 44 Da to *m/z* 161 and yielded a diagnostic reporter ion for doping control purposes (Fig. 3.19d).[228] Locating the hydroxyethyl residue at C-3, the ion at *m/z* 233 was consequently incremented by 44 Da to generate *m/z* 277, providing further evidence for the presence of HES in sports drug testing urine samples. The suggested dissociation pathways of HES-derived PMAAs under EI are illustrated in Scheme 3.15b.[227,228] Most abundant fragment ions are obtained from the fission of the C-2–C-3 bond yielding the complementary ions at *m/z* 161 and 233 in case of

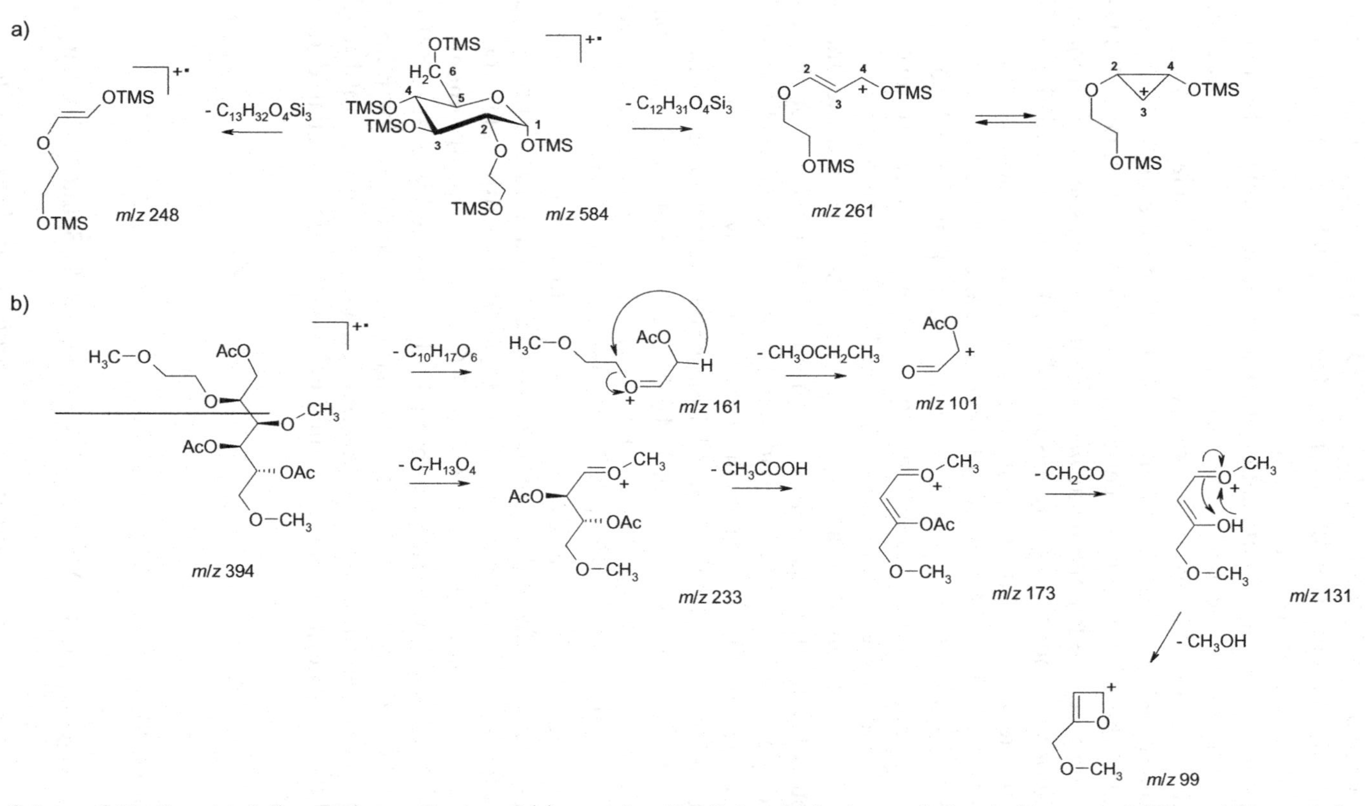

Scheme 3.15: Proposed dissociation pathways of (a) per-trimethylsilylated 2-hydroxyethylated glucose, and (b) partially methylated alditol acetate derived from 2-hydroxyethylated glucose after EI.[227,228]

the PMAA of 2-hydroxyethylated glucose. The first mentioned ion further released ethylmethyl ether (–60 Da) to generate the ion at *m/z* 101, which was supported by different deuterium labeling experiments introducing either trideuteromethyl or trideuteroacetyl residues into the analyte. Thus, the loss of acetic acid was excluded and the formation of the cation of acetic acid 2-oxo-ethyl ether (*m/z* 101) proposed. The ion at *m/z* 233 underwent several consecutive dissociation steps as substantiated by MS^n experiments and deuterium labeling, which demonstrated the elimination of acetic acid (–60 Da) and ketene (–42 Da) that produced the fragments at *m/z* 173 and 131, respectively. Finally, methanol was shown to be released (–32 Da) giving rise to the base peak of the EI mass spectrum at *m/z* 99 (Fig. 3.19d), which was suggested to comprise a 4-methoxymethyl-2H-oxete structure.

Dextrans consist of α-1,6-linked glucose monomers (Fig. 3.18, **3**), and their application as plasma volume expanders was initiated in the early 1940s.[229] Substances containing 1,6-linked glucose monomers occur naturally in human urine at low levels, which necessitated the unambiguous identification as well as quantitation of dextran for doping control purposes. Qualitative evidence for the presence of 1,6-linked glucose was obtained through the formation of PMAAs as reported above also for HES, and a typical EI mass spectrum is depicted in Fig. 3.19e. The dissociation routes of PMAAs and HES are highly comparable, and numerous ions such as those at *m/z* 233, 173, 117, and 99 were found in both cases; however, relative abundances of other fragments such as *m/z* 189, 129, and 87 were more prominent in EI spectra of PMAAs obtained from dextran, which was attributed to the preferred ionization of ether oxygens at C-3 and C-4. The resulting cleavage of the ionized molecule between C-3 and C-4 yielded the fragments at *m/z* 161 and 189 as well as the subsequently generated ions at *m/z* 101, 129, and 87 according to dissociation pathways described for HES.[230] For quantitative purposes, the PMAA-based approach was not found suitable, and alternatives based on LC-MS/MS were evaluated (see Chapter 6).

REFERENCES

1. Clasing, D. (2004) *Doping und seine Wirkstoffe*. Spitta Verlag GmbH, Köln.
2. McCrory, P. (2005) Last orders, gents.... *British Journal of Sports Medicine*, **39**, 879.

3. Gill, N.D., Shield, A., Blazevich, A.J., *et al.* (2000) Muscular and cardiorespiratory effects of pseudoephedrine in human athletes. *British Journal of Clinical Pharmacology*, **50**, 205–213.
4. Wyndham, C.H., Rogers, G.G., Benade, A.J.S., and Strydom, N.B. (1971) Physiological effects of the amphetamines during exercise. *South African Medical Journal*, **45**, 247–252.
5. Laties, V.G., and Weiss, B. (1981) The amphetamine margin in sports. *Federation Proceedings*, **40**, 2689–2692.
6. Ariens, E.J. (1965) Centrally-active drugs and performance in sports. *Schweizerische Zeitschrift für Sportmedizin*, **13**, 77–98.
7. Golding, L.A., and Barnard, J.R. (1963) The effect of d-amphetamine sulfate on physical performance. *Journal of Sports Medicine and Physical Fitness*, **44**, 221–224.
8. Karpovich, P.V. (1959) Effect of amphetamine sulfate on athletic performance. *Journal of the American Medical Association*, **170**, 558–561.
9. Smith, G.M., and Beecher, H.K. (1959) Amphetamine sulfate and athletic performance. I. Objective effects. *Journal of the American Medical Association*, **170**, 542–557.
10. Smith, G.M., and Beecher, H.K.(1960) Amphetamine, secobarbital, and athletic performance. II. Subjective evaluations of performance, mood states, and physical states. *Journal of the American Medical Association*, **172**, 1502–1514.
11. Smith, G.M., and Beecher, H.K. (1960) Amphetamine, secobarbital, and athletic performance. III. Quantiative effects on judgment. *Journal of the American Medical Association*, **172**, 1623–1629.
12. Magkos, F., and Kavouras, S.A. (2004) Caffeine and ephedrine: Physiological, metabolic and performance-enhancing effects. *Sports Medicine*, **34**, 871–889.
13. Bell, D.G., McLellan, T.M., and Sabiston, C.M. (2002) Effect of ingesting caffeine and ephedrine on 10-km run performance. *Medicine and Science in Sports and Exercise*, **34**, 344–349.
14. Hodges, K., Hancock, S., Currell, K., *et al.* (2006) Pseudoephedrine enhances performance in 1500-m runners. *Medicine and Science in Sports and Exercise*, **38**, 329–333.
15. Bohn, A.M., Khodaee, M., and Schwenk, T.L. (2003) Ephedrine and other stimulants as ergogenic aids. *Current Sports Medicine Reports*, **2**, 220–225.
16. Wadler, G.I., and Hainline, B. (1989) *Drugs and the Athlete*.F.A. Davis Company, Philadelphia.
17. George, A.J. (2000) Central nervous system stimulants. *Baillière's Best Practice and Research Clinical Endocrinology and Metabolism*, **14**, 79–88.
18. Avois, L., Robinson, N., Saudan, C., *et al.* (2006) Central nervous system stimulants and sport practice. *British Journal of Sports Medicine*, **40**, i16–i20.

19. Rasmussen, N. (2006) Making the first anti-depressant: Amphetamine in American Medicine, 1929–1950. *Journal of the History of Medicine*, **61**, 288–323.
20. Raza, M. (2006) A role for physicians in ethnopharmacology and drug discovery. *Journal of Ethno-Pharmacology*, **104**, 297–301.
21. Edeleano, L. (1887) Ueber einige Derivate der Phenylmethacrylsäure und der Phenylisobuttersäure. *Berichte der Deutschen Chemischen Gesellschaft*, **20**, 616–622.
22. Nagai, N. (1893) Kanyaku maou seibun kenkyuu seiseki (zoku). Yakugaku zasshi. *Journal of the Pharmaceutical Society of Japan*, **13**, 901–933.
23. Freudenmann, R.W., Öxler, F., and Bernschneider-Reif, S. (2006) The origin of MDMA (ecstasy) revisited: The true story reconstructed from the original documents. *Addiction*, **101**, 1241–1245.
24. Hartmann, M., and Seiberth, M. (1922) US 1,403,117, OF Chemical Industry SOC.
25. Hyde, J.F., Browning, E., and Adams, R. (1928) Synthetic homologs of *d,l*-ephedrine. *Journal of the American Chemical Society*, **50**, 2287–2292.
26. Shelton, R.S., and Van Campen, M.G. (1946) US 2,408,345, WM S Merry Company.
27. Heinzelman, R.V., and Aspergren, B.D. (1957) US 2,789,138, Upjohn Company.
28. McLafferty, F.W. (1956) Mass spectrometric analysis: Broad applicability to chemical research. *Analytical Chemistry*, **28**, 306–316.
29. Gohlke, R.S., and McLafferty, F.W. (1962) Mass spectrometric analysis: Aliphatic amines. *Analytical Chemistry*, **34**, 1281–1287.
30. McLafferty, F.W., and Turecek, F. (1993) *Interpretation of Mass Spectra*. University Science Books, Mill Valley.
31. Carpenter, W.R., Duffield, A.M., and Djerassi, C. (1967) Mass spectrometry in structural and stereochemical problems. CXLV. Factors governing the preferential loss of small vs. large radicals in ketones, Schiff bases, and ethers. *Journal of the American Chemical Society*, **89**, 6167–6170.
32. McLafferty, F.W. (1957) Mass spectrometric analysis: Aliphatic ethers. *Analytical Chemistry*, **29**, 1782–1789.
33. Brown, C.A., Duffield, A.M., and Djerassi, C. (1969) Mass spectrometry in structural and stereochemical problems—CLXXI: Factors governing the preferential loss of small *vs*. large radicals in the α-fission of aliphatic amines. *Organic Mass Spectrometry*, **2**, 625–630.
34. Hammerum, S., Norrman, K., Solling, T.I., *et al.* (2005) Competing simple cleavage reactions: The elimination of alkyl radicals from amine radical cations. *Journal of the American Chemical Society*, **127**, 6466–6475.
35. Hammerum, S. (1988) Distonic radical cations in gaseous and condensed phase. *Mass Spectrometry Reviews*, **7**, 123–202.

36. Hammerum, S., Christensen, J.B., Egsgaard, H., *et al.* (1983) Slow alkyl, alkene, and alkenyl loss from primary alkylamines. *International Journal of Mass Spectrometry and Ion Physics*, **47**, 351–354.
37. Hammerum, S. (1981) Rearrangement and hydrogen abstraction reactions of amine cation radicals: A gas-phase analogy to the Hofmann-Löffler-Freytag reaction. *Tetrahedron Letters*, **22**, 157–160.
38. Hammerum, S., and Nielsen, C.B. (2005) Intramolecular hydrogen bonding and hydrogen atom abstraction in gas-phase aliphatic amine radical cations. *Journal of Physical Chemistry A*, **109**, 12046–12053.
39. Reisch, J., Pagnucco, R., Alfes, H., *et al.* (1968) Mass spectra of derivatives of phenylalkylamines. *Journal of Pharmacy and Pharmacology*, **20**, 81–86.
40. Barry, T., and Petzinger, G. (1977) Structure and fragmentation mechanisms of some ions in the mass spectrum of ephedrine. *Biomedical Mass Spectrometry*, **4**, 129–133.
41. Baba, S., and Kawai, K. (1974) Studies on drug metabolism by use of isotopes. XI. Studies on synthesis and physicochemical features of ^{2}H-labeled *l*-ephedrine on benzene ring. Yakugaku zasshi. *Journal of the Pharmaceutical Society of Japan*, **94**, 783–787.
42. Valentine, J.L., Middleton, R. (2000) GC-MS identification of sympathomimetic amine drugs in urine: Rapid methodology applicable for emergency clinical toxicology. *Journal of Analytical Toxicology*, **24**, 211–222.
43. Thevis, M., Sigmund, G., Koch, A., and Schänzer, W. (2007) Determination of tuaminoheptane in doping control urine samples. *European Journal of Mass Spectrometry*, **13**, 213–221.
44. DeRuiter, J., Clark, C.R., and Noggle, F.T., Jr. (1990) Liquid chromatographic and mass spectral analysis of 1-(3,4-methylenedioxyphenyl)-1-propanamines: Regioisomers of the 3,4-methylenedioxyamphetamines. *Journal of Chromatographic Science*, **28**, 129–132.
45. Brownsill, R., Wallace, D., Taylor, A., and Campbell, B. (1991) Study of human urinary metabolism of fenfluramine using gas chromatography-mass spectrometry. *Journal of Chromatography*, **562**, 267–277.
46. Donike, M. (1973) Acylierung mit Bis(Acylamiden); N-Methyl-Bis(Trifluoracetamid) und Bis(Trifluoracetamid), zwei neue Reagenzien zur Trifluoracetylierung. *Journal of Chromatography*, **78**, 273–279.
47. Thurman, E.M., Pedersen, M.J., Stout, R.L., and Martin, T. (19920 Distinguishing sympathomimetic amines from amphetamine and methamphetamine in urine by gas chromatography/mass spectrometry. *Journal of Analytical Toxicology*, **16**, 19–27.
48. Wu, A.H., Onigbinde, T.A., Wong, S.S., and Johnson, K.G. (1992) Identification of methamphetamines and over-the-counter sympathometic amines by full-scan GC-ion trap MS with electron impact and chemical ionization. *Journal of Analytical Toxicology*, **16**, 137–141.

49. Gunnar, T., Ariniemi, K., and Lillsunde, P. (2005) Validated toxicological determination of 30 drugs of abuse as optimized derivatives in oral fluid by long column fast gas chromatography/electron impact mass spectrometry. *Journal of Mass Spectrometry*, **40**, 739–753.
50. Beckett, A.H., Tucker, G.T., and Moffat, A.C. (1967) Routine detection and identification in urine of stimulants and other drugs, some of which may be used to modify performance in sport. *Journal of Pharmacy and Pharmacology*, **19**, 273–294.
51. Donike, M., and Derenbach J. (1976) Die Selektive Derivatisierung Unter Kontrollierten Bedingungen: Ein Weg zum Spurennachweis von Aminen. *Zeitschrift für analytische Chemie*, **279**, 128–129.
52. Shin, H.S., and Donike, M. (1996) Stereospecific derivatization of amphetamines, phenol alkylamines, and hydroxyamines and quantification of the enantiomers by capillary GC/MS. *Analytical Chemistry*, **68**, 3015–3020.
53. Hemmersbach, P., and de la Torre, R. (1996) Stimulants, narcotics and β-blockers: 25 years of development in analytical techniques for doping control. *Journal of Chromatography B*, **687**: 221–238.
54. Kraft, M. (1990) Gleichzeitiger gas-chromatographisch/massenspektrometrischer Nachweis von Hydroxy- und Phenolalkylaminen (Stimulanzien, ß-Blocker, Narkotika) sowie synthetischen und endogenen anabol-androgenen Steroiden nach extraktiver Acylierung und gezielter Trimethylsilylierung. *Institut für Biochemie*, Deutsche Sporthochschule Köln.
55. Rosano, T.G., Hubbard, J.D., Meola, J.M., and Swift, T.A. (2000) Fatal strychnine poisoning: Application of gas chromatography and tandem mass spectrometry. *Journal of Analytical Toxicology*, **24**, 642–647.
56. Jackson, G., and Diggle, G. (1973) Strychnine-containing tonics. *British Medical Journal*, **2**, 176–177.
57. Pelletier, P.J., and Caventou, J.B. (1818) Note sur un Nouvel Alcali. *Annales de Chimie et de Physique*, **3**, 323–336.
58. Nicolaou, K.C., and Montagnon, T. (2008) *Molecules that changed the world*. Wiley-VCH, Weinheim.
59. Hoberman, J. (2007) History and prevalence of doping in the marathon. *Sports Medicine*, **37**, 386–388.
60. Biemann, K. (1962) *Mass spectrometry: Organic chemical applications*. McGraw-Hill, New York.
61. Galeffi, C., Nicoletti, M., Messana, I., and Marini-Bettolo, G.B. (1979) On the alkaloids of strychnos—XXXI: 15-Hydroxystrychnine, a new alkaloid from *Strychnos nux vomica* L. *Tetrahedron*, **35**, 2545–2549.
62. Galeffi, C., Miranda-dellle-Monache, E., and Marini-Bettolo, G.B. (1974) Strychnos alkaloids. XXVII. Separation and characterisation of isostrychnine in *Strychnos nux vomica* L. seeds. *Journal of Chromatography*, **88**, 416–418.

63. Marini-Bettolo, G.B., Ciasca, M.A., Galeffi, C., *et al.* (1972) The occurrence of strychnine and brucine in an American species of *Strychnos. Phytochemistry*, **11**, 381–384.
64. Budzikiewicz, H., Djerassi, C., and Williams, D.H. (1964) Structure elucidation of natural products by mass spectrometry—Volume I: Alkaloids. Holden-Day, Inc., San Francisco.
65. Gaedcke, F. (1855) Ueber das Erythroxylin. *Archiv der Pharmazie*, **132**, 141–150.
66. Niemann, A. (1860) Ueber eine neue organische base in den Cocablättern. *Archiv der Pharmazie*, **153**, 129–155.
67. Jindal, S.P., Lutz, T., and Vestergaard, P. (1978) Mass spectrometric determination of cocaine and its biologically active metabolite, Norcocaine, in human urine. *Biomedical Mass Spectrometry*, **5**, 658–663.
68. Curcuruto, O., Guidugli, F., Traldi, P., *et al.* Ion-trap mass spectrometry applications in forensic sciences. I. Identification of morphine and cocaine in hair extracts of drug addicts. *Rapid Communications in Mass Spectrometry*, **6**, 434–437.
69. Blossey, E.C., Budzikiewicz, H., Ohashi M., *et al.* Mass spectrometry in structural and stereochemical problems—XXXIX. Tropane alkaloids. *Tetrahedron*, **20**, 585–595.
70. Braun, H. (1946) *Pharmakologie des deutschen Arzneibuchs*. Wissenschaftliche Verlagsgesellschaft, Stuttgart.
71. Klockgether-Radke, A.P. (2002) F.W. Sertürner and the discovery of morphine. 200 years of pain therapy with opioids. *Anaesthesiologie, Intensivmedizin, Notfallmedizin, Schmerztherapie*, **37**, 244–249.
72. Sertürner, F.W. (1817) Ueber das Morphium, eine neue salzfähige Grundlage, und die mekonsäure, als Hauptbestandtheile des Opiums. *Annalen der Physik*, **55**, 56–89.
73. Gates, M., and Tschudi, G. (1956) The synthesis of morphine. *Journal of the American Chemical Society*, **78**, 1380–1393.
74. Wright, C.R.A. (1874) On the action of organic acids and their anhydrides on the natural alkaloids. Part I. *Journal of the Chemical Society*, **27**, 1031–1043.
75. von Eisleb O, and Schaumann, O. (1939) Dolantin, ein neuartiges Spasmolytickum und Analgetikum. *Deutsche medizinische Wochenschrift*, **55**.
76. Latta, K.S., Ginsberg, B., and Barkin, R.L. (2002) Meperidine: A critical review. *American Journal of Therapeutics*, **9**, 53–68.
77. Eddy, N.B., and May, E.L. (1973) The search for a better analgesic. *Science*, **181**, 407–414.
78. Bentley, K.W. (1969) US 3433791, Reckitt & Sons Ltd.
79. Mather, L. (1983) Clinical pharmacokinetics of fentanyl and its newer derivatives. *Clinical Pharmacokinetics*, **8**, 422–446.

80. World Anti-Doping Agency (2009) The 2009 Prohibited List. Available at http://www.wada-ama.org/rtecontent/document/2009_Prohibited_List_ENG_Final_20_Sept_08.pdf. Accessed 02-01-2009.
81. Wheeler, D., Kinstle, T., Rinehart, K. Jr. (1967) Mass spectral studies of alkaloids related to morphine. *Journal of the American Chemical Society* **89**, 4494–4501.
82. Audier, H., Fetzion, M., Ginsburg, D.,*et al.* (1965) Mass spectrometry of the morphine alkaloids. *Tetrahedron Letters*, **1**, 13–22.
83. Thevis, M. (1998) Untersuchungen zur quantitativen gas-chromatographisch / massenspektrometrischen Bestimmung von Morphin, insbesondere in Humanurin nach Konsum von Mohn. Institute of Biochemistry, German Sport University Cologne. Master thesis.
84. Feng, L., Xuying, H., and Yi, L. (1994) Investigation of meperidine and its metabolites in urine of an addict by gas chromatography-flame ionization detection and gas chromatography-mass spectrometry. *Journal of Chromatography B*, **658**, 375–379.
85. Alha, A., Karlsson, M., and Korte T. (1975) Fatal combined anileridine-pethidine poisoning. A gas chromatography, thin layer chromatography and mass spectrometry investigation. *Zeitschrift für Rechtsmedizin*, **75**, 293–298.
86. Blomquist, M., Bonnichsen, R., Fri, C-G., *et al.* Gas chromatography-mass spectrometry in forensic chemistry for identification of substances isolated from tissue. *Zeitschrift für Rechtsmedizin*, **69**, 52–61.
87. Lindberg, C., Bondesson, U., and Hartvig, P. (1980) Investigation of the urinary excretion of pethidine and five of its metabolites in man using selected ion monitoring. *Biomedical Mass Spectrometry*, **7**, 88–92.
88. Harvey, S.C., Toussaint, C.P., Coe, S.E., *et al.* Stability of meperidine in an implantable infusion pump using capillary gas chromatography-mass spectrometry and a deuterated internal standard.*Journal of Pharmaceutical and Biomedical Analysis*, **21**, 577–583.
89. de Kruif, P. (1945) *The Male Hormone*.Harcourt, Brace and Company, New York.
90. Maisel, A.Q. (1965) *The Hormone Quest*. Random House New York.
91. Butenandt, A., Hanisch, G. (1935) Über Testosteron. Umwandlung des Dehydroandrosterons in Androstendiol und Testosteron, ein Weg zur Darstellung des Tesotosterons aus Cholesterin. *Hoppe-Seyler's Zeitschrift für Physiologische Chemie*, **237**, 89–97.
92. Ruzicka, L., and Wettstein, A. (1935) Sexualhormone VII. Über die künstliche Herstellung des Testikelhormons Testosteron (Androsten-3-on-17-ol). *Helvetica Chimica Acta*, **18**, 1264–1275.
93. Kochakian, C.D. (1976) Metabolic effects of anabolic-androgenic steroids in experimental animals, in *Anabolic-Androgenic Steroids* (ed C.D. Kochakian), Springer, Berlin-Heidelberg-New York, pp. 5–44.

94. Ruzicka, L., Goldberg, M.W., and Rosenberg, H.R. (1935) Sexualhormone X. Herstellung des 17-Methyltestosteron und anderer Androsten- und Androstanderivate. Zusammenhänge zwischen chemischer Konstitution und männlicher Hormonwirkung. *Helvetica Chimica Acta*, **18**, 1487–1498.
95. Rommerts, F.F.G. (2004) Testosterone: An overview of biosynthesis, transport, metabolism and non-genomic actions, in *Testosterone: Action, Deficiency, Substitution* (eds E. Nieschlag and H.M. Behre),Cambridge University Press, Cambridge, pp. 1–38.
96. Gooren, L.J., and Bunck, M.C. (2004) Androgen replacement therapy: Present and future. *Drugs*, **64**, 1861–1891.
97. Chen, J., Kim, J., and Dalton, J.T. (2005) Discovery and therapeutic promise of selective androgen receptor modulators. *Molecular Interventions*, **5**, 173–188.
98. Bagatell, C.J., and Bremner, W.J. (1996) Androgens in men—Uses and abuses. *New England Journal of Medicine*, **334**, 707–714.
99. Negro-Vilar, A. (1999) Selective androgen receptor modulators (SARMs): A novel approach to androgen therapy for the new millennium. *Journal of Clinical Endocrinology and Metabolism*, **84**, 3459–3462.
100. Cadilla, R., and Turnbull, P. (2006) Selective androgen receptor modulators in drug discovery: Medicinal chemistry and therapeutic potential. *Current Topics in Medicinal Chemistry*, **6**, 245–270.
101. Todd, J., and Todd, T. (2001) Significant events in the history of drug testing and the Olympic movement: 1960–1999, in *Doping in Elite Sport* (eds W. Wilson and E. Derse), Human Kinetics, Champaign, pp. 63–128.
102. Todd, T. (1987) Anabolic steroids: The gremlins of sport. *Journal of Sports History*, **14**, 87–107.
103. Donike, M. (1969) N-Methyl-N-trimethylsilyl-trifluoracetamide, ein neues Silylierungsmittel aus der Reihe der silylierten Amide. *Journal of Chromatography*, **42**, 103–104.
104. Donike, M., and Zimmermann, J. (1980) Zur Darstellung von Trimethylsilyl-, Triethylsilyl- und tert.- Butyldimethylsilyl- enoläthern von Ketosteroiden für gas- chromatographische und massenspektrometrische Untersuchungen. *Journal of Chromatography*, **202**, 483–486.
105. Björkhem, I., Gustafsson, J-A., and Sjövall, J. (1972) A novel fragmentation of trimethylsilyl ethers of 3β-hydroxy-δ^5-steroids. *Organic Mass Spectrometry*, **7**, 277–281.
106. Brooks, C.J.W., Harvey, D.J., Middleditch, B.S., and Vouros, P. (1973) Mass spectra of trimethylsilyl ethers of some δ^5-3β-hydroxy C_{19} steroids. *Organic Mass Spectrometry*, **7**, 925–948.
107. Diekman, J., and Djerassi, C. (1967) Mass spectrometry in structural and stereochemical problems. CXXV. Mass spectrometry of some steroid trimethylsilyl ethers. *Journal of Organic Chemistry*, **32**, 1005–1011.
108. Gaskell, S.J., Smith, A.G., and Brooks, C.J.W. (1975) Gas chromatography mass spectrometry of trimethylsilyl ethers of sidechain hydroxylated δ^4-3-

ketosteroids. Long range trimethylsilyl group migration under electron impact. *Biomedical Mass Spectrometry*, **2**, 148–155.

109. Gustafsson, J-A., Ryhage, R., Sjövall, J., and Moriarty, R.M. (1969) Migrations of the trimethylsilyl group upon electron impact in steroids. *Journal of the American Chemical Society*, **91**, 1234–1236.
110. Harvey, D.J., and Vouros, P. (1979) Influence of the 6-trimethylsilyl group on the fragmentation of the trimethylsilyl derivatives of some 6-hydroxy- and 3,6-dihydroxy-steroids and related compounds. *Biomedical Mass Spectrometry*, **6**, 135–143.
111. Masse, R., Laliberte, C., and Tremblay, L. (1985) Gas chromatography-mass spectrometry of epimeric 19-norandrostan-3-ol-17-ones as the trimethylsilyl ether, methoxime-trimethylsilyl ether and trimethylsilyl-enol trimethylsilyl derivatives. *Journal of Chromatography*, **339**, 11–23.
112. McCloskey, J.A., Stillwell, R.N., and Lawson, A.M. (1968) Use of deuterium-labeled trimethylsilyl derivatives in mass spectrometry. *Analytical Chemistry*, **40**, 233–236.
113. Vouros, P., and Harvey, D.J. (1973) Method for selective introduction of trimethylsilyl and perdeuterotrimethylsilyl groups in hydroxy steroids and its utility in mass spectrometric interpretations. *Analytical Chemistry*, **45**, 7–12.
114. Vouros, P., and Harvey, D.J. (1980) The electron impact induced cleavage of the C-17–C-20 bond D-ring in trimethylsilyl derivatives of C_{21} steroids. Reciprocal exchange of trimethylsilyl groups. *Biomedical Mass Spectrometry*, **7**, 217–225.
115. Budzikiewicz, H., Djerassi, C., and Williams, D.H. (1964) *Structure elucidation of natural products by mass spectrometry—Volume II: Steroids, terpenoids, sugars, and miscellaneous classes*. Holden-Day, Inc., San Francisco.
116. Friedland, S.S., and Lane, G.H. (1959) Mass spectra of steroids. *Analytical Chemistry*, **31**, 169–174.
117. Reed, R.I. (1958) Electron impact and molecular dissociation. Part I. Some steroids and triterpenoids. *Journal of the Chemical Society*, 3432–3436.
118. Biemann, K., and Seibl, J. (1959) Application of mass spectrometry to structure problems. II. Stereochemistry of epimeric, cyclic alcohols. *Journal of the American Chemical Society*, **81**, 3149–3150.
119. Fales, H.M., and Luukkainen, T. (1965) O-Methyloximes as carbonyl derivatives in gas chromatography, mass spectrometry, and nuclear magnetic resonance. *Analytical Chemistry*, **37**, 955–957.
120. Shapiro, R.H., and Djerassi, C. (1964) Mass spectrometry in structural and stereochemical problems. L. Fragmentation and hydrogen migration reactions of α,β-unsaturated 3-keto steroids. *Journal of the American Chemical Society*, **86**, 2825–2832.

121. Brown, F.J., and Djerassi, C. (1980) Elucidation of the course of the electron impact induced fragmentation of α,β-unsaturated 3-keto steroids. *Journal of the American Chemical Society*, **102**, 807–817.
122. Karliner, J., Budzikiewicz, H., and Djerassi, C. (1966) Mass spectrometry in structural and stereochemical problems. XCI. The electron impact induced elimination of water from 3-hydroxy steroids. *Journal of Organic Chemistry*, **31**, 710–713.
123. Macdonald, C.G., Shannon, J.S., and Sugowdz, G. (1963) Studies in mass spectrometry stereochemistry and elimination reactions of hydroxy and acetoxy compounds. *Tetrahedron Letters*, **4**, 807–814.
124. Djerassi, C., and Kielczewski, M.A. (1963) The introduction of deuterium into the C-19 angular methyl group. *Steroids*, **2**, 125–134.
125. Horning, E.C., Brooks, C.J.W., and Vanden Heuvel, W.J.A. (1968) Gas phase analytical methods for the study of steroids, in *Advances in Lipid Research* (eds R. Paoletti, and D. Kritchevsky), Academic Press, New York, pp. 273–392.
126. Djerassi, C. (1964) Isotope labelling and mass spectrometry of natural products. *Pure and Applied Chemistry*, **9**, 159–178.
127. Egger, H., and Spiteller, G. (1966) Massenspektren und Stereochemie von Hydroxyverbindungen, 1. Mitt.: Hydroxysteroide. *Monatshefte für Chemie*, **97**, 579–601.
128. Budzikiewicz, H., and Djerassi, C. (1962) Mass spectrometry in structural and stereochemical problems I. Steroid ketones. *Journal of the American Chemical Society*, **84**, 1430–1439.
129. Biemann, K., Bommer, P., and Desiderio, D.M. (1964) Element-mapping, a new approach to the interpretation of high resolution mass spectra. *Tetrahedron Letters*, **5**, 1725–1731.
130. Thevis, M., and Schänzer, W. (2007) Mass spectrometry in sports drug testing: Structure characterization and analytical assays. *Mass Spectrometry Reviews*, **26**, 79–107.
131. Donike, M., Zimmermann, J., Bärwald, K.R., *et al.* (1984) Routinebestimmung von Anabolika in Harn. *Dtsch Z Sportmed*, **35**, 14–24.
132. Ayotte, C., Goudreault, D., and Charlebois, A. (1996) Testing for natural and synthetic anabolic agents in human urine. *Journal of Chromatography B, Biomedical Sciences and Applications*, **687**, 3–25.
133. Bowers, L.D., and Borts, D.J. (1996) Separation and confirmation of anabolic steroids with quadrupole ion trap tandem mass spectrometry. *Journal of Chromatography B: Biomedical Sciences and Applications*, **687**, 69–78.
134. Kazlauskas, R., and Trout, G. (2000) Drugs in sports: Analytical trends. *Therapeutic Drug Monitoring*, **22**, 103–109.
135. Marcos, J., Pascual, J.A., de la Torre, X., and Segura, J. (2002) Fast screening of anabolic steroids and other banned doping substances in human

urine by gas chromatography/tandem mass spectrometry. *Journal of Mass Spectrometry*, **37**, 1059–1073.

136. Diekman, J., Thomson, J.B., and Djerassi, C. (1968) Mass spectrometry in structural and stereochemical problems. CLV. Electron impact induced fragmentations and rearragements of some trimethylsilyl ethers of aliphatic glycols and related compounds. *Journal of Organic Chemistry*, **33**, 2271–2284.
137. Harvey, D.J., Horning, M.G., and Vouros, P. (1971) Some stereochemical factors in the formation of rearrangement ions in the mass spectra of trimethylsilyl derivatives of steroidal phosphates. *Tetrahedron*, **27**, 4231–4243.
138. Rendic, S., Nolteernsting, E, and Schänzer, W. (1999) Metabolism of anabolic steroids by recombinant human cytochrome P450 enzymes: Gas chromatographic–mass spectrometric determination of metabolites. *Journal of Chromatography B*, **735**, 73–83.
139. Durbeck, H.W., and Buker, I. (1980) Studies on anabolic steroids. The mass spectra of 17 alpha-methyl-17 beta-hydroxy-1,4-androstadien-3-one (Dianabol) and its metabolites. *Biomedical Mass Spectrometry*, **7**, 437–445.
140. Schänzer, W., Horning, S., Opfermann, G., and Donike, M. (1996) Gas chromatography/mass spectometry identification of long-term excreted metabolites of the anabolic steroid 4-chloro-1,2-dehydro-17alpha-methyltestosterone in humans. *Journal of Steroid Biochemistry and Molecular Biology*, **57**, 363–376.
141. Schänzer, W., and Donike, M. (1992) Metabolism of boldenone in man: Gas chromatographic/mass spectrometric identification of urinary excreted metabolites and determination of excretion rates. *Biological Mass Spectrometry*, **21**, 3–16.
142. Donike, M., Ueki, M., Kuroda, Y., *et al.* (1995) Detection of dihydrotestosterone (DHT) doping: Alterations in the steroid profile and reference ranges for DHT and its 5 alpha-metabolites. *Journal of Sports Medicine and Physical Fitness*, **35**, 235–250.
143. Schänzer, W., and Donike, M. (1994) Synthesis of deuterated steroids for GC/MS quantification of endogenous steroids, in *Recent Advances in Doping Analysis* (eds M. Donike, H. Geyer, A. Gotzmann, and U. Mareck-Engelke) Sport und Buch Strauß, Cologne, pp. 93–112.
144. Schänzer, W., Geyer, H., and Donike, M. (1991) Metabolism of metandienone in man: Identification and synthesis of conjugated excreted urinary metabolites, determination of excretion rates and gas chromatographic-mass spectrometric identification of bis-hydroxylated metabolites. *Journal of Steroid Biochemistry and Molecular Biology*, **38**, 441–464.
145. Brown, F.J., and Djerassi, C. (1980) Elucidation of the course of the electron impact induced fragmentation of α,β-unsaturated 3-keto steroids. *Journal of the American Chemical Society*, **102**, 807–811.
146. Shapiro, R.H., and Djerassi C. (1964) Mass spectrometry in structural and stereochemical problems. L. Fragmentation and hydrogen migration

reactions of α,β-unsaturated 3-keto steroids. *Journal of the American Chemical Society*, **86**, 2825–2832.

147. Schänzer, W., Opfermann, G., and Donike, M. (1992) 17-Epimerization of 17 alpha-methyl anabolic steroids in humans: Metabolism and synthesis of 17 alpha-hydroxy-17 ß-methyl steroids. *Steroids*, **57**, 537–549.
148. Thevis, M., and Schänzer, W. (2008) Mass spectrometry of selective androgen receptor modulators. *Journal of Mass Spectrometry*, **43**, 865–876.
149. Dalton, J.T., Mukherjee, A., Zhu, Z., *et al.* (1998) Discovery of nonsteroidal androgens. *Biochemical and Biophysical Research Communications*, **244**, 1–4.
150. Gao, W., and Dalton, J.T. (2007) Expanding the therapeutic use of androgens via selective androgen receptor modulators (SARMs). *Drug Discovery Today*, **12**, 241–248.
151. Miner, J.N., Chang, W., Chapman, M.S., *et al.* (2007) An orally active selective androgen receptor modulator is efficacious on bone, muscle, and sex function with reduced impact on prostate. *Endocrinology*, **148**, 363–373.
152. Rosen, J., and Negro-Vilar, A. (2002) Novel, non-steroidal, selective androgen receptor modulators (SARMs) with anabolic activity in bone and muscle and improved safety profile. *Journal of Musculoskeletal & Neuronal Interactions*, **2**, 222–224.
153. Ostrowski, J., Kuhns, J.E., Lupisella, J.A., *et al.* (2006) Pharmacological and x-ray structural characterization of a novel selective androgen receptor modulator: potent hyperanabolic stimulation of skeletal muscle with hypostimulation of prostate in rats. *Endocrinology*, **48**, 4–12.
154. Sun, C., Robl, J.A., Wang, T.C., *et al.* (2006) Discovery of potent, orally-active, and muscle-selective androgen receptor modulators based on an N-aryl-hydroxybicyclohydantoin scaffold. *Journal of Medicinal Chemistry*, **49**, 7596–7599.
155. Thevis, M., Kohler, M., Schlörer, N., *et al.* (2008) Screening for two selective androgen receptor modulators using gas chromatography-mass spectrometry in doping control analysis. *European Journal of Mass Spectrometry*, **14**, 153–161.
156. Van Langenhove, A., Costello, C.E., Biller, J.E., et al. (1980) A mass spectrometric method for the determination of stable isotope labeled phenytoin suitable for pulse dosing studies. *Biomedical Mass Spectrometry*, **7**, 576–581.
157. Locock, R.A., and Coutts, R.T. (1970) The mass spectra of succinimides, hydantoins, oxazolidinediones and other medicinal anti-epileptic agents. *Organic Mass Spectrometry*, **3**, 735–745.
158. Rau, S. (2006) Die Geschichte der Diuretika. *Pharmazie in unserer Zeit*, **35**, 286–292.
159. Sneader, W. (2005) *Drug Discovery: A History*. John Wiley and Sons, Chichester.

160. Schultz, E.M., Bicking, J.B., DeSolms, S.J., and Stokker, G.E. (1971) Structure of the diuretic merbaphen. *Journal of Medicinal Chemistry*, **14**, 998–999.
161. von Schroeder, W. (1887) Ueber die Wirkung des Coffeins als Diureticum. *Archiv für Experimentelle Pathologie und Pharmakologie*, **22**, 39–61.
162. Roblin, R.O., and Clapp, J.W. (1950) The preparation of heterocyclic sulfonamides. *Journal of the American Chemical Society*, **72**, 4890–4892.
163. Werner, L.H., Halamandaris, A., Ricca, S., *et al.* (1960) Dihydrobenzothiadiazine 1,1-dioxides and their diuretic properties. *Journal of the American Chemical Society*, **82**, 1161–1166.
164. Sturm, K., Siedel, W., and Weyer, R. (1962) US 3058882, *Hoechst AG.*
165. Pachter, I.J. (1963) Pteridines. II. Synthesis of 6-substituted 7-aminopteridines from aldehydes. *Journal of Organic Chemistry*, **28**, 1191–1196.
166. Cragoe, E.J. (1967) US 3313813, Merck & Co, Inc.
167. Casy, A.F. (1987) Electron impact mass spectrometry of diuretic agents. *Journal of Pharmaceutical and Biomedical Analysis*, **5**, 247–257.
168. Yoon, C.N., Lee, T.H., and Park, J. (1990) Mass spectrometry of methyl and methyl-d3 derivatives of diuretic agents. *Journal of Analytical Toxicology*, **14**, 96–101.
169. Walter, M.J., and Holtzman, M.J. (2005) A centennial history of research on asthma pathogenesis. *American Journal of Respiratory Cell and Molecular Biology*, **32**, 483–489.
170. Zeile, K., Thomä, O., and Mentrup, A. (1967) US 3341593, *Boehringer Ingelheim.*
171. Lunts, L.H.C., and Toon, P. (1972) US 3644353, Allen and Hanburys, Ltd..
172. Keck, J., Krüger, G., Machleidt, H., *et al.* (1970) US 3536712, Boehringer Ingelheim.
173. Keck, J., Krüger, G., Noll, K., and Machleidt, H. (1972) Synthesen von neuen Amino-Halogen-substituierten Phenyl-aminoäthanolen [Synthesis of new amino-halogen substituted phenyl-amino ethanols]. *Arzneimittel-Forschung*, **22**, 861–869.
174. Kim, Y.S., and Sainz, R.D. (1992) Beta-adrenergic agonists and hypertrophy of skeletal muscles. *Life Sciences*, **50**, 397–407.
175. Maltin, C.A., Reeds, P.J., Delday, M.I., *et al.* (1986) Inhibition and reversal of denervation-induced atrophy by the beta-agonist growth promoter, clenbuterol. *Bioscience Reports*, **6**, 811–818.
176. Zeman, R.J., Ludemann, R., and Etlinger, J.D. (1987) Clenbuterol, a beta 2-agonist, retards atrophy in denervated muscles. *American Journal of Physiology*, **252**, E152–155.
177. von Deutsch, D.A., Abukhalaf, I.K., Wineski, L.E., *et al.* (2002) Distribution and muscle-sparing effects of clenbuterol in hindlimb-suspended rats. *Pharmacology*, **65**, 38–48.

178. Polettini, A. (1996) Bioanalysis of β_2-agonists by hyphenated chromatographic and mass spectrometric techniques. *Journal of Chromatography B*, **687**, 27–42.
179. Damasceno, L., Ventura, R., Ortuno, J., and Segura, J. (2000) Derivatization procedures for the detection of beta(2)-agonists by gas chromatographic/mass spectrometric analysis. *Journal of Mass Spectrometry*, **35**, 1285–1294.
180. Clare, R.A., Davies, D.S., and Baillie, T.A. (1979) The analysis of terbutaline in biological fluids by gas chromatography electron impact mass spectrometry. *Biomedical Mass Spectrometry*, **6**, 31–37.
181. Henze, M.K., Opfermann, G., Spahn-Langguth, H., and Schänzer, W. (2001) Screening of beta-2 agonists and confirmation of fenoterol, orciprenaline, reproterol and terbutaline with gas chromatography-mass spectrometry as tetrahydroisoquinoline derivatives. *Journal of Chromatography B, Biomedical Sciences and Applications*, **751**, 93–105.
182. Dumasia, M.C., and Houghton, E. (1991) Screening and confirmatory analysis of beta-agonists, beta-antagonists and their metabolites in horse urine by capillary gas chromatography-mass spectrometry. *Journal of Chromatography*, **564**, 503–513.
183. Blanchflower, W.J., Hewitt, S.A., Cannavan, A., *et al.* (1993) Detection of clenbuterol residues in bovine liver, muscle, retina and urine using gas chromatography/mass spectrometry. *Biological Mass Spectrometry*, **22**, 326–330.
184. Amendola, L., Colamonici, C., Rossi, F., and Botre, F. (2002) Determination of clenbuterol in human urine by GC-MS-MS-MS: Confirmation analysis in antidoping control. *Journal of Chromatography B, Analytical Technologies in the Biomedical and Life Sciences*, **773**, 7–16.
185. Henze, M.K. (2001) Screening auf β_2-Sympathomimetika in der Dopinganalytik. Institute for Pharmaceutical Chemistry, Martin-Luther University Halle-Wittenberg. Dissertation.
186. Bourne, G.R. (1981) The metabolism of β-adrenoreceptor blocking drugs. *Progress in Drug Metabolism*, **6**, 77–110.
187. Powell, C.E., and Slater, I.H. (1958) Blocking of inhibitory adrenergic receptors by a dichloro analog of isoproterenol. *Journal of Pharmacology and Experimental Therapeutics*, **122**, 480–488.
188. Crowther, A.F., and Smith, L.H. (1967) US 3337628, ICI Ltd.
189. Barrett, A.M., Carter, J., Hull, R., *et al.* (1972) US 3663607, ICI Ltd.
190. Jonas, R. Becker, K-H., Enenkel, H-J., *et al.* (1981) US 4258062, Merck.
191. Brändstöm, A.E., Carlsson, P.A.E., Carlsson, S.A.I., *et al.* (1976) US 3998790, Hassle.
192. Donike, M. (1975) N-Trifluoracetyl-O-trimethylsilyl-phenolalkylamine—Darstellung und massenspezifischer gaschromatographischer Nachweis. *Journal of Chromatography*, **103**, 91–112.
193. Maurer, H., and Pfleger, K. (1986) Identification and differentiation of beta-blockers and their metabolites in urine by computerized gas

chromatography—Mass spectrometry. *Journal of Chromatography*, **382**, 147–165.

194. Garteiz, D.A., and Walle, T. (1972) Electron impact fragmentation studies of b-blocking drugs and their metabolites by GC-mass spectroscopy. *Journal of Pharmaceutical Sciences*, **61**, 1728–1731.
195. Amendola, L., Molaioni, F., and Botre, F. (2000) Detection of beta-blockers in human urine by GC-MS-MS-EI: Perspectives for the anti-doping control. *Journal of Pharmaceutical and Biomedical Analysis*, **23**, 211–221.
196. Delbeke, F.T. (1996) Disposition of human drug preparations in the horse. V. Orally administered oxprenolol. *Biomedical Chromatography*, **10**, 172–178.
197. Zamecnik, J. (1990) Use of cyclic boronates for GC/MS screening and quantitation of beta-adrenergic blockers and some bronchodilators. *Journal of Analytical Toxicology*, **14**, 132–136.
198. Walle, T., Morrison, J., Walle, K., and Conradi, E. (1975) Simultaneous determination of propranolol and 4-hydroxypropranolol in plasma by mass fragmentography. *Journal of Chromatography*, **114**, 351–359.
199. Vu, V.T., and Abramson, F.P. (1978) Quantitative analysis of propranolol and metabolites by a gas chromatograph mass spectrometer computer technique. *Biomedical Mass Spectrometry*, **5**, 686–691.
200. Hermann, P., Fraisse, J., Allen, J., *et al.* (1984) Determination of betaxolol, a new beta-blocker, by gas chromatography mass spectrometry: Application to pharmacokinetic studies. *Biomedical Mass Spectrometry*, **11**, 29–34.
201. Lee, C.R., Coste, A.C., and Allen, J. (2988) Determination of the beta-blocker betaxolol and labelled analogues by gas chromatography/mass spectrometry with selected ion monitoring of the alpha-cleavage fragment (m/z 72). *Biomedical and Environmental Mass Spectrometry*, **16**, 387–392.
202. Upthagrove, A.L., Hackett, M., and Nelson, W.L. (1999) Fragmentation pathways of selectively labeled uropranolol using electrospray ionization on an ion trap mass spectrometer and comparison with ions formed by electron impact. *Rapid Communications in Mass Spectrometry*, **13**, 534–541.
203. Kaneko, N. (1994) New 1,4-benzothiazepine derivative, K201, demonstrates cardioprotective effects against sudden cardiac cell death and intracellular calcium blocking action. *Drug Development Research*, **33**. 429–438.
204. Kaneko, N., Oosawa, T., Sakai, T., and Oota, H. (1995) US 5416066, Kirin Brewery Co. Ltd.
205. Bellinger, A.M., Reiken, S., Dura, M., *et al.* (2008) Remodeling of ryanodine receptor complex causes "leaky" channels: A molecular mechanism

for decreased exercise capacity. *Proceedings of the National Academy of Sciences of the United States of America*, **105**, 2198–2202.

206. Bellinger, A.M., Mongillo, M., and Marks, A.R. (2008) Stressed out: The skeletal muscle ryanodine receptor as a target of stress. *Journal of Clinical Investigation*, **118**, 445–453.
207. Thevis, M., Beuck, S., Thomas, A., *et al.* (2009) Electron ionization mass spectrometry of the ryanodine-based Ca2+-channel stabilizer S-107 and its Implementation into routine doping control. *Rapid Communications in Mass Spectrometry*, **23**, 2363–2370.
208. Reynolds, J.E.F. (2002) *Martindale: The Extra Pharmacopoeia*. The Pharmaceutical Press, London.
209. Petitt, B.R., King, G.S., and Blau, K. (1980) The analysis of hexitols in biological fluid by selected ion monitoring. *Biomedical Mass Spectrometry*, **7**, 309–313.
210. Guddat, S., Thevis, M., and Schänzer, W. (2008) Identification and quantification of the osmodiuretic mannitol in urine for sports drug testing using gas chromatography-mass spectrometry. *European Journal of Mass Spectrometry*, **14**, 127–133.
211. Niwa, T., Yamamoto, N., Maeda, K., and Yamada, K. (1983) Gas chromatographic—mass spectrometric analysis of polyols in urine and serum of uremic patients. Identification of new deoxyalditols and inositol isomers. *Journal of Chromatography*, **277**, 25–39.
212. Renner, F., Schmitz, A., and Gehring, H. (1998) Rapid and sensitive gas chromatography-mass spectroscopy method for the detection of mannitol and sorbitol in serum samples. *Clinical Chemistry*, **44**, 886–887.
213. Frank, M.S., Nahata, M.C., and Hilty, M.D. (1981) Glycerol: A review of its pharmacology, pharmacokinetics, adverse reactions, and clinical use. *Pharmacotherapy*, **1**, 147–160.
214. Wald, S.L., and McLaurin, R.L. (1982) Oral glycerol for the treatment of traumatic intracranial hypertension. *Journal of Neurosurgery*, **56**, 323–331.
215. Meyer, J.S., Charney, J.Z., Rivera, V.M., and Mathew, N.T. (1971) Treatment with glycerol of cerebral oedema due to acute cerebral infarction. *Lancet*, **298**, 993–997.
216. Coutts, A., Reaburn, P., Mummery, K., and Holmes, M. (2002) The effect of glycerol hyperhydration on olympic distance triathlon performance in high ambient temperatures. *International Journal of Sport Nutrition and Exercise Metabolism*, **12**, 105–119.
217. Hitchins, S., Martin, D., Burke, L., *et al.* (1999) Glycerol hyperhydration improves cycle time trial performance in hot humid conditions. *European Journal of Applied Physiology and Occupational Physiology*, **80**, 494–501.

218. Ackermans, M.T., Ruiter, A.F.C., and Endert, E. (1998) Determination of glycerol concentrations and glycerol isotopic enrichments in human plasma by gas chromatography/mass spectrometry. *Analytical Biochemistry*, **258**, 80–86.
219. Thevis, M., Guddat, S., Flenker, U., and Schänzer W. (2008) Quantitative analysis of urinary glycerol levels for doping control purposes using gas chromatography-mass spectrometry. *European Journal of Mass Spectrometry*, **14**, 117–125.
220. Magni, F., Arnoldi, L., Monti, L., *et al.* Determination of plasma glycerol isotopic enrichment by gas chromatography-mass spectrometry: An alternative glycerol derivative. *Analytical Biochemistry*, **211**, 327–328.
221. Flakoll, P.J., Zheng, M., Vaughan, S., and Borel, M.J. (2000) Determination of stable isotopic enrichment and concentration of glycerol in plasma via gas-chromatography-mass spectrometry for the estimation of lipolysis in vivo. *Journal of Chromatography B*, **744**, 47–54.
222. Gilker, C.D., Pesola, G.R., and Matthews, D. (1992) A mass spectrometric method for measuring glycerol levels and enrichments in plasma using 13C and 2H stable isotopic tracers. *Analytical Biochemistry*, **205**, 172–178.
223. Dellacherie, E. (1996) Polysaccharides in oxygen-carrier blood substitutes, in *Polysaccharides in Medicinal Applications* (ed S. Dumitriu), Marcel Dekker, Inc., New York, pp. 525–544.
224. Nitsch, E. (1998) Volumenersatz mit künstlichen Kolloiden. *Anaesthesiologie, Intensivmedizin, Notfallmedizin, Schmerztherapie*, **33**, 255–260.
225. Czerny, A. (1894) Versuche über Bluteindickung und ihre Folgen. *Naunyn-Schmiedeberg's Archives of Pharmacology*, **34**, 268–280.
226. Moffitt, E.A. (1975) Blood substitutes. *Canadian Anaesthetists' Society Journal*, **22**, 12–19.
227. Thevis, M., Opfermann, G., and Schänzer, W. (2000) Detection of the plasma volume expander hydroxyethyl starch in human urine. *Journal of Chromatography B*, **744**, 345–350.
228. Thevis, M., Opfermann, G., and Schänzer, W. (2000) Mass spectrometry of partially methylated alditol acetates derived from hydroxyethyl starch. *Journal of Mass Spectrometry*, **35**, 77–84.
229. de Belder, A.N. (1996) Medical applications of dextran and its derivatives, in *Polysaccharides in Medicinal Applications* (ed S. Dumitriu), Marcel Dekker, Inc., New York, pp. 505–523.
230. Guddat, S. (2006) Identifizierung und quantitative Bestimmung des Plasmavolumenexpanders Dextran und des Diuretikums Mannitol in Humanurin mittels Flüssigkeitschromatographie / Massenspektrometrie und Gaschromatographie / Massenspektrometrie zu Dopingkontrollzwecken. Institute of Biochemistry, German Sport University Cologne. Dissertation.
231. Thevis, M., Opfermann, G., and Schänzer, W. (2004) *N*-methyl-*N*-trimeth ylsilyltrifluoroacetamide promoted synthesis and mass spectrometric

characterization of deuterated ephedrines. *European Journal of Mass Spectrometry*, **10**, 673–681.

232. Leloux, M.S., and Maes, R.A. (1990) The use of electron impact and positive chemical ionization mass spectrometry in the screening of beta blockers and their metabolites in human urine. *Biomedical and Environmental Mass Spectrometry*, **19**, 137–142.

4 Structure Characterization of Low Molecular Weight Target Analytes: Electrospray Ionization

The use of atmospheric pressure ionization such as electrospray ionization (ESI) in connection with liquid chromatography has become routine in doping controls since the late 1990s.[1] Although several reports of its utility in the field of sports drug testing were described earlier,[2–6] ESI or atmospheric pressure chemical ionization (APCI) were not employed on a daily basis until 2000/2001.[7–9] With the constantly increasing number of analytes and applications of LC-MS(/MS), detailed knowledge of mass spectrometric behavior and dissociation pathways of target compounds was required, and numerous studies were dedicated to the elucidation of fragmentation routes initiated by protonation or deprotonation followed by collision-induced dissociation (CID). In contrast to most EI-MS(/MS) studies described in Chapter 3, which were commonly conducted at 70 eV, the dissociation of precursor ions using collisional activation after soft ESI yields considerably different product ion mass spectra depending on whether low, medium, or high collision offset voltages are applied. Moreover, the type of mass spectrometer and the used MS/MS mode, i.e., in-space or in-time, have been shown to produce significantly deviating ESI-MS/MS spectra. Hence, the analyzers used to generate the product ion mass spectra shown in the following pages will be included in respective figure captions.

Mass Spectrometry in Sports Drug Testing: Characterization of Prohibited Substances and Doping Control Analytical Assays, By Mario Thevis
Copyright © 2010 John Wiley & Sons, Inc.

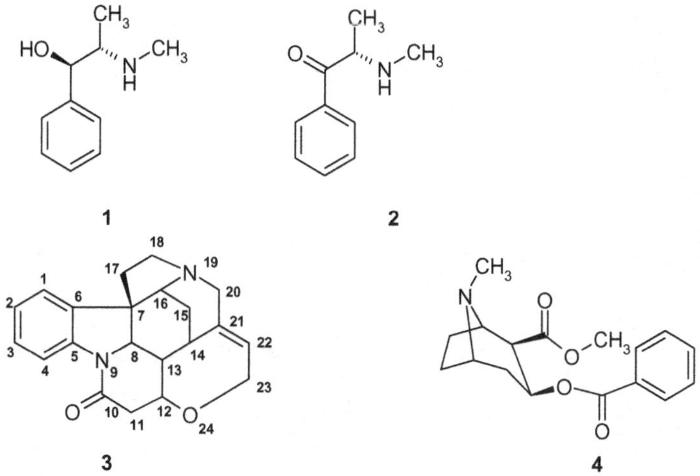

Figure 4.1: Structures of selected stimulants: ephedrine (**1**), methcathinone (**2**), strychnine (**3**), and cocaine (**4**).

4.1 STIMULANTS

As mentioned in Chapters 1 and 3, stimulants have been considered as relevant for doping controls for more than a century and were subject of very early anti-doping research. A variety of stimulating agents contains common structural features such as the phenylethylamine core, e.g., ephedrine and methcathinone (Fig. 4.1, **1** and **2**, respectively), which are complemented by several alkaloids such as strychnine and cocaine (Fig. 4.1, **3** and **4**, respectively), which have also demonstrated stimulating properties. The considerable proton affinity of amines in general enabled efficient ionization of stimulants using positive ESI, and CID yielded informative product ion mass spectra as shown for example with ephedrine, strychnine, and cocaine (Figure 4.2, a–c).

The protonated molecules of phenylpropanolamines were shown to dissociate in a distinct way as outlined in detail for ephedrine (Scheme 4.1a).[10,12,14,15] In one route, the precursor ion at *m/z* 166 eliminated water (–18 Da) to generate the ion at *m/z* 148, which subsequently expelled methylamine (–31 Da) to yield the product ions at *m/z* 117 and 115, representing the cations of allyl-benzene and prop-2-ynyl-benzene, respectively. Alternatively, a second dissociation route starting also

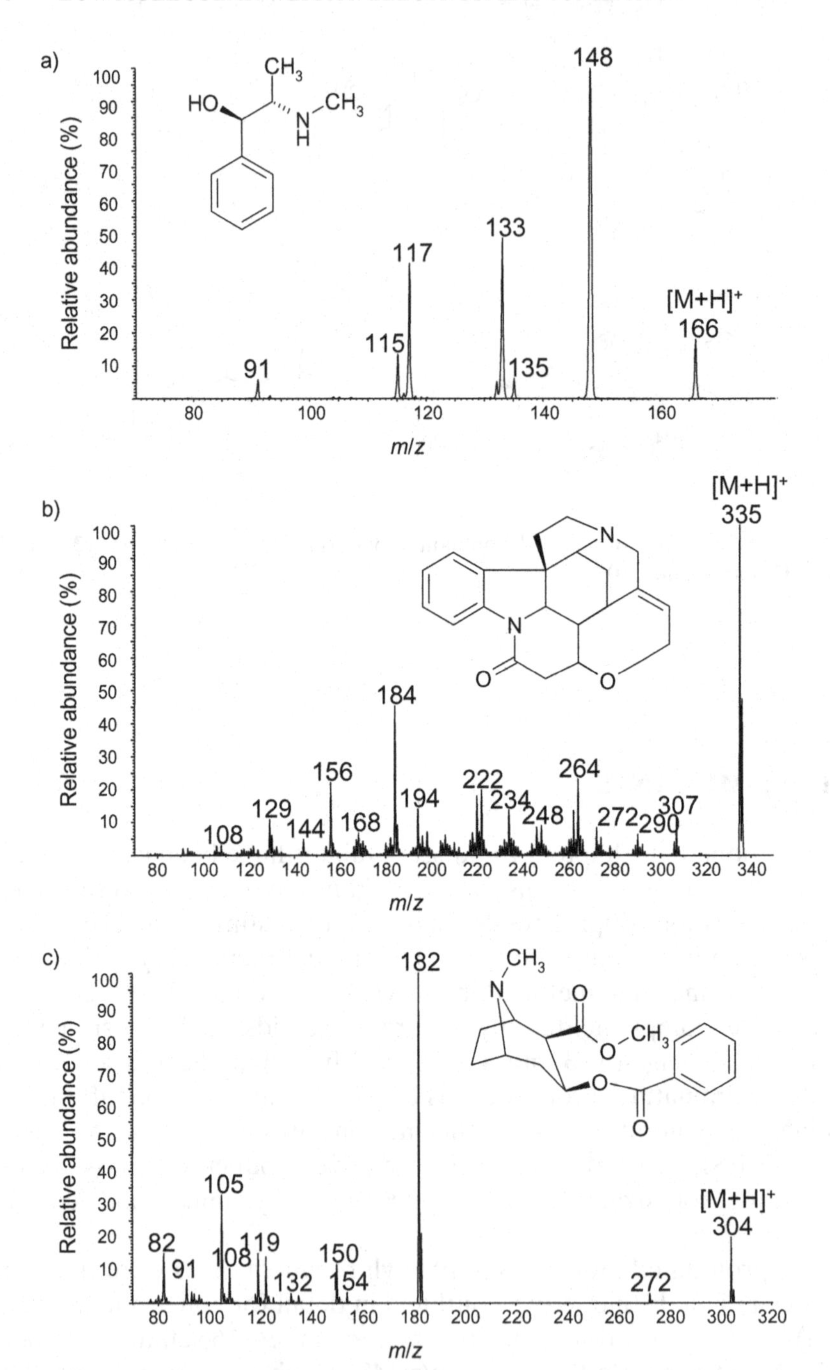

Figure 4.2: ESI product ion mass spectra of protonated molecules $[M+H]^+$ of (a) ephedrine (mol wt = 165), (b) strychnine (mol wt = 334), and (c) cocaine (mol wt = 303), all recorded on a QTrap 2000 system.

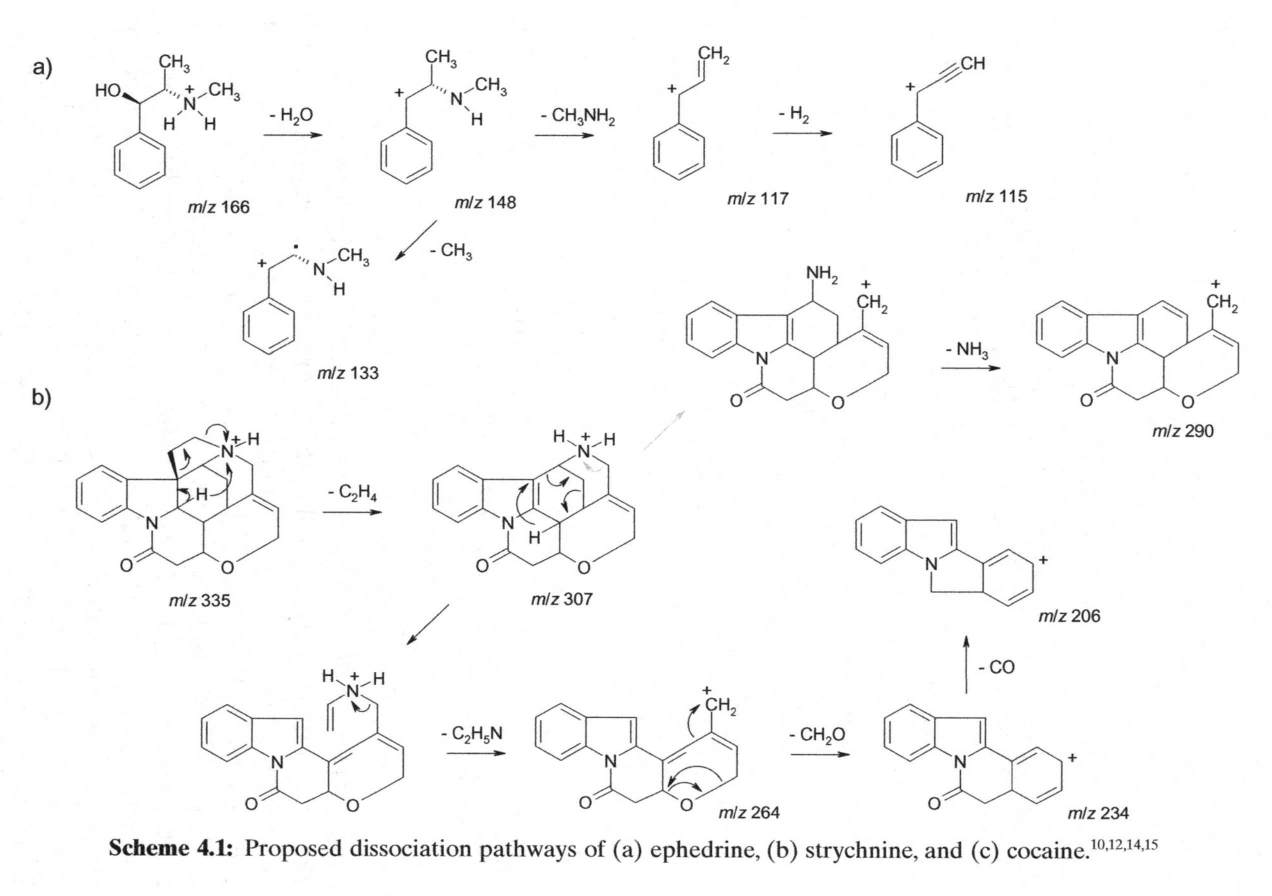

Scheme 4.1: Proposed dissociation pathways of (a) ephedrine, (b) strychnine, and (c) cocaine.[10,12,14,15]

c)

Scheme 4.1: (*Continued*)

from $[M+H]^+$-H_2O at *m/z* 148 was determined, which included the elimination of a methyl radical (–15 Da) originating from either of the methyl functions as demonstrated by MS^3 and deuterium labeling experiments and produced the ion at *m/z* 133.[10] Oxidation of the hydroxyl function to the corresponding carbonyl residue, as in case of methcathinone (Fig. 4.1, **2**) did not alter the dissociation behavior significantly; however, the simultaneous modification of the secondary amine to a tertiary amine as present in metamfepramone favored a different fragmentation process with the release of the alkylated amine followed by carbon monoxide rather than the loss of water followed by the alkylated amine.[11] An overview of characteristic product ions generated from protonated molecules of selected stimulants is listed in Table 4.1.

TABLE 4.1: Characteristic Product Ions of Selected Stimulants Generated by ESI and CID

Compound	Mol Wt (Da)	Precursor Ion (m/z)	Product Ions (m/z)		
Amfepramone	205	206	133	105	100
Amphetamine	135	136	119	91	90
Benzphetamine	239	240	148	119	91
Cathine	151	152	134	117	115
Cocaine	303	304	272	182	82
Dimethylamphetamine	163	164	119	91	90
Ephedrine	165	166	148	133	117
Etamivan	223	224	151	123	108
Etilefrine	181	182	164	135	91
Fencamphamine	215	216	171	129	91
Fenfluramine	231	232	187	159	109
Fenproporex	188	189	158	119	91
p-hydroxyephedrine	181	182	164	149	133
Metamfepramone	177	178	160	133	105
Methylenedioxyamphetamine	179	180	163	135	105
Methylenedioxymethamphetamine	193	194	163	135	105
Methoxyphenamine	179	180	149	121	91
Methylephedrine	179	180	162	146	115
Methylphenidate	233	234	174	129	84
Nikethamide	178	179	151	108	80
Pethidine	247	248	220	174	131
Phendimetrazine	191	192	174	148	91
Phenmetrazine	177	178	160	134	117
Phentermine	149	150	133	105	91
Sibutramine	279	280	153	139	125
Strychnine	334	335	307	264	184
Tuaminoheptane	115	116	99	57	43

With the early recognition of the misuse of strychnine in sports (Chapter 1), the difficulty of its analysis was also observed; however, the use of LC-MS/MS has significantly facilitated and improved the analytical options and detection limits as, in contrast to EI-MS (Chapter 3), abundant and diagnostic product ions are formed under ESI-CID conditions (Fig. 4.2b). With the use of high resolution/high accuracy MS^n, dissociation pathways and potential structures of major product ions were proposed (Scheme 4.1b). The initial loss of ethylene (−28 Da) was identified as the starting point of different dissociation routes, yielding the intermediately formed product ion at m/z 307. Here, the carbons C-17 and C-18 were suggested to be removed from the protonated molecule, necessitating the migration of a hydrogen that presumably originated from C-8. Subsequently, either ammonia (−17 Da) or vinylamine (−43 Da) was eliminated to generate the product ions at *m/z* 290 or 264, respectively. The latter required the opening of the ring structure composed of C-7, C-8, C-13–16 to yield a cation comprising four rings that allowed the loss of C_2H_5N including C-15, C-16, and N-19. The resulting ion at *m/z* 264 further released formaldehyde (−30 Da) and carbon monoxide (−28 Da) consecutively, producing the ions observed at *m/z* 234 and 206.[12] The use of high resolution/high accuracy MS^n to support the suggested dissociation routes was of utmost importance because of the complexity of CID spectra obtained from *strychnos* alkaloids. Earlier attempts of assigning product ions to possible structures failed most likely due to the lack of sufficient information about elemental compositions.[13]

Cocaine is commonly referred to being one of the first (mis)used ergogenic and stimulating drugs, and the options on how to detect cocaine in various clinical, forensic, and doping control arenas have been studied extensively. Besides EI-MS, cocaine and its main metabolites were characterized and analyzed using ESI-MS/MS by means of diagnostic product ions derived from protonated molecules, which were elucidated in-depth using stable-isotope, H/D-exchange, and MS^n experiments.[14–16] Within these studies, several dissociation pathways were described, explaining the formation of the abundant and characteristic first and second generation product ions of cocaine (Fig. 4.2c). Initial protonation of cocaine is suggested to appear at the nitrogen atom, but CID of the activated molecule allows proton transfer reactions between the functional groups; hence, the dissociation routes outlined in Scheme 4.1c represent only selected options of several possibilities. The protonated precursor ion at *m/z* 304 yielded a low abundant product ion at *m/z* 272 due to the elimination of methanol (−32 Da), which was shown to include the proton introduced during the ionization process. The protonation of the bridgehead nitrogen of cocaine was suggested to be followed by a chair to boat conformation transformation, which requires a comparably low energy of 7 kcal/mol, in order to allow the release of benzoic acid (−122 Da) yielding the base peak of the product ion mass spectrum at *m/z* 182. The proposed structure of *m/z* 182, the cation of 8-methyl-8-aza-bicyclo[3.2.1]octane-2-carboxylic acid methyl ester, further dissociates via two pathways yielding the ions at *m/z* 150, 122, 119, 82, and 91. Except for the ion at *m/z* 82, which was suggested to comprise the cation of 1-methyl-2,3-dihydro-1H-pyrrole formed by the loss of but-3-enoic acid methyl ester (−100 Da), all second generation product ions were suggested to be generated via *m/z* 150, which was shown to result from the loss of methanol (–32 Da). The proposed structure of *m/z* 150 represents an 8-methyl-8-aza-bicyclo[3.2.1]oct-2-en-2-ylmethylidyne-oxonium or 8-methyl-8-aza-bicyclo[3.2.1]oct-2-ene-2-carbaldehyde cation, which subsequently eliminates carbon monoxide (–28 Da), methylamine (–31 Da), and methylamine plus carbon monoxide to give rise to the product ions at *m/z* 122, 119, and 91, respectively. Alternatively, the precursor ion at *m/z* 304 was shown to expel 3-hydroxy-8-methyl-8-aza-bicyclo[3.2.1] octane-2-carboxylic acid methyl ester (–199 Da) to produce the ion at *m/z* 105 (benzaldehyde or phenylmethylidyne-oxonium cation).

4.2 NARCOTICS

Although the ergogenic properties of narcotics in elite sport are rather limited, their formerly widespread use (see Chapters 1 and 3) and the fact that modern elite sport represents enormous physiological demands and risks of injuries, the misuse of narcotics has been prohibited in competition since systematic doping controls were initiated in the 1960s.

The high proton affinity of most narcotics such as morphine and related compounds (e.g., heroin, buprenorphine) or the synthetic opioids pethidine, fentanyl, and its derivatives (Fig. 4.3, **1–5**), allows the efficient ionization by means of ESI and, thus, the sensitive detection

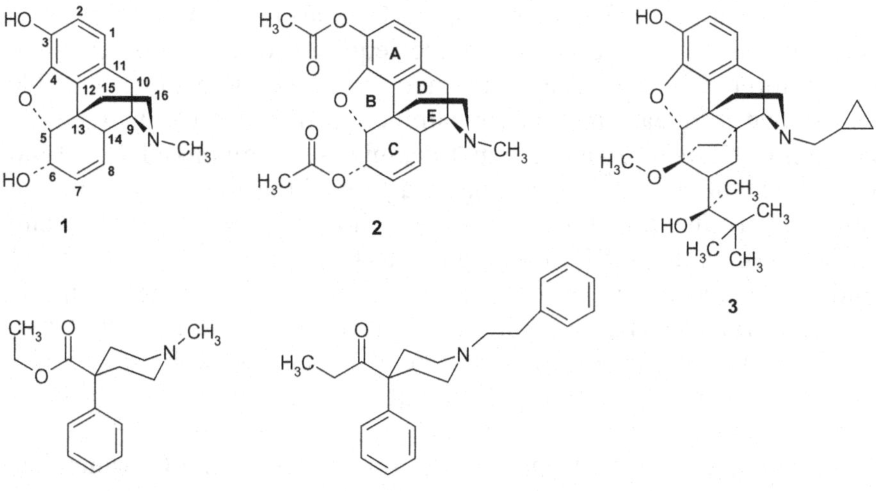

Figure 4.3: Structures of selected narcotics: morphine (**1**), heroin (**2**), buprenorphine (**3**), pethidine (**4**), and fentanyl (**5**).

TABLE 4.2: Characteristic Product Ions of Selected Narcotics Generated by ESI and CID

Compound	Mol Wt (Da)	Precursor Ion (*m/z*)	Product Ions (*m/z*)		
Alfentanil	416	417	385	268	197
Buprenorphine	467	468	414	396	55
Fentanyl	336	337	216	188	105
Hydromorphone	285	286	227	185	157
Heroin	369	370	328	310	268
Methadone	309	310	265	219	105
Morphine	285	286	181	165	153
Oxycodone	315	316	298	256	241
Pentazocine	285	286	218	175	159
Pethidine	247	248	220	174	131
Remifentanil	376	377	317	285	228
Sufentanil	386	387	355	238	224

of the active drugs as well as major metabolites in doping control samples. Depending on the principle structure, i.e., whether it is a morphine-like alkaloid or a fully synthetic substitute such as pethidine or fentanyl, the dissociation pathways of protonated molecules under CID conditions were shown to be considerably different (Table 4.2) and at least occasionally of enormous complexity. The product ion mass spectrum of morphine is depicted in Fig. 4.4a, which illustrates the comprehensive fragmentation of the protonated molecule after collisional activation, and high resolution/high accuracy MS^n as well as selective stable isotope labeling were required to rationalize possible dissociation routes.[17–19] The protonated morphine molecule at *m/z* 286 was shown to release methyl-vinylamine (–57 Da), which included the carbons C-15 and C-16 and the methylamine function linked to C-9 and C-16 (Scheme 4.2a),[17–19] resulting in a loss of the E-ring system. The obtained product ion at *m/z* 229 further eliminated water (–18 Da) and carbon monoxide (–28 Da), giving rise to the ions observed at *m/z* 211 and 201, respectively, with the latter lacking C-5 and the formerly present furan ring (B). In addition, structures of various abundant and/or diagnostic product ions were proposed, for instance for *m/z* 185, 183, and 181 (Scheme 4.2a). While the ion at *m/z* 185 was shown to be composed of intact A, B, and D-rings, the product ions at *m/z* 183 and 181 comprised the A- and D-ring plus the remainder of ring C (which lost carbon C-5), forming a cyclopentanaphthalene nucleus. The release of carbon monoxide (–28 Da) from *m/z* 185 and 181 yielded the characteristic ions at *m/z* 157 and 153, and the loss of water (–18 Da) from

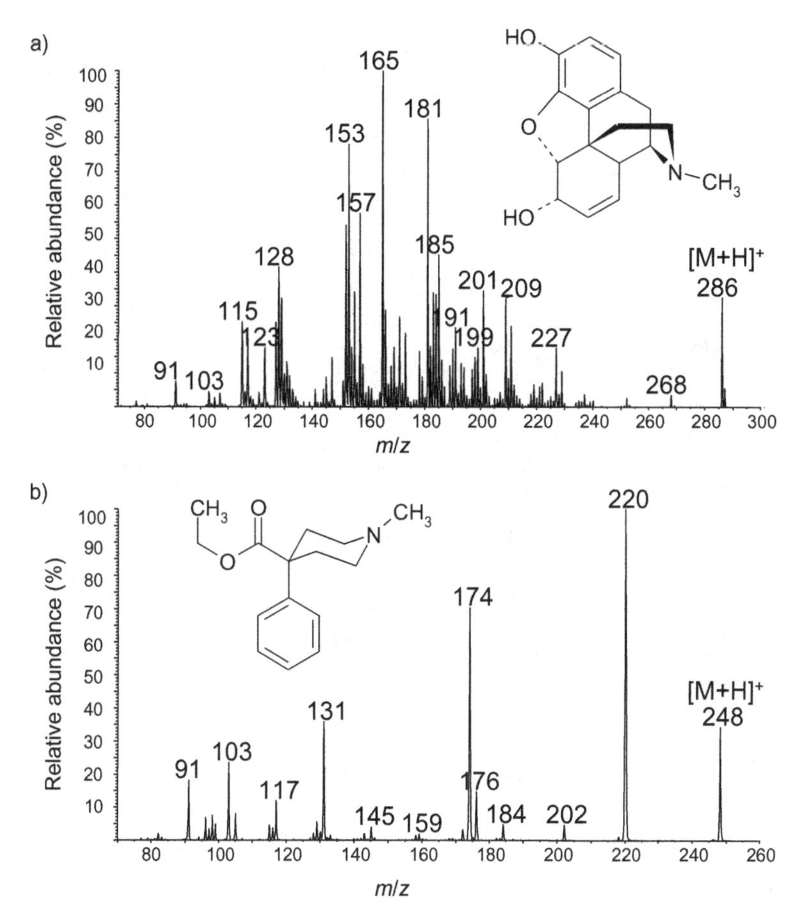

Figure 4.4: ESI product ion mass spectra of protonated molecules $[M+H]^+$ of (a) morphine (mol wt = 285), and (b) pethidine (mol wt = 247), recorded on a QTrap 2000 system.

m/z 183 resulted in the base peak ion (3Ah-cyclopenta[a]naphthalene) of the product ion mass spectrum of morphine found at *m/z* 165 (Fig. 4.4a).

The dissociation behavior of the protonated molecule of pethidine (Fig. 4.3, **4**) was significantly different and yielded less but still informative product ions that characterize the synthetic opioid (Fig. 4.4b).[20] The precursor ion at *m/z* 248 was predominantly de-esterified by the elimination of ethylene (–28 Da) to yield the ion at *m/z* 220, which subsequently released carbon dioxide (–44 Da) or water (–18 Da) and carbon monoxide (–28 Da) to produce *m/z* 176 and 174, respectively. The consecutive losses resulting in the product ion at *m/z* 174 were

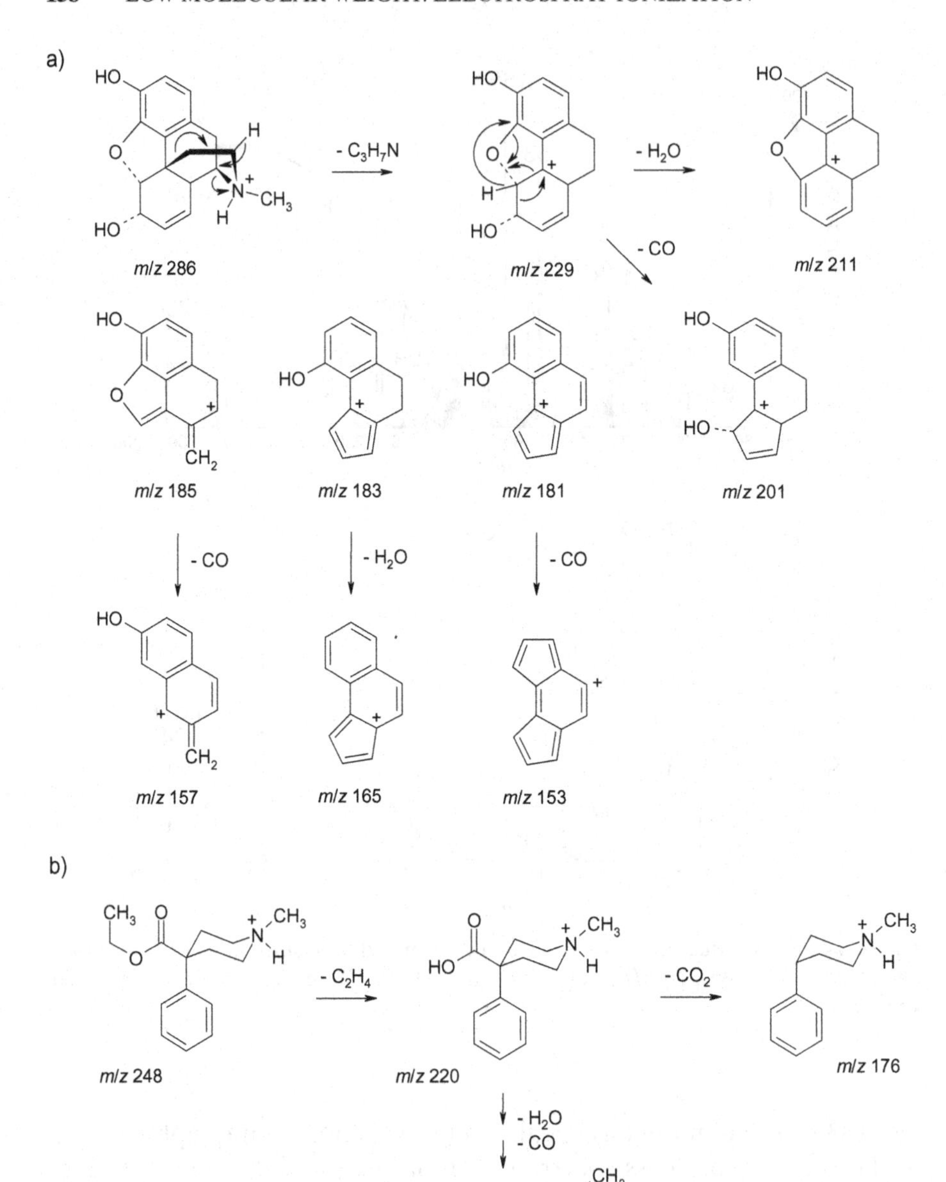

Scheme 4.2: Proposed dissociation pathways of (a) morphine, and (b) pethidine.[17–19]

further followed by the elimination of methyl-methyleneamine (–43 Da) leading to the formation of the (1-methyl-allyl)-benzene cation at *m/z* 131 (Scheme 4.2b).

4.3 ANABOLIC ANDROGENIC STEROIDS

The class of anabolic androgenic steroids has been a particular challenge for ESI-CID studies due to various aspects. The great variety of analytes with similar or even identical elemental compositions, the complex product ion spectra resulting from the cyclopenta[α]phenanthrene skeleton, and last but not least the limited proton affinity of numerous representatives has impeded but not stopped studies elucidating the intricate dissociation pathways of anabolic androgenic steroids and their metabolites.[21] Due to the comprehensiveness of this class of compounds, categories comparable to those used in Chapter 3 were employed to structure the discussion of their ESI-CID properties.

4.3.1 α,β-Saturated 3-Keto-Steroids

Steroidal agents such as 5α-androstan-17β-ol-3-one (Fig. 4.5, **1**) and selected analogues (Fig. 4.5, **2–4**) dissociate efficiently upon ESI and CID as shown exemplarily for 5α-androstan-17β-ol-3-one (Fig. 4.6a). Commonly, the saturated scaffold of cyclopenta[α]phenanthrene-derived compounds results in product ion mass spectra predominantly containing ions of low specificity, which are separated by methylene units (14 Da). Although α,β-unsaturated 3-keto-steroids also yield such product ions, the presence of double bonds commonly directs the dissociation pathway to specific and more characteristic fragments (*vide infra*).

A diagnostic product ion obtained from α,β-saturated 3-keto-steroids is commonly found at *m/z* 215 (Fig. 4.6a), which was suggested to result from the elimination of water (–18 Da) and acetone (–58 Da) from the A-ring of the protonated molecule (Scheme 4.3a).[22,131] The analysis of analogues bearing methyl residues at C-17 or C-1 yielded a fragment ion at *m/z* 229, corresponding to *m/z* 215, which indicated the preservation of these sites within the product ion after elimination of a total of 76 Da from the precursor ion (Table 4.3). Moreover, deuterium labeling at C-16 and C-17 caused a shift of *m/z* 215 to 218, which further substantiated the presence of these carbons in the respective product ion. In contrast, four deuterium atoms introduced at C-2 and C-4 were eliminated from steroids with the above reported loss of

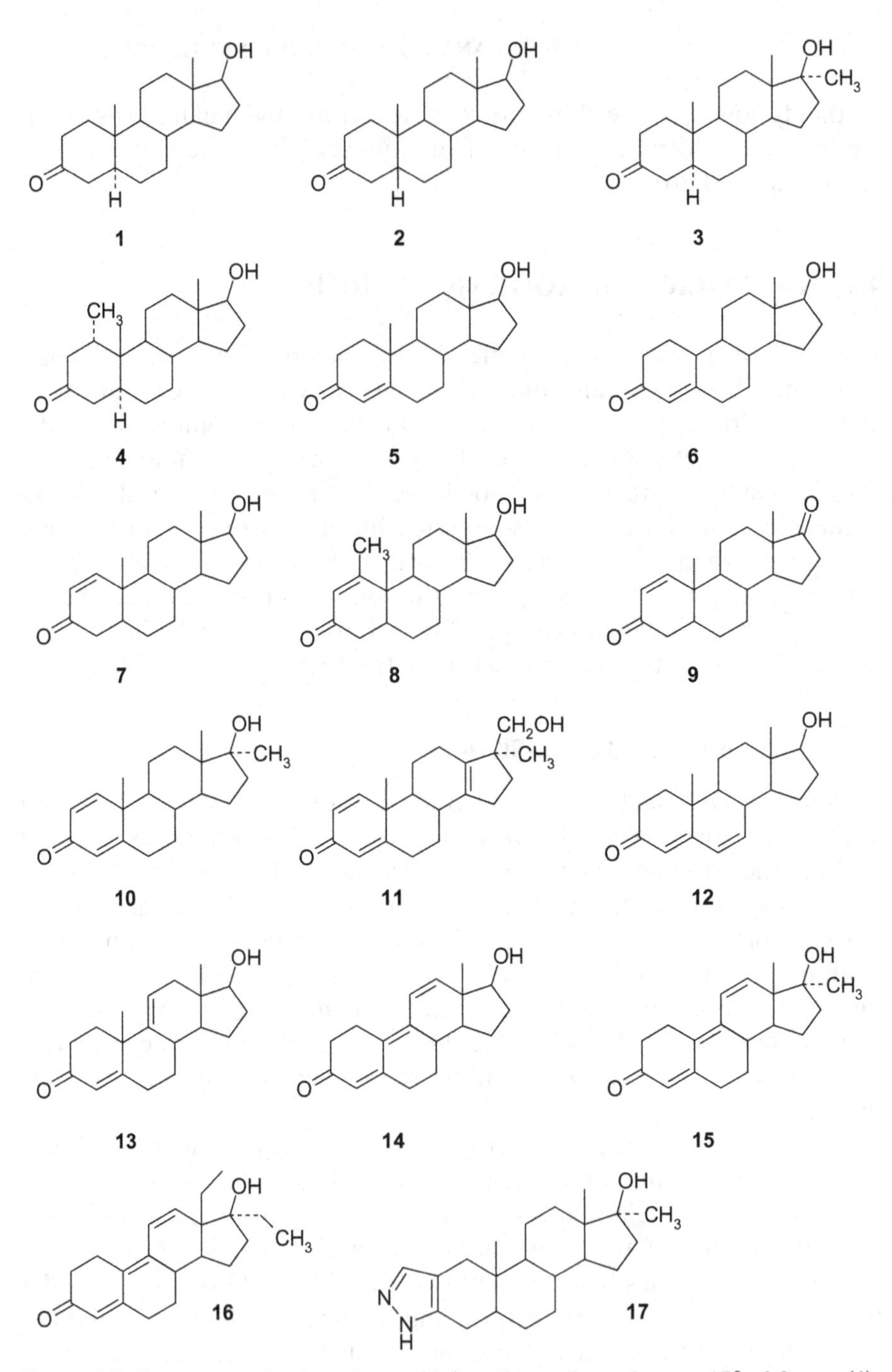

Figure 4.5: Structures of selected steroidal analytes: 5α-androstan-17β-ol-3-one (**1**), 5β-androstan-17β-ol-3-one (**2**), 17α-methyl-5α-androstan-17β-ol-3-one (**3**), 1α-methyl-5α-androstan-17β-ol-3-one (**4**), testosterone (**5**), nandrolone (**6**), 1-testosterone (**7**), metenolone (**8**), 5α-androst-1-ene-3,17-dione (**9**), metandienone (**10**), 18-nor-17β-hydroxymethyl,17α-methyl-androst-1,4,13-trien-3-one (metandienone metabolite) (**11**), androsta-4,6-dien-17β-ol-3-one (**12**), androsta-4,9-dien-17β-ol-3-one (**13**), trenbolone (**14**), methyltrienolone (**15**), tetrahydrogestrinone (THG, **16**), and stanozolol (**17**).

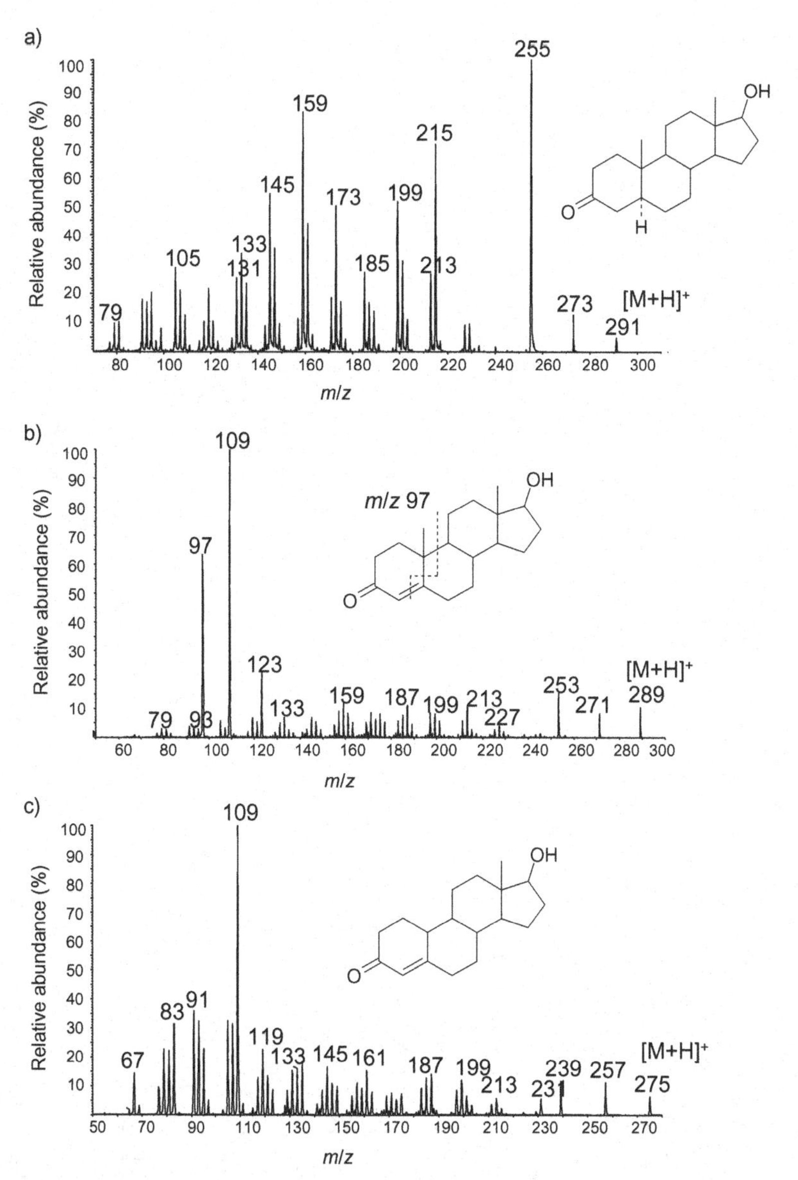

Figure 4.6: ESI product ion mass spectra of protonated molecules $[M+H]^+$ of (a) 5α-androstan-17β-ol-3-one (mol wt = 290), (b) testosterone (mol wt = 288), and (c) nandrolone (mol wt = 274) recorded on a QTrap 2000 system.

acetone.[22] Also 2- and 3-keto-steroids were described to release acetone molecules after EI in earlier studies by Budzikiewicz, Djerassi, and co-workers.[23–25] Selective labeling experiments proved the participation of the hydrogen atom located at C-6 in the required rearrangement, and dissociation route to *m/z* 215 was shown to be strongly influenced

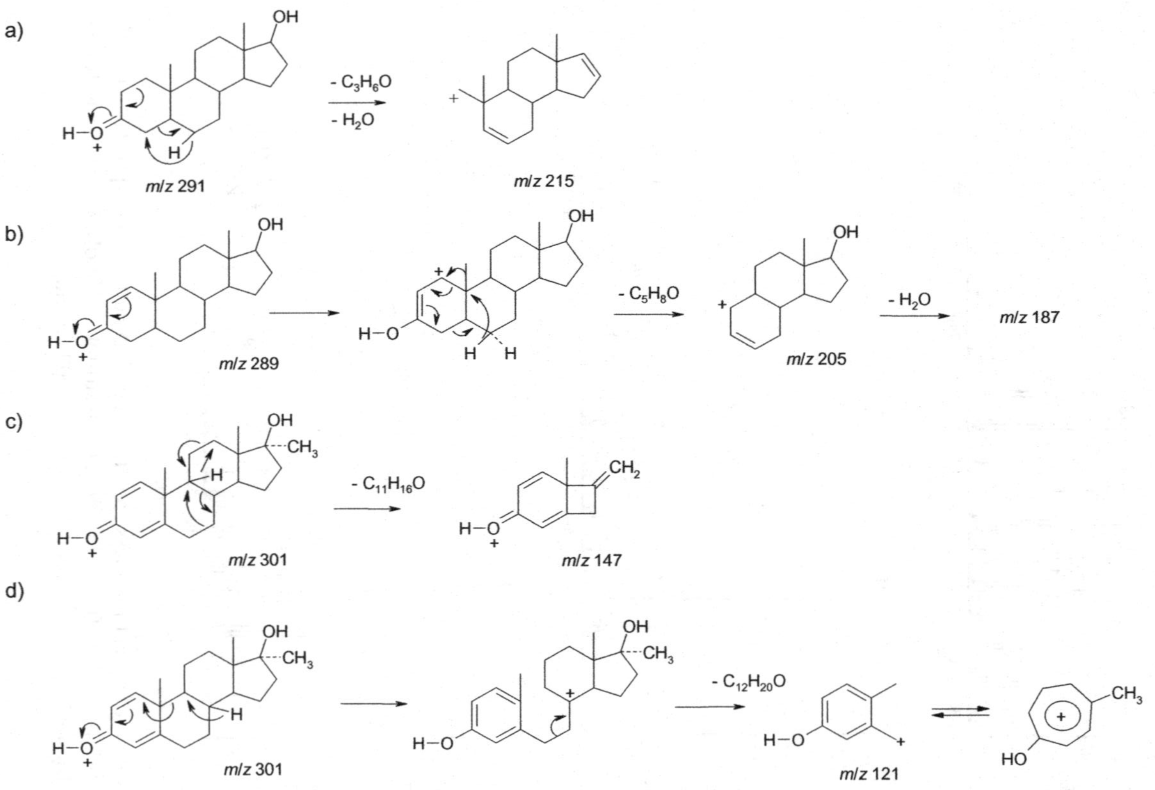

Scheme 4.3: Suggested dissociation pathways of (a) 5α-androstan-17β-ol-3-one, (b) 1-testosterone, and (c) and (d) metandienone.[22,131]

TABLE 4.3: Characteristic Product Ions of Selected Steroidal Agents Generated by ESI and CID[a]

Steroid Nucleus	Representative Compound	Mol Wt (Da)	Precursor Ion (m/z) $[M+H]^+$	Product Ions (m/z) $[M+H]^+$-18				
3-keto	5α-dihydrotestosterone	290	291	273	255[b]	215	159	145
	5β-dihydrotestosterone	290	291	273	255[b]	215	173	159
	17α-methyl-5α-androstane-17β-ol-3-one	304	305	287	269[b]	229	173	159
	1α-methyl-5α-androstane-17β-ol-3-one	304	305	287	269[b]	229	213	159
3-keto-4-ene	Testosterone	288	289	271	253	123	109[a]	97
	Nandrolone	274	275	257	239	119	109[a]	91
3-keto-1-ene	1-testosterone	288	289	271	253	187[b]	145	131
	Metenolone	302	303	285	267	187[b]	145	131
	Androst-1-en-3,17-dione	286	287	269	203	185[b]	143	129
3-keto-1,4-diene	1-dehydrotestosterone	286	287	269	173	147	135	121[b]
	Metandienone	300	301	283	149	147	135	121[b]
3-keto-4,6-diene	6-dehydrotestosterone	286	287	269	173	151	133[b]	97
3-keto-4,9-diene	Androsta-4,9(11)-diene-17β-ol-3-one	286	287	269	254	239	147[b]	145
3-keto-4,9,11-triene	Trenbolone	270	271	253	238	227	199[b]	157
	Gestrinone	308	309	291	262	241	199[b]	179
	Tetrahydrogestrinone	312	313	295	266	241[b]	199	159
	Propyltrenbolone	312	313	295	252	227[b]	199	183

[a] Refs. 21 and 131.
[b] Base peak.

by stereochemical properties of the androstan-17β-ol-3-ones. The orientation of the hydrogen at C-5 plays a key role as relative abundances of *m/z* 215 differed significantly from 5α-androstan-17β-ol-3-one (71%) to 5β-androstan-17β-ol-3-one (9%) under identical mass spectrometric conditions, which was considered a means to distinguish between these stereoisomers using mass spectrometric approaches.

4.3.2 3-Keto-4-ene and 3-Keto-1-ene Steroids

Typical representatives of α,β-unsaturated 3-keto-steroids are testosterone, nandrolone, and 1-testosterone (Fig. 4.5, **5–7**). These analytes possess a comparably high proton affinity and are more efficiently ionized using ESI than their saturated derivatives.[26] Characteristic dissociation pathways of protonated precursor ions were extensively studied, and in particular for testosterone and its hydroxylated analogues, stable-isotope labeled analogues were measured to provide substantial information on the origin and composition of diagnostic product ions.[27] Major product ions of testosterone obtained by positive ESI and CID are observed at *m/z* 97 and 109, which characterize the general 3-keto-androst-4-ene nucleus as detected in various analogous compounds (Table 4.3). The generation of *m/z* 109 was suggested to include cleavages of the bonds between C-1 and C-10, C-5 and C-10 as well as C-7 and C-8 accompanied by hydrogen migrations from C-2 to C-5 and C-19 to C-11, which was supported by extensive labeling studies including the hydrogen atoms at C-19, C-1, C-2, C-4, and C-6 in addition to ^{13}C-labeling of carbon 3.[27,28] In accordance, the structure of *m/z* 97 was investigated, providing evidence for the presence of carbons 1–4 and 19. A complex rearrangement is required that includes the cleavage of the C-C bonds between carbons 9 and 10, 5 and 10, as well as 4 and 5, and a possible route was recently suggested[29] as indicated in the product ion mass spectrum of testosterone (Fig. 4.6b).

Nandrolone (19-nortestosterone, Fig. 4.5, **6**) lacks the C-19 methyl residue of testosterone, a fact that causes a considerable change of the dissociation behavior of protonated molecules compared to the 3-keto-androst-4-ene-derived steroids. While the ion at *m/z* 109 is still observed, the fragmentation of the positively charged species yields many additional abundant product ions, but the diagnostic fragment ion at *m/z* 97 is not detected due to the lack of the C-19 residue (Fig. 4.6c).

Relocating the C-4–C-5 double bond to C-1–C-2, which yields a compound commonly referred to as 1-testosterone (5α-androst-1-en-17β-ol-3-one, Fig. 4.5, **7**), also initiates a significantly different dissociation route compared to testosterone. An intense product ion at *m/z* 187 is found (Table 4.3) that is presumably composed by the A- and B-ring of the steroid nucleus. Evidence for a proposed dissociation route was obtained by deuterium labeling and analysis of structurally related drugs such as metenolone and 5α-androst-1-ene-3,17-dione (Fig. 4.5, **8–9**),[30] and the dissociation of the protonated 1-testosterone to *m/z* 187 was suggested to result from an elimination of the carbons C-1–C-4 and C-19 followed by a loss of water (Scheme 4.3b). Indicators for the suggested fragmentation route were the detection of *m/z* 187 in product ion spectra of C-1-substituted 1-testosterone (metenolone), which proved the release of C-1 (and its substituent) from the steroid, and the elimination of the entire steroidal A-ring (C-1–C-4) during the fragmentation process as demonstrated by means of H/D-exchange. However, the postulated generation of *m/z* 187 requires a Wagner-Meerwein rearrangement,[31] i.e., the migration of C-19 from C-10 to C-1 of the leaving group (Scheme 4.3b), for which several examples were described in steroid biochemistry.[32–34]

4.3.3 3-Keto-1,4-diene Steroids

A typical representative of 3-keto-1,4-diene steroids is metandienone (Fig. 4.5, **10**), one of the most characteristic product ions found (though of limited abundance) at *m/z* 147 (Fig. 4.7a). This ion is commonly observed in ESI product ion spectra of steroids with cross-conjugated π-electron systems established by 1,4-diene-3-one structures such as metandienone, boldenone, and many synthetic corticosteroids, e.g., dexamethasone, isoflupredone, prednisolone, and prednylidene. The product ion at *m/z* 147 is proposed to result from the steroidal A-, B-, and C-rings by fissions of the linkages between C-6 and C-7, C-8 and C-9, and C-11 and C-12 (Scheme 4.3c). It was suggested that the protonated species undergoes a hydrogen transfer from C-9 to C-12 that generates a double bond between C-9 and C-11, and the subsequent formation of a 4-membered ring structure requiring the rearrangement of bonds between C-6 and C-7 as well as C-8 and C-9 yields the postulated product ion structure at *m/z* 147. Another characteristic product ion is observed at *m/z* 121, the proposed generation of which is also initiated by the protonation of the 3-keto function followed by the formation of a 3-linked 4-methylphenol residue. The migration of the C-8 hydrogen was proposed to accompany the cleavage of the bond between C-9 and C-10 as supported by deuterium labeling experiments.[35] A temporary charge transfer to C-8 initiates the fission of the linkage between C-5 and C-6 yielding the fragment ion at *m/z* 121 (Scheme 4.3d).[22]

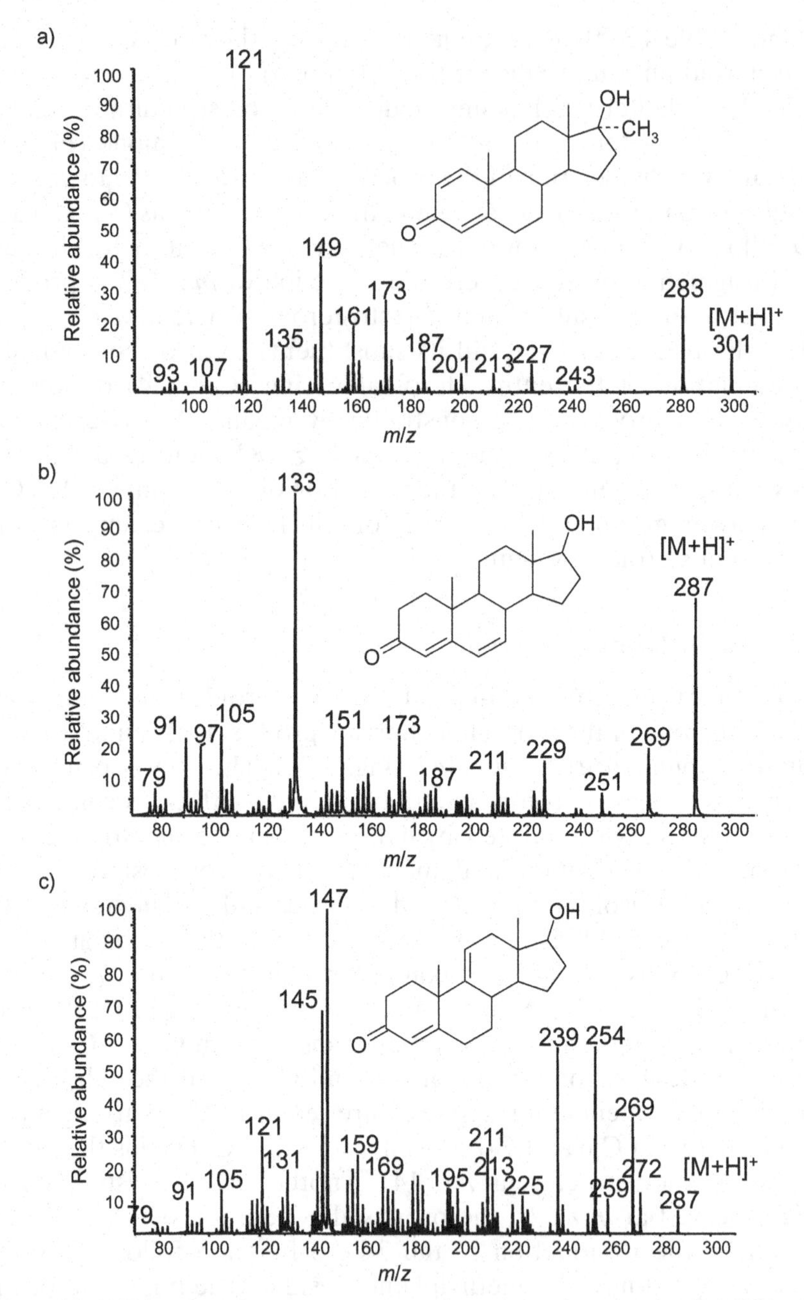

Figure 4.7: ESI product ion mass spectra of protonated molecules $[M+H]^+$ of (a) metandienone (mol wt = 300), (b) androsta-4,6-dien-17β-ol-3-one (mol wt = 286), and (c) androst-4,9-dien-17β-ol-3-one (mol wt = 286) recorded on a QTrap 2000 system.

The utility of the detailed knowledge on the mass spectrometric behavior of steroidal agents was recently demonstrated when the combined information obtained from EI and ESI-MS(/MS) experiments led to the successful identification of a long-term metabolite of metandienone, namely 18-nor-17β-hydroxymethyl,17α-methyl-androst-1,4,13-trien-3-one (Fig. 4.5, **11**),[36] which subsequently enabled the detection of more than 60 doping rule violations by elite athletes with metandienone.[37]

4.3.4 3-Keto-4,6-diene Steroids

Locating two double bonds between carbons 4 and 5 as well as 6 and 7, as found for instance in the case of androstadien-17β-ol-3-one (Fig. 4.5, **12**) commonly results in a specific dissociation pattern, which yields a predominant product ion at *m/z* 133 (Fig. 4.7b). It is proposed to be composed of the steroidal C- and D-rings, and postulated to be generated by cleavages of the linkages between C-9 and C-10 as well as C-7 and C-8. Here, the migration of the hydrogen located at C-14 to the leaving group was suggested, which allowed the formation of an intermediate product ion at *m/z* 151 that further eliminated water (–18 Da) to produce *m/z* 133.[22]

4.3.5 3-Keto-4,9-diene Steroids

Steroids with 3-keto-4,9-diene nucleus such as androst-4,9-dien-17β-ol-3-one (Fig. 4.5, **13**) were shown to eliminate methyl radicals (Fig. 4.7c), which was attributed to carbon 19 being in a favorable position since the resulting radical cation is stabilized by a conjugated electron system. Additionally, androst-4,9-dien-17β-ol-3-one generated a base peak at *m/z* 147, the suggested dissociation pathway of which included the elimination of the entire D-ring (–98 Da) by means of a retro-Diels-Alder rearrangement and the release of ketene (–42 Da) from the A-ring including carbons 2 and 3.[38]

4.3.6 3-Keto-4,9,11-triene Steroids

The category of 3-keto-4,9,11-triene steroids has gained considerable attention due to the findings of numerous doping rule violations with compounds such as trenbolone, methyltrienolone, or tetrahydrogestrinone (Fig. 4.5, **14–16**, respectively). Their principal fragmentation behavior is outlined by means of the product ion mass spectrum of tetrahydrogestrinone (THG, Fig. 4.5, **16**), a designer steroid that was discovered in 2003.[39] A variety of product ions is obtained from the protonated molecule $[M+H]^+$ *m/z* 313 upon CID (Fig. 4.8a), including more and less characteristic eliminations. The neutral loss of water (−18 Da), presumably originating from C-17, yielded the product ion at *m/z* 295, and, the subsequent release of an ethyl radical gave rise to the ion at *m/z* 266. This exemption from the commonly accepted "even-electron-rule"[40] was observed with several synthetic analogues such as gestrinone, dihydrogestrinone, or deuterated THG, obtained from respective protonated molecules after elimination of water. The released ethyl radical was suggested to be composed of the C-13-linked alkyl residue that is located in allylic position to the large conjugated 8-π-electron system, explaining the stable character of the generated radical cation.

The product ions at *m/z* 241 and 199 were detected in all product ion spectra of THG analogues (Table 4.3) and, thus, required a structure independent from the steroidal D-ring. Hence, a neutral loss of C-16 and C-17 including their substituents was postulated giving rise to *m/z* 241 as demonstrated in Scheme 4.4a.[3,42,43] While the ion at *m/z* 241 was obtained immediately from the protonated molecules, the secondary ion at *m/z* 199 resulted from a neutral loss of 42 Da from *m/z* 241, which was assigned to the elimination of propene composed of the carbons C-15, C-18, and C-19 (Scheme 4.4a). The ion at *m/z* 241 proved to be diagnostic for steroids bearing a 4,9,11-triene nucleus and an ethyl residue at C-13. The substitution of the ethyl group by a methyl function gave rise to a fragment at *m/z* 227 instead of *m/z* 241 due to the involved C-13-linked alkyl residue. In contrast, the generation of *m/z* 199 includes the removal of this particular part of the molecule and was not influenced by any modification of C-13. Exchanging the ethyl function of gestrinone by a methyl group results in a decrement of *m/z* 241 by 14 Da yielding *m/z* 227 as detected in product ion spectra of compounds such as trenbolone and methyltrienolone (Fig. 4.8b). The latter was subject of various adverse analytical findings in the pre-Olympic doping control program of the Beijing 2008 Olympic Games,[41] which stopped numerous athletes from participating in this event.

4.3.7 Stanozolol

Stanozolol, 17β-hydroxy-17α-methyl-5α-androst-2-eno(3,2-c)-pyrazole (Fig. 4.5, **17**), represents one of the most frequently misused anabolic androgenic steroids with analytical peculiarities due to the presence of

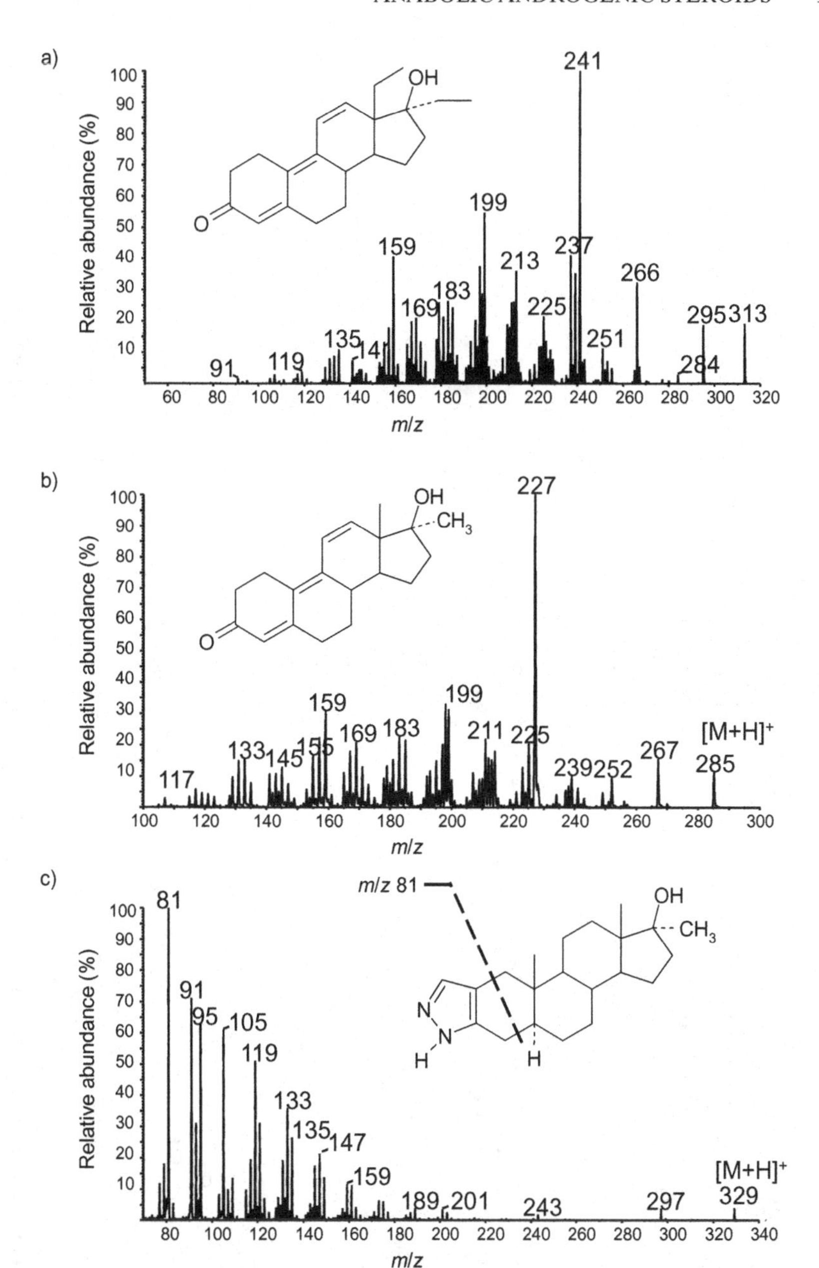

Figure 4.8: ESI product ion mass spectra of protonated molecules $[M+H]^+$ of (a) tetrahydrogestrinone (mol wt = 312), (b) methyltrienolone (mol wt = 284), and (c) stanozolol (mol wt = 328) recorded on a QTrap 2000 system.

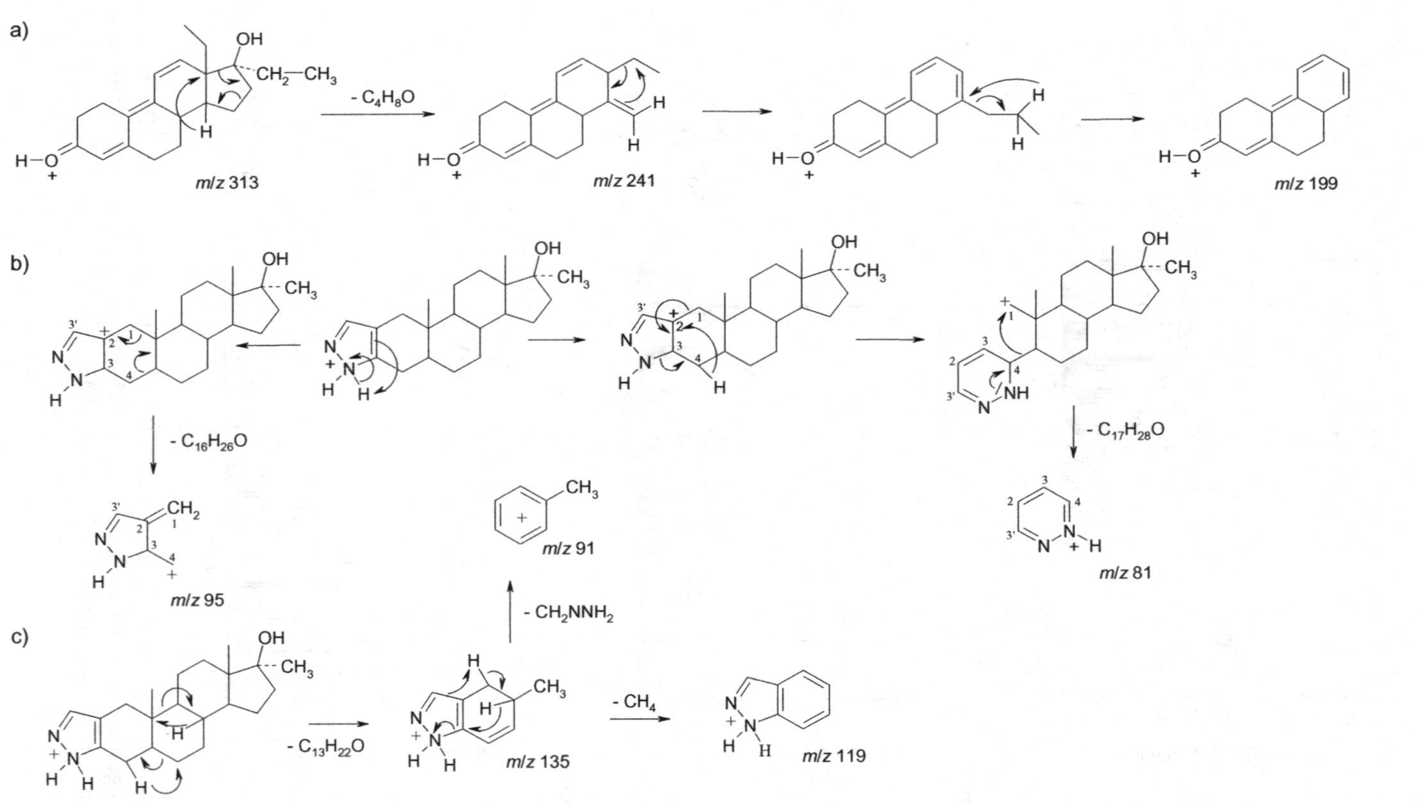

Scheme 4.4: Suggested dissociation pathways of (a) tetrahydrogestrinone, and (b) and (c) stanozolol.[3,42,43]

the pyrazole structure attached to the steroidal A-ring. The product ion mass spectrum of $[M+H]^+$ *m/z* 329 of stanozolol is illustrated in Fig. 4.8c, and major product ions were the subject of detailed studies using deuterium labeling, derivatization, and high resolution/high accuracy mass spectrometry to elucidate the complex dissociation pathways.[3,42,43] The most abundant product ion of stanozolol was observed at *m/z* 81, the origin of which was suggested to be the pyrazole ring including C-2–C-4 of the steroidal A-ring after cleavage of the C-1–C-2 and C-4–C-5 bonds[3,43] as obtained by charge-driven rearrangement and dissociation mechanisms[44] (Scheme 4.4b). Starting with a protonation of *N*-2, the charge is presumably relocated at C-2, triggering the formation of a 6-membered ring structure. With the cleavage of the C-4–C-5 linkage, the product ion at *m/z* 81 is liberated comprising a stable heterocyclic structure with conjugated π-electron system. The product ion at *m/z* 95 was suggested to originate from the same protonated species as *m/z* 81 by a rearrangement of the bonds between C-10 and C-1 as well as C-4 and C-5 (Scheme 4.4b). In contrast to these fragments immediately generated from the protonated intact molecule, the ion at *m/z* 91 was proposed to originate, at least partially, from the ion at *m/z* 135 as generated by cleavage of the bonds between C-9 and C-10 as well as C-5 and C-6 and subsequent elimination of methylene-hydrazine ($H_2N\text{-}N{=}CH_2$, −44 Da). Also, the fragment ion at *m/z* 119 was identified to result from *m/z* 135 by a methane elimination (−16 Da) to yield the conjugated π-electron system of *m/z* 119, which was considered the driving force in the dissociation route (Scheme 4.4c).

4.4 SELECTIVE ANDROGEN RECEPTOR MODULATORS (SARMs)

Alternatively to anabolic androgenic steroids, SARMs have emerged into the pharmaceutical market as drug candidates potentially useful for the treatment of debilitating diseases, osteoporosis, muscle wasting, or male contraception.[45–51] They represent a heterogeneous group of substances including arylpropionamides, quinolines, tetrahydroquinolines, and bicyclic hydantoins (Fig. 4.9), and compounds of the first mentioned category have successfully completed advanced clinical trials. Due to the prohibition of SARMs in sports since January 2008, the necessity of mass spectrometry-based detection methods was evident, and various studies aiming the structural characterization of target analytes were conducted.[52] A summary of characteristic product ions derived from various SARMs is presented in Table 4.4.

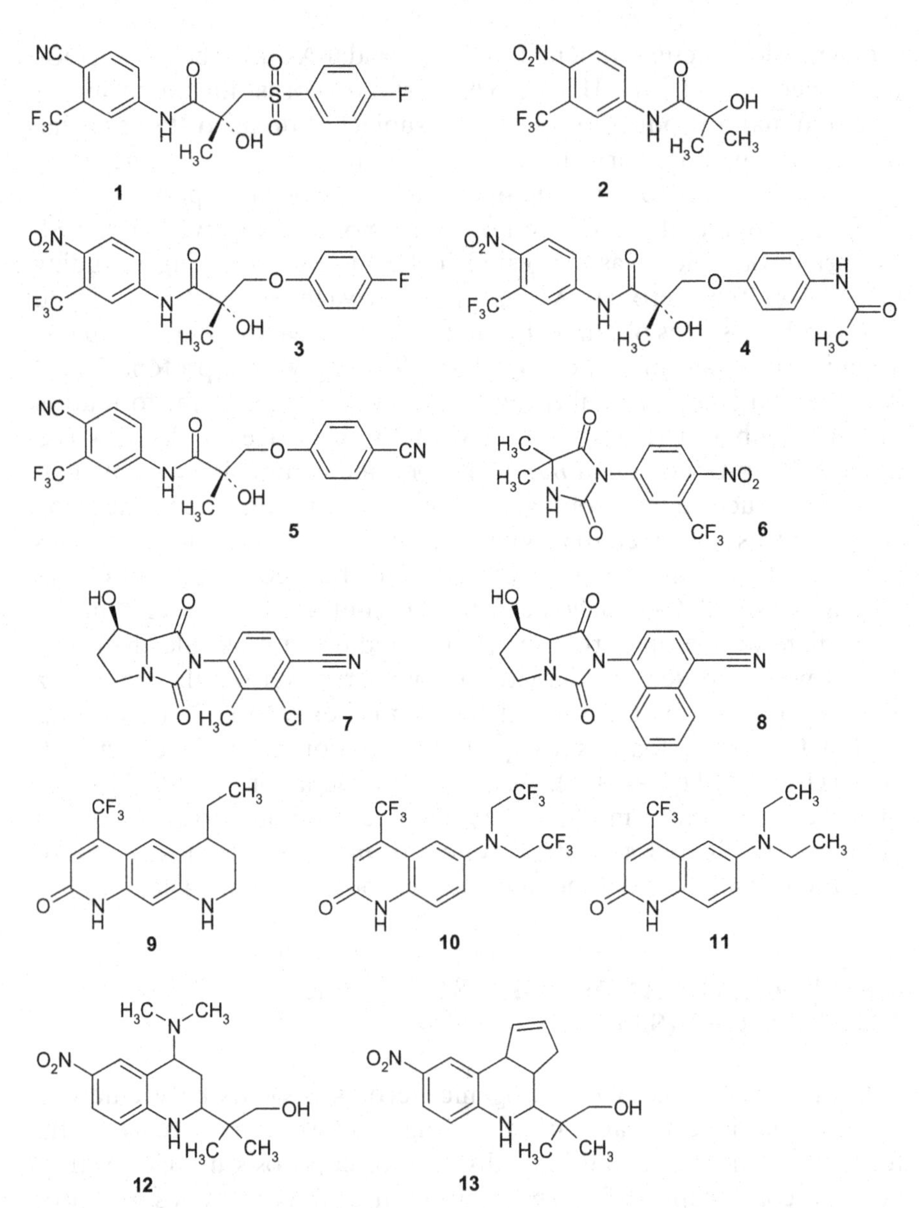

Figure 4.9: Structures of selected SARMs: androgen receptor antagonists bicalutamide (**1**) and flutamide (**2**); arylpropionamide-based agonists S-1 (**3**), S-4 (Andarine, **4**), S-22 (**5**); bicyclic hydantoin-derived antagonist nilutamide (**6**); agonists BMS-564929 (**7**), and 4-(7-hydroxy-1,3-dioxo-tetrahydro-pyrrolo[1,2-c]imidazol-2-yl)-naphthalene-1-carbonitrile (**8**); quinolinone-based SARMs LG-121071 (**9**), LGD-2226 (**10**), and bisethylamino-4-trifluoromethylquinolin-2(1H)-one (**11**); and tetrahydroquinoline-derived SARMs S-40503 (**12**), and 2-methyl-2-(8-nitro-3a,4,5,9b-tetrahydro-3H-cyclopenta[c] chinolin-4-yl)propan-1-ol (**13**).

TABLE 4.4: Characteristic Product Ions of Selected SARMs Generated by ESI and CID

SARMs Nucleus	Representative Compound	Mol Wt (Da)	Precursor Ion (*m/z*)	Product Ions (*m/z*)			
Arylpropionamide	S-1	402	401 [M-H]$^-$	289	261[a]	205	111
	S-4	441	440 [M-H]$^-$	289	261[a]	205	150
	S-9	418	417 [M-H]$^-$	289	261[a]	205	127
	S-22	389	388 [M-H]$^-$	269	241	185	118[a]
	S-23	416	415 [M-H]$^-$	269	241	185	145[a]
	S-24	382	381 [M-H]$^-$	269	241[a]	185	111
Bicyclic hydantoin	BMS-564929	305	306 [M-H]$^+$	288	278[a]	260	193
	BMS-564929	305	304 [M-H]$^-$	286[a]	260	248	193
	SARM (**8**)[b]	307	308 [M-H]$^+$	290	280[a]	264	262
Quinolinone	LGD-2226	392	393 [M-H]$^+$	375	310	241	—
	SARM (**11**)[c]	284	285 [M-H]$^+$	267	257[a]	256	241
Tetrahydroquinoline	S-40503	293	294 [M-H]$^+$	249[a]	231	219	177
	SARM (**13**)[d]	288	289 [M-H]$^+$	272[a]	259	223	199

[a] Base peak.
[b] 4-(7-hydroxy-1,3-dioxo-tetrahydro-pyrrolo[1,2-c]imidazol-2-yl)-naphthalene-1-carbonitrile.
[c] Bisethylamino-4-trifluoromethylquinolin-2(1H)-one.
[d] 2-Methyl-2-(8-nitro-3a,4,5,9b-tetrahydro-3H-cyclopenta[c]chinolin-4-yl)propan-1-ol.

4.4.1 Arylpropionamide-Derived SARMs

The first non-steroidal androgen receptor agonists were derived from bicalutamide and flutamide (Fig. 4.9, **1** and **2**, respectively), both of which contain an arylpropionamide nucleus. The scientific breakthrough of preparing the first SARM was accomplished in 1998, and major advantages of these SARMs over steroids in replacement therapies have been the considerably reduced undesirable effects such as hepatic toxicity, decreased levels of HDL cholesterol, gynecomastia, and negative influences on prostate and cardiovascular systems.[45] In addition, SARMs have demonstrated full anabolic activity in target tissues such as muscles and bones, as well as a considerable gain in lean body mass concomitant with a dose-dependent increase in functional performance.[53]

The class of arylpropionamide-derived SARMs includes a variety of promising drug candidates,[54] which differ mainly in number and nature of ring substituents as outlined with selected examples in Figure 4.9 (**3**–**5**). Although ionized in positive and negative mode by ESI, these compounds are preferably deprotonated yielding abundant [M-H]$^-$ ions.[55] Employing CID, diagnostic product ions are generated that allow the characterization of both aromatic ring systems (A and B) as illustrated with the product ion mass spectrum of compound **5** (S-22,

Fig. 4.10a). The site of deprotonation is suggested to occur at the amide nitrogen of the propionanilide-derived nucleus, attributed to the acidity of the respective hydrogen, which results from significant electron-withdrawing inductive effects (–I-effects) exerted by substituents such as the trifluoromethyl- and nitro-functions. The deprotonated molecule of **5** was shown to eliminate 4-hydroxybenzonitrile (–119 Da) yielding the product ion at *m/z* 269, which corresponds to *m/z* 289 in case of analytes with a nitro residue located at R_1 (para-position in A-ring),[56] and the loss of the B-ring is followed by the release of carbon monoxide giving rise to the product ion at *m/z* 241 (corresponding to *m/z* 261 with $R_1 = NO_2$).[57] The formation of product ions corresponding to bisubstituted and deprotonated anilines, e.g., 4-cyano-3-trifluoromethyl-aniline at *m/z* 185 in case of **5** (Fig. 4.10a), further characterizes the SARM A-ring, and structurally related derivatives yielded diagnostic fragments accordingly.[52] Product ions resulting from the B-ring commonly represent the deprotonated and substituted hydroxyphenyl residue,[56,57] which yields the base peak product ion at *m/z* 118 in case of **5** (Fig. 4.10a). The principle dissociation routes of arylpropionamide-derived SARMs are summarized in Scheme 4.5.[57] Using a few but diagnostic product ions, modifications of either ring system can be detected, allowing the identification of metabolic products[56,58,59] as well as modified designer analogues.

4.4.2 Hydroxybicyclic Hydantoin-Derived SARMs

SARMs bearing a hydroxybicyclic hydantoin core are structurally related to hydantoin-based androgen receptor (AR) antagonists such as nilutamide (Fig. 4.9, **6**). In contrast to these AR-blocking agents, compounds such as BMS-564929 (Fig. 4.9, **7**), comprise a hydroxylated five-membered ring structure, which enables excellent AR binding affinities with activating properties and high muscle-tissue selectivity.[60–62] Drug candidates with bicyclic hydantoin nuclei (Fig. 4.9, **7** and **8**) are readily protonated as well as deprotonated using ESI, and product ion mass spectra contain a variety of characteristic information (Fig. 4.10b-c).

Hydantoins contain different protonation sites including the nitrogens or the carbonyl oxygens[63] with a thermodynamically favored initial *O*-protonation.[64] The complex structure of SARMs such as compound **7**, however, significantly influences the proton affinity of *N*-3 resulting in a privileged protonation at *N*-1 and carbonyl residues as substantiated by density functional theory (DFT) calculations.[65] Still, also here the above mentioned mobile nature of protons,[66,67] in particular after

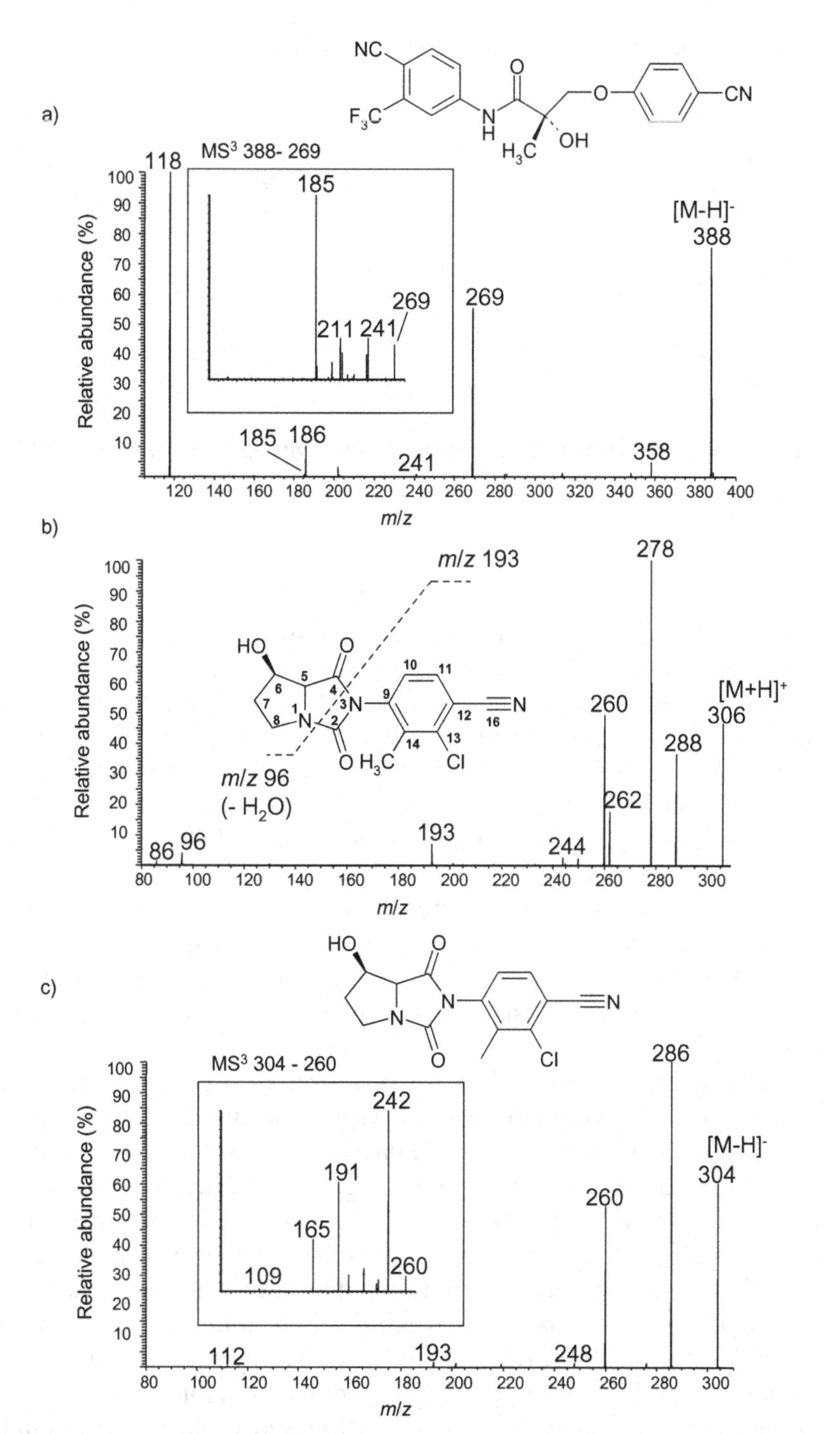

Figure 4.10: ESI product ion mass spectra of (a) the deprotonated molecule $[M-H]^-$ of S-22 (mol wt = 389), (b) the protonated molecule $[M+H]^+$ of BMS-564929 (mol wt = 305), and (c) the deprotonated molecule $[M-H]^-$ of BMS-564929 (mol wt = 305) recorded on an LTQ-Orbitrap system.

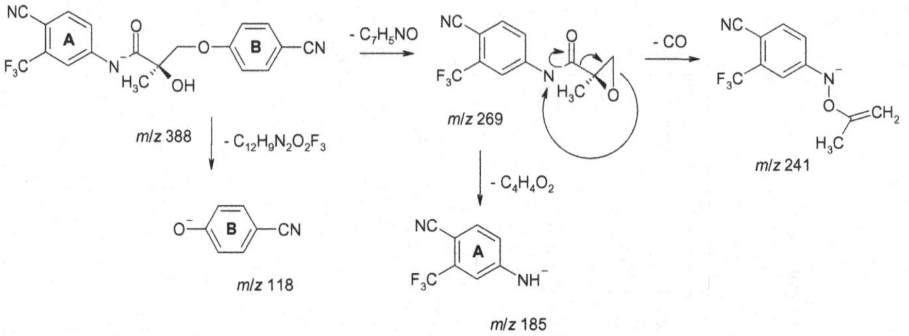

Scheme 4.5: Suggested dissociation pathways of the arylpropionamide-derived SARM S-22.[57]

excitation of ionized molecules by CID, allows dissociation pathways starting from both options. The protonated molecule of BMS-564929 (**7**) at *m/z* 306 eliminates water (–18 Da) and carbon monoxide (–28 Da) in either sequence to generate product ions at *m/z* 288, 278, and 260 (Fig. 4.10b). A more characteristic feature of 6-hydroxylated bicyclic hydantoins such as **7** was shown to be the loss of acetaldehyde as structural analogues, e.g., with 7-hydroxylation, are lacking this particular fragment in CID spectra.[65] Additional product ions found at *m/z* 193 and 96 are complementary fragments originating from a cleavage of the hydantoin core following the fission of the linkages between *N*-1 and *C*-2 as well as *N*-3 and *C*-4.[63] Evidence for this dissociation route was obtained by the analysis of stable-isotope labeled analogues to **7** (Scheme 4.6a)[65].

Deprotonation of **7** using ESI was proposed to occur predominantly at C-5 or the hydroxyl function at C-6. Upon collisional activation, only few but informative ions specifically characterizing the structure of the lead drug candidate **7** were found (Fig. 4.10c). The loss of carbon dioxide (–44 Da) from $[M\text{-}H]^-$ (*m/z* 304) yielding the product ion at *m/z* 260 requires a complex rearrangement involving an intermediate six-membered ring structure, with the hydroxyl function located at C-6 playing a key role as analogues lacking this residue do not show the loss of CO_2. The newly formed 2,3-dihydro-1H-pyrrole residue of *m/z* 260 is subsequently released (–69 Da) producing the ion at *m/z* 191 (Fig. 4.10c, inset). Complementary, the deprotonated molecule of **7** eliminates water (–18 Da), and consecutive losses of imino-methanone (–43 Da) and but-1-en-3-yne (–52 Da) also give rise to *m/z* 191 as summarized in Scheme 4.6b.

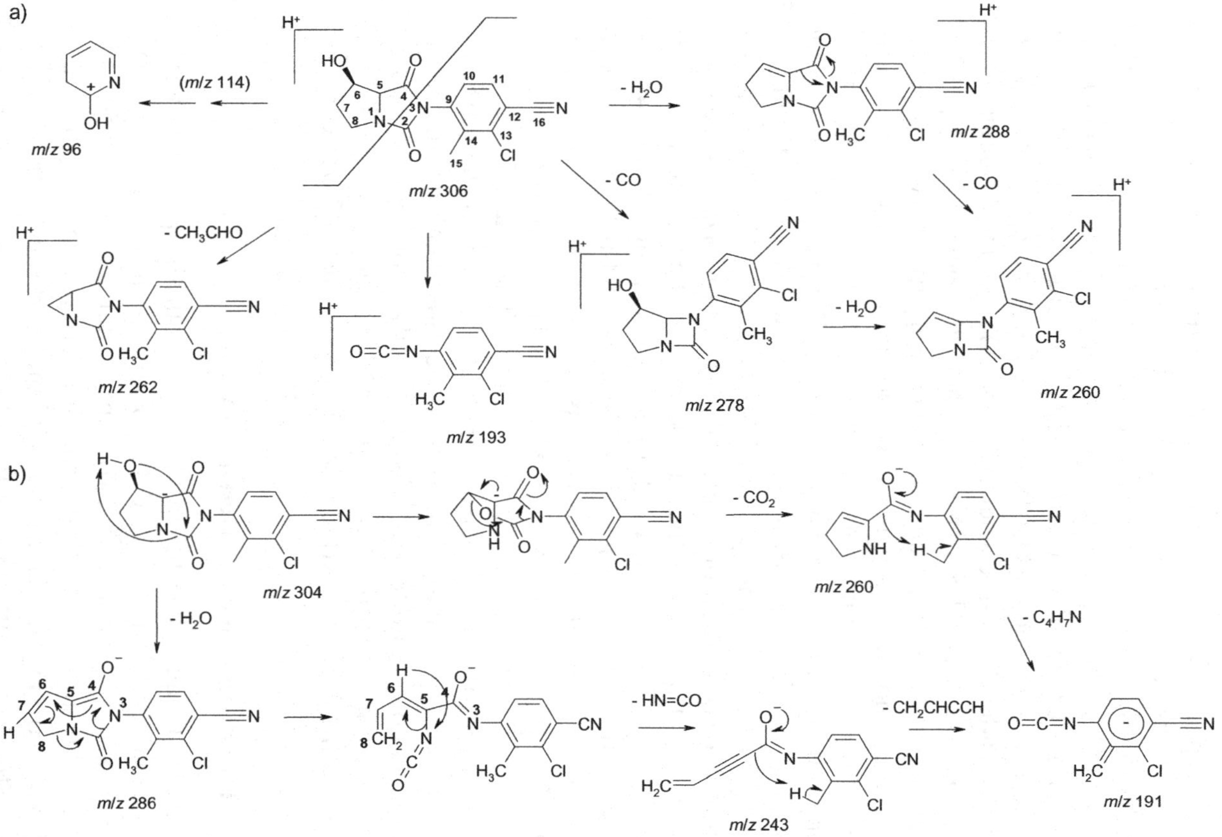

Scheme 4.6: Proposed dissociation pathways of (a) the protonated and (b) deprotonated bicyclic hydantoin-based SARM BMS-564929.[65]

4.4.3 2-Quinolinone-Derived SARMs

Agonistic SARM-like activity was reported for several bi- and tricyclic quinoline derivatives, including LG 121071 and LGD 2226 (Fig. 4.9, **9** and **10**, respectively).[68–73] Those include a 4-trifluoromethyl-2-quinolinone nucleus and either C-ring substituents or a 6-located bis(trifluoroethyl)amine residue that enable activation of the androgen receptor (AR). The 2-oxo residue of the A-ring and the ethyl group at the C-ring (or the bis(trifluoroethyl)amine moiety) were found to mimic the 3-keto- and 17-OH-functions of testosterone[54] and demonstrated considerable tissue selectivity and AR binding affinities.[74,75]

Quinolinone-derived SARMs are efficiently ionized using positive ESI, and common as well as unique dissociation pathways were observed for these compounds (e.g., substance **10**, Fig. 4.9)[40,76]. The protonated molecule (*m/z* 393) eliminates a trifluoroethyl residue (–83 Da) yielding the product ion at *m/z* 310 (Fig. 4.11a). The driving force for the loss of a radical from an even-electron precursor ion is attributed to the conjugated π-electron system of 2-quinolinones that promotes the generation of radical cations under CID conditions. The resulting odd-electron ion at *m/z* 310 further dissociates by eliminating a trifluoromethyl radical (–69 Da) to yield a core product ion at *m/z* 241 that represents the common nucleus of bisalkylated 4-trifluoro-2-quinolinones.[77] In subsequent MS^3 experiments, the even-electron product ion at *m/z* 241 released a methyleneamine radical (–28 Da) giving rise to the product ion at *m/z* 213 (Scheme 4.7a).[77,81] The unusual alternation between even- and odd-electron ions may be due to the particular properties of 2-quinolinones to form stable radical cations. In contrast to bisalkylated 2-quinolinones, monoalkylated analogues were reported to yield a common product ion at *m/z* 228 (instead of *m/z* 241), which represents the 6-amino-4-trifluoromethyl-1H-quinolin-2-one core.[77]

4.4.4 Tetrahydroquinoline-Derived SARMs

In addition to quinoline-based SARMs, tetrahydroquinoline-derived drug candidates were reported to possess tissue-selective androgen receptor agonist activity.[78,79] Two representatives are S-40503 and 2-methyl-2-(8-nitro-3a,4,5,9b-tetrahydro-3H-cyclopenta[c]chinolin-4-yl)propan-1-ol (Fig. 4.9, **12** and **13**), and the mass spectrometric behavior was studied for both compounds.[80,81] The protonated molecule of **13** at *m/z* 289 dissociates under CID conditions by the loss of a hydroxyl radical (–17 Da) originating from the nitro function.[76] The proton affinity of 1,2,3,4-tetrahydroquinoline was determined as 225 kcal/mol[82] and

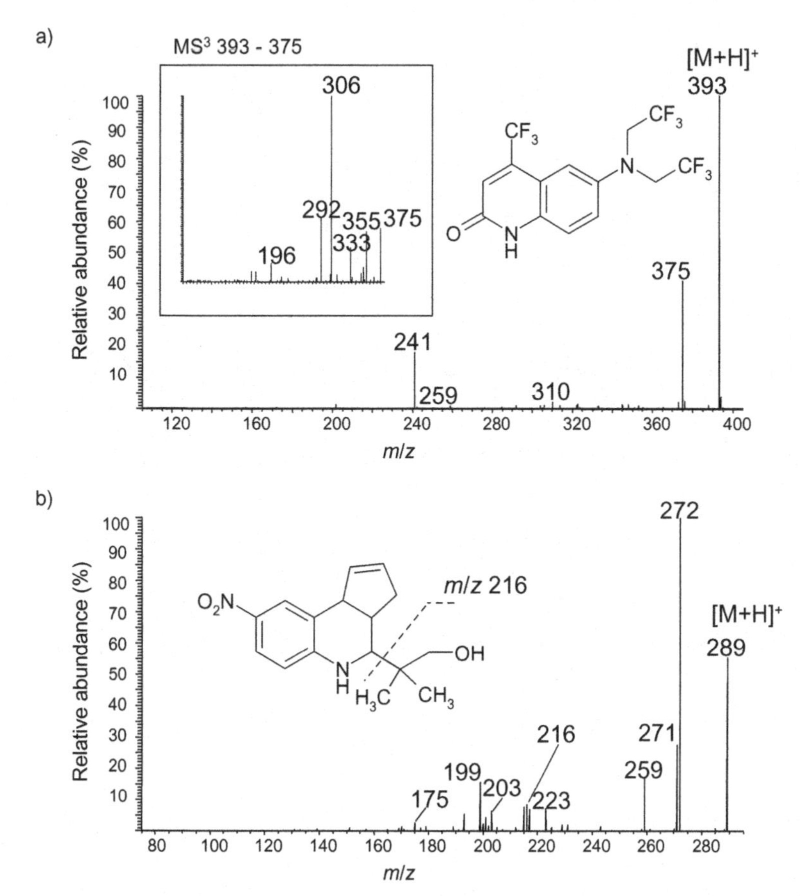

Figure 4.11: ESI product ion mass spectra of the protonated molecules $[M+H]^+$ of (a) LGD 2226 (mol wt = 392), and (b) 2-methyl-2-(8-nitro-3a,4,5,9b-tetrahydro-3H-cyclopenta[c]chinolin-4-yl)propan-1-ol (mol wt = 288) recorded on an LTQ-Orbitrap system.

is, thus, higher than the corresponding affinity of nitrobenzene (164 kcal/mol)[83] without accounting for the –I-effect caused by the NO_2-residue. Consequently, the site of initial protonation is not clear, and also here CID can induce proton migration and trigger dissociation processes at various sites of the molecule.[66] Hence, the loss of the OH-radical might result from immediate protonation at the nitro function or after proton transfer. In addition to the loss of •OH, losses of water (–18 Da), formaldehyde (–30 Da), and the 2-linked side chain with homolytic or heterolytic cleavages are observed yielding the product ions at *m/z* 271, 259 and 217, 216, and 215, respectively (Fig. 4.11b). The elimination of the hydroxyl radical (*m/z* 272) is followed by the loss of a 2-methyl

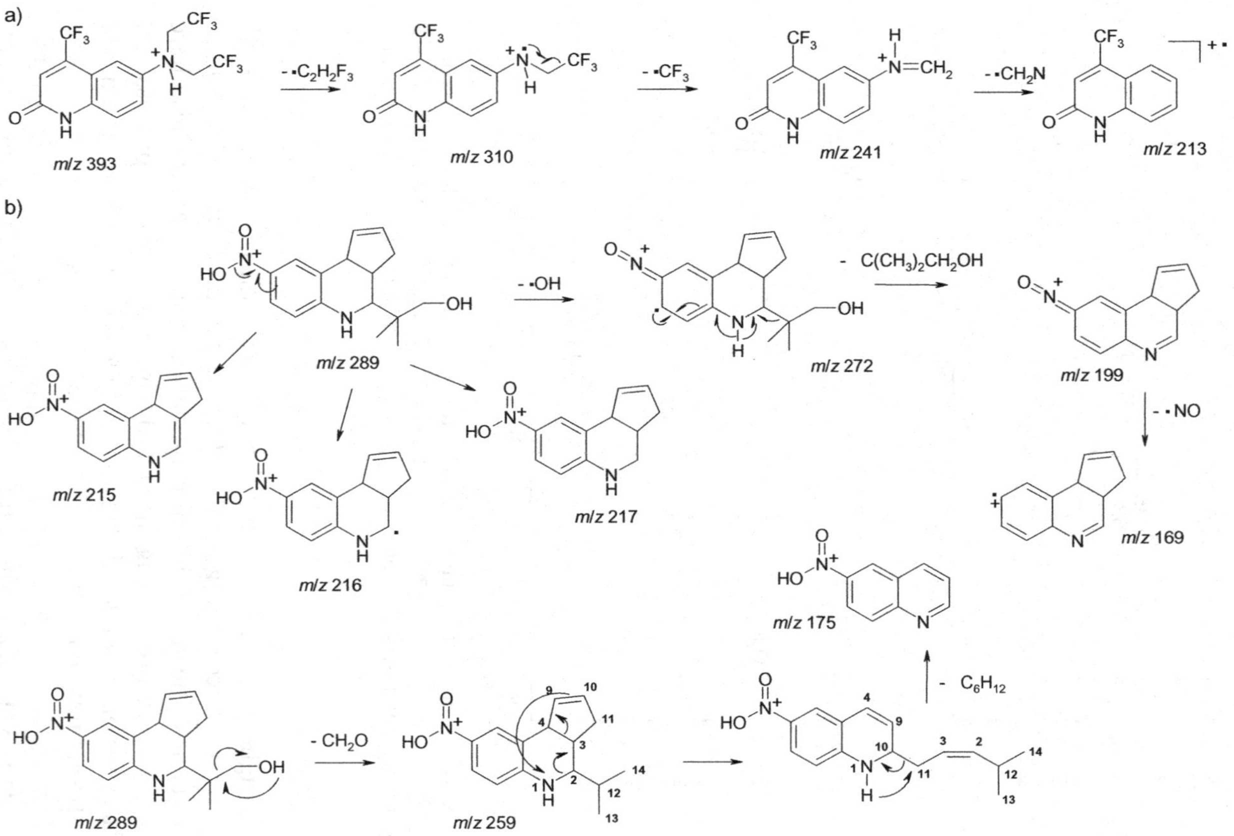

Scheme 4.7: Suggested dissociation pathways of (a) LGD-2226, and (b) 2-methyl-2-(8-nitro-3a,4,5,9b-tetrahydro-3H-cyclopenta[c]chinolin-4-yl)propan-1-ol.[77,81]

propanol radical (−73 Da) yielding *m/z* 199, and the release of formaldehyde from the precursor ion resulting in the fragment at *m/z* 259 is followed by the loss of 4-methylpent-2-ene (−84 Da) yielding 6-nitroquinoline (*m/z* 175), which necessitates an intramolecular rearrangement (Scheme 4.7b).

4.5 DIURETICS

Due to the considerable structural diversity and the polar nature of many diuretic agents, which is responsible also for mostly limited GC properties (see Chapter 3), ESI-MS(/MS) was greatly appreciated for characterizing and analyzing diuretics in sports drug testing.[8,84–87] The majority of diuretic agents contains acidic (e.g., carboxyl or sulfonamide residues) or basic (amino, pyrazine, pteridine, or guanidine moieties) sites, which allow an efficient deprotonation or protonation, respectively. CID of quasimolecular ions further results in diagnostic product ions that are used to identify the active drug, its metabolites, and/or degradation products as shown by means of selected examples including hydrochlorothiazide, furosemide, bumetanide, and amiloride (Fig. 4.12, **1–4**).

4.5.1 Thiazide-Derived Drugs

Hydrochlorothiazide (Fig. 4.12, **1**) is one of the most frequently prescribed drugs to correct hypertension and has been frequently detected in doping controls for many years. Its deprotonated molecule at *m/z* 296 gives rise to a number of product ions that are characteristic for the thiazide-typical 7-sulfamoyl-3,4-dihydro-1,2,4-benzothiadiazine 1,1-dioxide nucleus. Those include the ions at *m/z* 269, 205, 126, and 78 (Fig. 4.13a), which were suggested to result from an initial deprotonation of the sulfonamide residue followed by a cascade of rearrangement and elimination processes (Scheme 4.8a);[84–86,88] however, the predominant site of ionization has yet to be determined conclusively.[85] The generation of *m/z* 269 was proposed to be initiated by the fission of the C-3–N-4 bond and a subsequent elimination of HCN (−27 Da).[86] Here, the presence of a single bond between the atoms 3 and 4 and the following cleavage of this linkage was found essential for the existence of *m/z* 269 as two structural analogues, chlorothiazide and benzthiazide (Table 4.5), which contain a double bond at this particular position, do not generate the common fragment at *m/z* 269 or any other corresponding ion (Table 4.5). The elimination of sulfur dioxide (−64 Da) from *m/z*

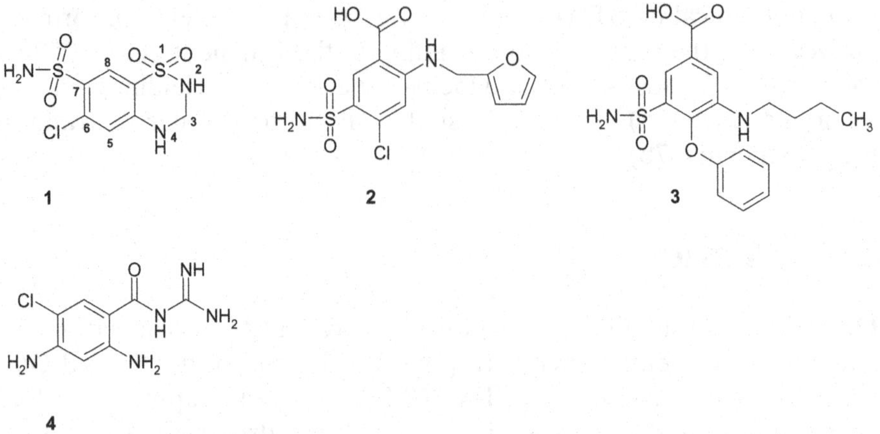

Figure 4.12: Structures of selected diuretic agents: hydrochlorothiazide (**1**), furosemide (**2**), bumetanide (**3**), and amiloride (**4**).

269 yielded the ion at *m/z* 205, and the subsequent loss of SO_2NH produces the common fragment at *m/z* 126 representing the deprotonated 3-chloroaniline.[88] In case of 6-trifluoromethylated thiazides such as hydroflumethiazide and bendroflumethiazide, the counterparts to *m/z* 269, 205, and 126 were detected at *m/z* 303, 239, and 160, respectively (Table 4.5). Consequently, the substituent at C-6 (-Cl or $-CF_3$) does not essentially influence the observed dissociation process. The composition of the product ion at *m/z* 78 was proposed to be SO_2N^- since the ion shifts to *m/z* 79 with ^{15}N-labeling of the sulfonamide function.[85] Alternatively to the above suggested dissociation route, product ions of comparably low abundance indicate a second pathway to the ion at *m/z* 205. The initial loss of sulfur dioxide (–64 Da) from the deprotonated molecule at *m/z* 296 yielded the product ion at *m/z* 232 that was proposed to comprise a five-membered ring structure (Scheme 4.8a). The subsequent elimination of HCN (–27 Da), which exclusively includes the nitrogen N-2, also gave rise to the central product ion at *m/z* 205.

4.5.2 Benzoic Acid-Derived Loop Diuretics

The loop diuretics furosemide (Fig. 4.12, **2**), bumetanide (Fig. 4.12, **3**), and piretanide include a benzoic acid residue, which suggests a favored negative ionization of these drugs by deprotonation of the carboxyl function. The product ion mass spectrum of *m/z* 363 of bumetanide (3-butylamino-4-phenoxy-5-sulfamoylbenzoic acid, Fig. 4.13b), contains

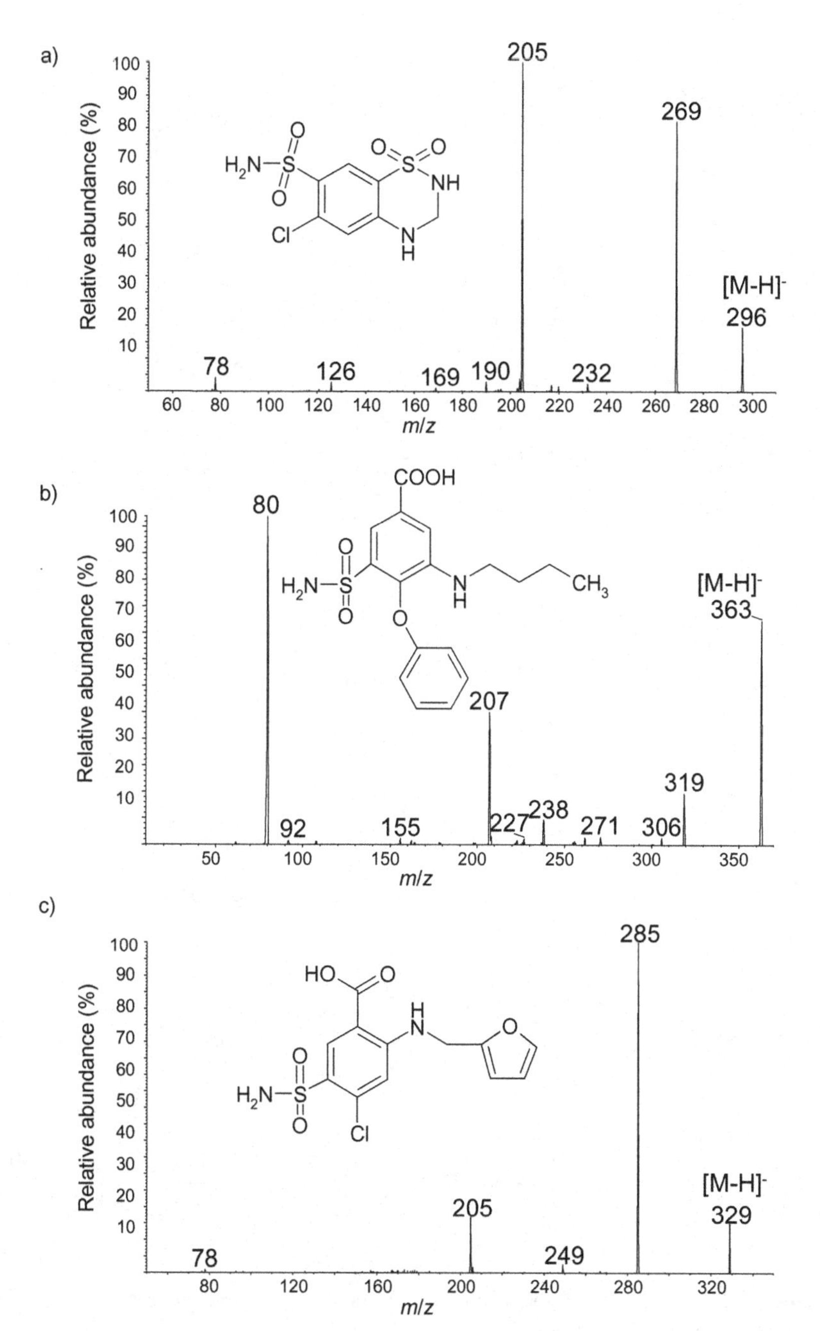

Figure 4.13: ESI product ion mass spectra of the deprotonated molecules $[M\text{-}H]^-$ of (a) hydrochlorothiazide (mol wt = 297), (b) bumetanide (mol wt = 364), and (c) furosemide (mol wt = 330) recorded on an API2000 QTrap system.

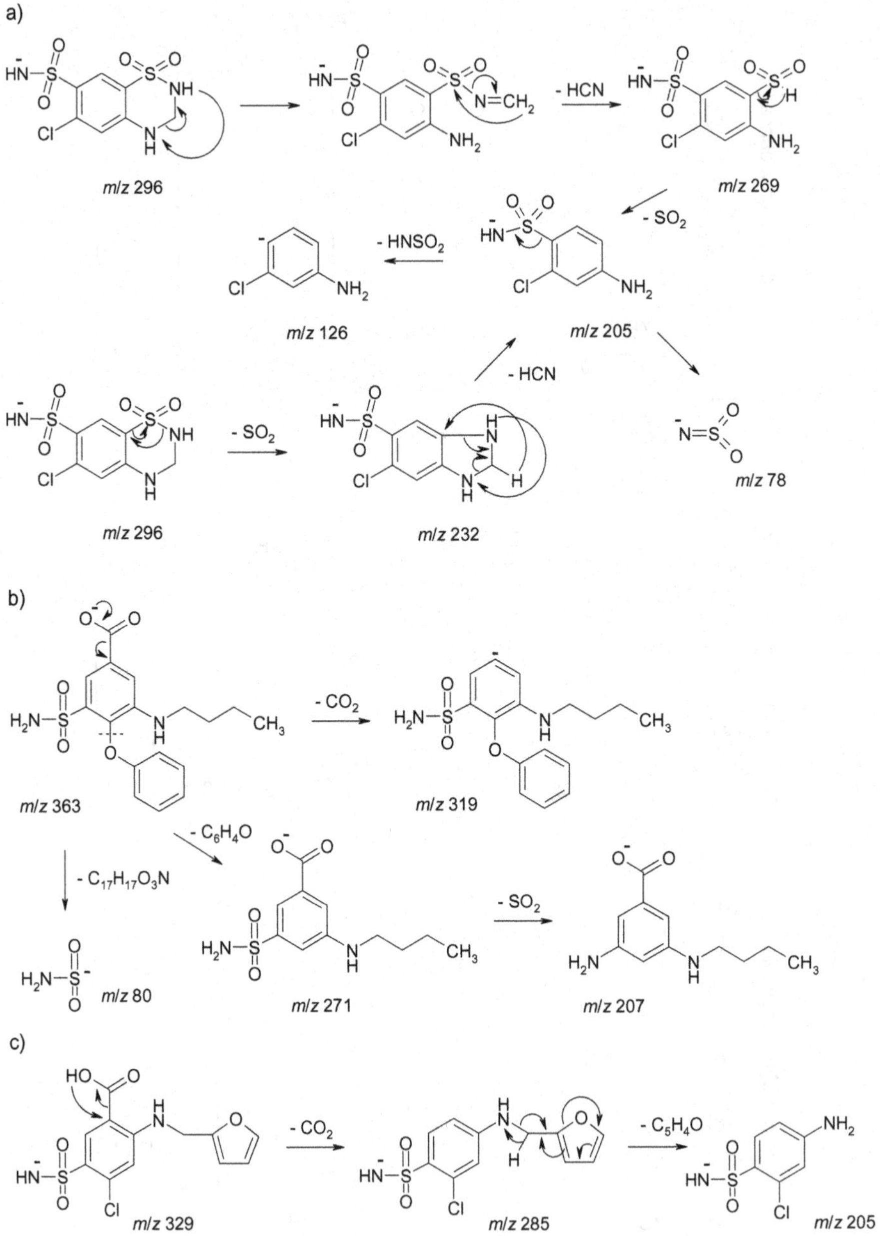

Scheme 4.8: Suggested dissociation pathways of (a) hydrochlorothiazide, (b) bumetanide, and (c) furosemide.[84–86,88]

TABLE 4.5: Characteristic Product Ions of Selected Diuretic Agents Generated by ESI and CID[a]

Substance Class	Representative Compound	Mol Wt (Da)	Precursor Ion (*m/z*)	Product Ions (*m/z*)					
Androst-4,6-dien-3-one	Canrenone	340	341 $[M+H]^+$	173	119	107	105	97	95
Thiazide	Althiazide	383	382 $[M-H]^-$	341	269	205	—	126	78
	Bemetizide	401	400 $[M-H]^-$	294	230	204	195	124	78
	Bendroflumethiazide	421	420 $[M-H]^-$	328	289	197	160	113	78
	Benzthiazide	431	430 $[M-H]^-$	308	228	193	175	124	113
	Buthiazide	353	352 $[M-H]^-$	261	269	205	190	126	78
	Chlorothiazide	295	294 $[M-H]^-$	214	179	149	115	88	78
	Cyclopenthiazide	379	378 $[M-H]^-$	287	269	205	190	126	78
	Cyclothiazide	389	388 $[M-H]^-$	322	269	205	—	126	78
	Epithiazide	425	424 $[M-H]^-$	300	269	205	190	126	78
	Ethiazide	325	324 $[M-H]^-$	233	269	205	190	126	78
	Hydrochlorothiazide	297	296 $[M-H]^-$	—	269	205	190	126	78
	Hydroflumethiazide	331	330 $[M-H]^-$	—	269	205	190	126	78
	Polythiazide	439	438 $[M-H]^-$	398	324	204	—	124	78
Phenoxyacetic acid	Ethacrynic acid	302	301 $[M-H]^-$	243	207	192	160	69	—
Pteridine	Triamterene	253	254 $[M+H]^+$	237	210	195	168	116	104
Pyrazine	Amiloride	229	230 $[M+H]^+$	171	161	143	116	101	60
Sulfamoyl benzoic acid	Bumetanide	364	363 $[M-H]^-$	319	306	271	238	207	80
	Furosemide	330	329 $[M-H]^-$	285	249	205	—	126	78
	Piretanide	362	361 $[M-H]^-$	317	—	269	225	205	80
Sulfamoyl-benzamide	Clopamide	345	344 $[M-H]^-$	—	308	280	189	80	78
	Indapamide	365	364 $[M-H]^-$	—	233	216	189	80	78
Sulfamoylbenzoyl-aniline	Xipamide	354	353 $[M-H]^-$	—	274	273	206	127	78

[a] Refs. 84 and 85.

an abundant ion at *m/z* 80, which was proposed to be composed of the sulfamoyl side $SO_2NH_2^-$. The generation of this ion requires the presence of a sulfamoyl group comprising two hydrogens, which indicated ionization remote from this residue and, thus, the mechanism of charge-remote fragmentation. Evidence for this postulation was obtained by the analysis of the methyl ester derivative of **3**, which did not yield the ion at *m/z* 80 but an intense fragment ion at *m/z* 78 with a composition of SO_2N^- obtained from charge-driven dissociation comparable to that described above for thiazide-based diuretics.[84] In addition to the ion resulting from the sulfamoyl residue, neutral losses of carbon dioxide (−44 Da), the phenoxy group (accompanied by the migration of one hydrogen) and sulfur dioxide (−64 Da) from the deprotonated molecule, contribute to the characteristic product ion mass spectrum of **3** (Fig. 4.13b), the proposed dissociation routes of which are summarized in Scheme 4.8b. Furosemide (Fig. 4.12, **2**) also comprises the 3-sulfamoylbenzoic acid nucleus, but instead of a 5-positioned butylamino residue, it bears a furfurylamino group, which is ortho-located to the carboxyl function. Moreover, the para phenoxy group of **3** is replaced by a chlorine atom. As a consequence, the product ion mass spectra of the deprotonated molecules of **2** and **3** differ significantly as **2** does not give rise to the intense product ion at *m/z* 80 but only a product ion at *m/z* 78 of low abundance (Fig. 4.13c). According to the data discussed above, this fact indicates a preferred ionization at the sulfamoyl residue and not at the carboxyl group, owing presumably to the presence of a strong hydrogen bond with the secondary amino group present at position 6. The entire fragmentation pattern of **2** (Scheme 4.8c) is consistent with a hydrogen abstraction at the sulfamoyl residue followed by subsequent eliminations of CO_2 (−44 Da to *m/z* 285), C_5H_4O of the furfuryl group (−80 Da to *m/z* 205), and SO_2NH (−79 Da to *m/z* 126), which is in close accordance to the dissociation of the above described thiazidic diuretics.

4.5.3 Potassium-Sparing Diuretics

In contrast to thiazides and benzoic acid derivatives, the pyrazine-based drug amiloride (Fig. 4.12, **4**) represents a potassium-sparing diuretic agent, which is one of the few diuretics that are preferably analyzed as protonated species under ESI conditions. Assuming an initial protonation of **4** at the guanidine moiety, a precursor ion at *m/z* 230 is obtained, which predominantly eliminates guanidine (−59 Da) followed by the release of carbon dioxide (−28 Da) and HCN (−27 Da) to produce the diagnostic ions at *m/z* 171, 143, and 116, respectively (Table 4.5), all of which comprise the 3-chloropyrazine-2,6-diamine nucleus. In addition, a product ion representing the guanidine function is found at *m/z* 60.

4.6 β_2-AGONISTS

Due to the close structural relation to catecholamines, β_2-agonists exhibit marginal GC properties if analyzed without adequate derivatization (see Chapter 3). In contrast, their polar nature and mostly alkaline properties allow efficient ionization using ESI, and various diagnostic product ions were obtained under CID conditions. The vast majority of β_2-agonists comprises a common nucleus consisting of a singly or multiply substituted phenyl residue bearing an ethanolamine side chain that terminates in either a *tert.*-butyl residue, an isopropyl group, or functionalities different from these common structures such as clenbuterol, cimaterol, and reproterol, respectively (Fig. 4.14, **1–3**).

Protonation of clenbuterol and analogues is likely to occur at the nitrogen atom of the ethanolamine side chain, which can initiate charge-driven as well as charge remote fragmentation.[89] The product ion mass spectrum of *m/z* 277 of clenbuterol (Fig. 4.15a) contains a series of product ions that are typical for β_2-agonists bearing a *tert.*-butyl residue. At a moderate collision energy, the neutral loss of a water molecule (−18 Da) is observed, generating the product ion at *m/z* 259, which is followed by the elimination of isobutene (−56 Da) giving rise to the ion at *m/z* 203.[90–92] A subsequent release of either a chlorine radical or

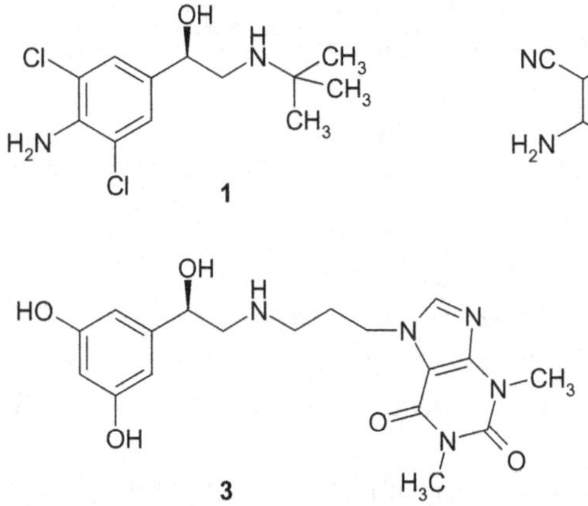

Figure 4.14: Structures of selected β_2-agonists: clenbuterol (**1**), cimaterol (**2**), and reproterol (**3**).

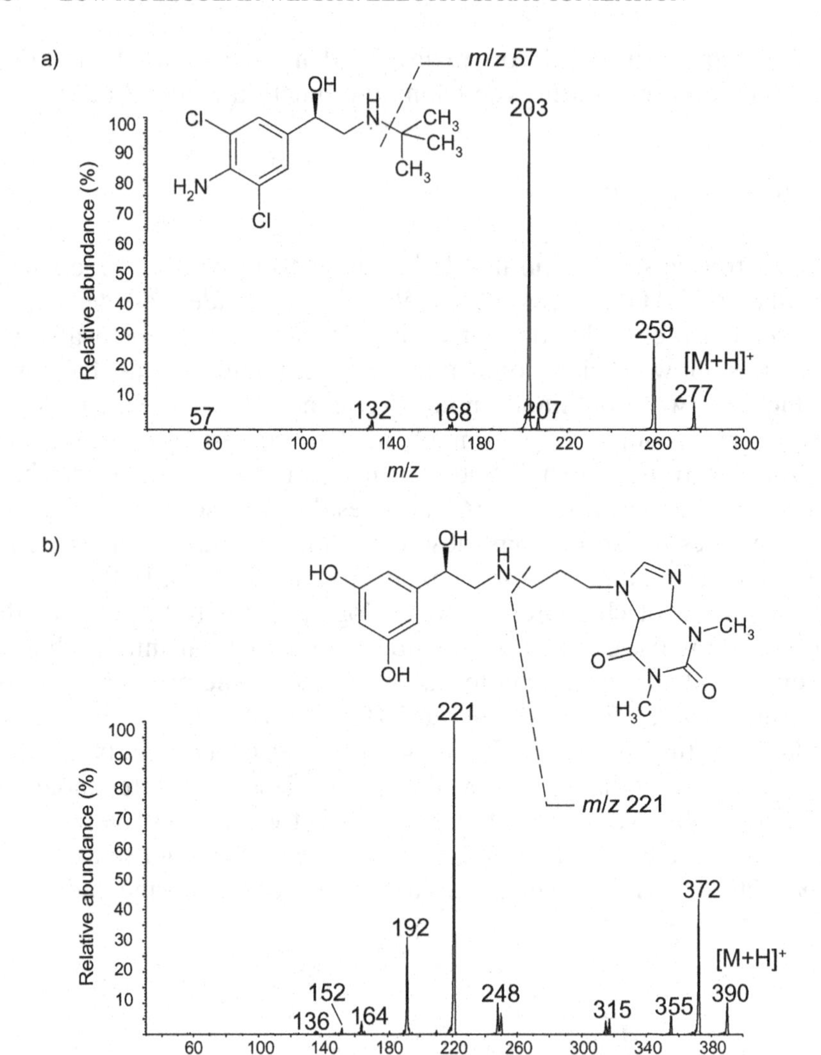

Figure 4.15: ESI product ion mass spectra of the protonated molecules $[M+H]^+$ of (a) clenbuterol (mol wt = 276), and (b) reproterol (mol wt = 389) recorded on an API2000 QTrap system.

hydrogen chloride yields the product ions at *m/z* 168 or 167, respectively, each of which expels the remaining chlorine atom complementary as hydrogen chloride or chlorine radical to *m/z* 132 (Scheme 4.9a).[90–92]

In all product ion spectra of β_2-agonists with an ethanolamine side chain terminating in either a *tert.*-butyl or an isopentyl group such as clenbuterol, mabuterol, clenpenterol, and mapenterol, the product ions

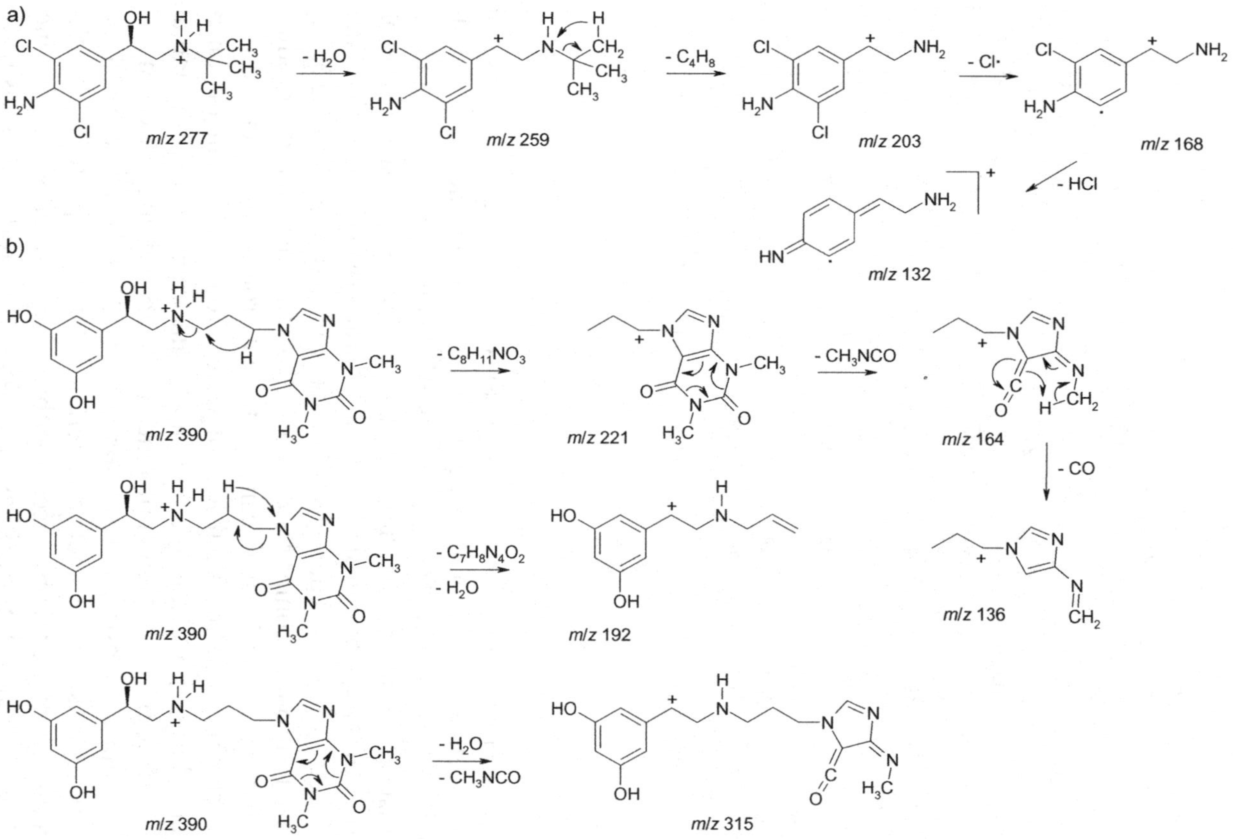

Scheme 4.9: Suggested dissociation pathways of (a) clenbuterol, and (b) reproterol.[90–92]

TABLE 4.6: Characteristic Product Ions of Selected β_2-agonists Generated by Positive ESI and CID

Compound	Mol Wt (Da)	Precursor Ion (*m/z*)	Product Ions (*m/z*)					
Bambuterol	367	368	350	312	294	277	249	72
Brombuterol	364	365	347	—	291[a]	212	132	57
Cimaterol	219	220	202	—	160[a]	143	121	43
Cimbuterol	233	234	216	178	160[a]	152	143	57
Clenbuterol	276	277	259	—	203[a]	168	132	57
Clenpenterol	290	291	273	—	—	203[a]	168	71
Fenoterol	303	304	286	—	—	152	135[a]	107
Formoterol	344	345	327	—	—	179	149[a]	121
Isoxsuprine	301	302	284[a]	—	—	150	135	107
Mabuterol	310	311	293	—	237[a]	217	202	181
Mapenterol	324	325	307	—	237[a]	217	202	—
Procaterol	290	291	273[a]	—	231	—	162	130
Ractopamine	301	302	284	—	—	164[a]	121	107
Reproterol	389	390	372	—	—	315	221[a]	192
Ritodrine	287	288	270	—	—	150	121[a]	93
Salmeterol	415	416	398[a]	—	—	380	248	232
Salbutamol	239	240	222	—	—	166	148[a]	57
Terbutaline	225	226	208	170	152[a]	125	107	57
Tulobuterol	227	228	210	172	154[a]	119	118	57

[a] Base peak.

composed by the nitrogen-linked functions are observed at *m/z* 57 and 71, respectively (Table 4.6), the abundance of which increases significantly with elevated collision energies. With compounds bearing more extensive substituents at the secondary amino group, e.g., fenoterol, reproterol, and salmeterol, additional characteristic product ions are produced upon CID as demonstrated with reproterol (compound **3**, Fig. 4.15b). Reproterol comprises a theophylline structure and the product ion mass spectrum is significantly different from those obtained e.g., from clenbuterol. An initial loss of water from the precursor ion is observed at *m/z* 372, and the most abundant fragment ion is found at *m/z* 221, the proposed origin of which is a theophyllinylpropylium ion. Origin and composition of the ion at *m/z* 221 were confirmed by a variety of mass spectrometric experiments using reproterol and the structurally related substances theobromine and ethyltheophylline. CID of *m/z* 221 in MS^3 analyses gave rise to the product ions resulting from elimination of 57, 28, and 27 Da, the origins of which were proposed to be releases of methyl isocyanate, carbon monoxide, and HCN, respectively, indicating the presence of the theophylline structure.

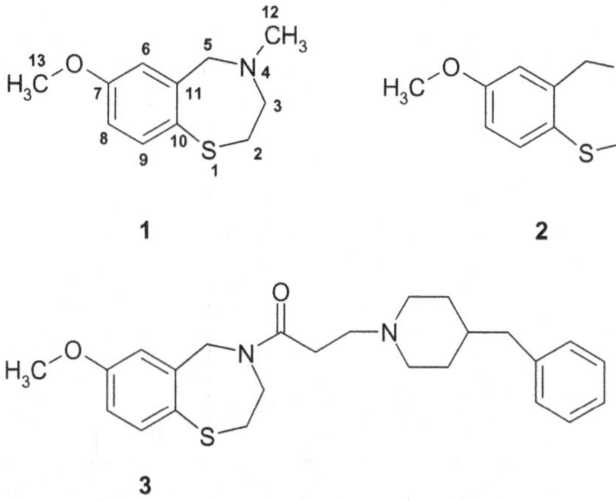

Figure 4.16: Structures of the rycals S-107 (1), its desmethylated metabolite (2), and JTV-519 (3).

Moreover, the losses of theophylline (–180 Da) or methyl isocyanate (–57 Da) after release of a water molecule from the protonated molecule of reproterol were observed, yielding the product ions at *m/z* 192 or 315, respectively (Scheme 4.9b).[92,93]

4.7 CALCIUM-CHANNEL MODULATORS (RYCALS)

The beneficial effects of rycals such as S-107 and JTV-519 (Fig. 4.16, **1** and **3**, respectively) on cardiac as well as skeletal muscles with regard to arrhythmia and fatigue have alerted doping control authorities as outlined in Chapter 3. While S-107 and potential metabolites are amenable for GC-MS analyses, JTV-519 is hardly suitable for gas chromatography; consequently, ESI-MS/MS studies on both compounds were conducted to characterize these potentially performance-enhancing drugs, although none of these is currently banned according to the regulations of the World Anti-Doping Agency.

Due to the presence of various heteroatoms, protonation of S-107 and JTV-519 via ESI may occur at different sites of the molecules; however, the proton affinities of sulfur and nitrogen atoms within the analytes as estimated from related compounds suggested an initial protonation at the tertiary amino function and the piperidine residue of **1** and **3**, respectively.[94] S-107 (**1**) yielded a comprehensive dissociation pattern after protonation and CID, which characterized the benzothiazepine core structure due the ionization of the bicyclic nucleus

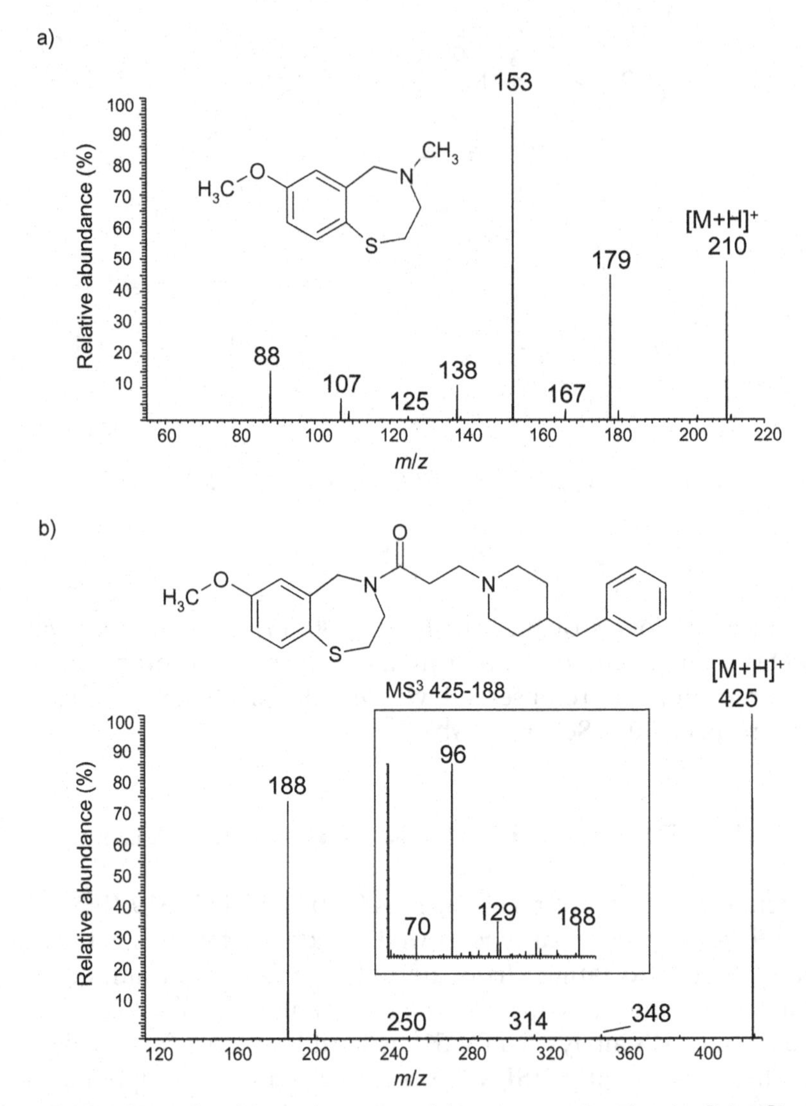

Figure 4.17: ESI product ion mass spectra of the protonated molecules $[M+H]^+$ of (a) S-107 (mol wt = 209), and (b) JTV-519 (mol wt = 424) recorded on an LTQ-Orbitrap system.

(Fig. 4.17a). By means of deuterium labeling experiments and analyses of structurally related compounds,[94] concise information on fragmentation pathways was obtained (Scheme 4.10a).[94] Ring opening processes of the protonated molecule (*m/z* 210) accompanied by the losses of methylamine (–31 Da) or 1-methyl-aziridine (–57 Da) yielded major product ions at *m/z* 179 and 153, which were assigned to a charged

4-methoxy-1-vinylsulfanyl-cycloheptatriene and dehydrogenated 4-methoxy-2-methyl-benzenethiol structure, respectively (Scheme 4.10a). The latter was shown to be of considerable diagnostic value as it was observed as a common product ion of **1** and its *N*-desalkylated analogue (Fig. 4.16, **2**) and might be useful for future metabolite screening purposes in precursor ion scan experiments. The ions at *m/z* 138 and 107 were shown to result from *m/z* 153 in MS^3 studies, and a product ion at *m/z* 88 was suggested to consist of a 2-methyl-4H[1,2]-thiazete structure derived from the protonated molecule at *m/z* 210 (Scheme 4.10a).

The protonated molecule of **3** $[M+H]^+$ at *m/z* 425 primarily generated one abundant product ion at *m/z* 188 (Fig. 4.17b). The elimination of 4-acetyl-7-methoxy-2,3,4,5-tetrahydro-1,4-benzothiazepine (–237 Da) was suggested to yield the product ion with the elemental composition of $C_{13}H_{18}N$ (*m/z* 188), presumably consisting of a 4-benzyl-1-methylene piperidine residue (Scheme 4.10b). Subsequently, neutral losses of toluene (–92 Da) or ethyl-methylamine (–59) from *m/z* 188 were observed in MS^3 experiments, which gave rise to product ions at *m/z* 96 and 129, respectively, the structures of which were assigned to 1-methylene-tetrahydropyridine and buta-1,3-dienyl benzene (Scheme 4.10b). All abundant product ions derived from **3** originated from the 4-benzyl-1-methyl-piperidine portion of the molecule, presumably due to the charge-driven dissociation processes initiated by the protonation at the piperidine nitrogen; however, a characteristic though low abundance product ion was detected at *m/z* 250, which resulted from the release of 4-benzylpiperidine (–175 Da) from the protonated molecule of **3** (Scheme 4.10b) and, thus, complements the information obtained from the product ion mass spectrum of the target analyte.

4.8 PEROXISOME-PROLIFERATOR ACTIVATED RECEPTOR-δ (PPARδ) AND ADENOSINE MONOPHOSPHATE ACTIVATED PROTEIN KINASE (AMPK) AGONISTS

The importance of the peroxisome-proliferator-activated receptor (PPAR)δ as a key regulator of fat utilization[95,96] and its influence on the regulation of muscle fiber type and physical performance has become evident in studies concerning potential targets for the treatment of obesity and metabolic disorders.[97] In animal experiments, the pharmacological activation of PPARδ with 2–5 mg/kg/day of the lead drug candidate GW1516 (also referred to as GW501516, Fig. 4.18, **1**)

a)

m/z 210 → → $-NH_2CH_3$ → m/z 179 ⇌

m/z 210 → → $-C_3H_7N$ → m/z 153 → m/z 138, m/z 107

m/z 210 → → $-C_8H_{10}O$ → m/z 88

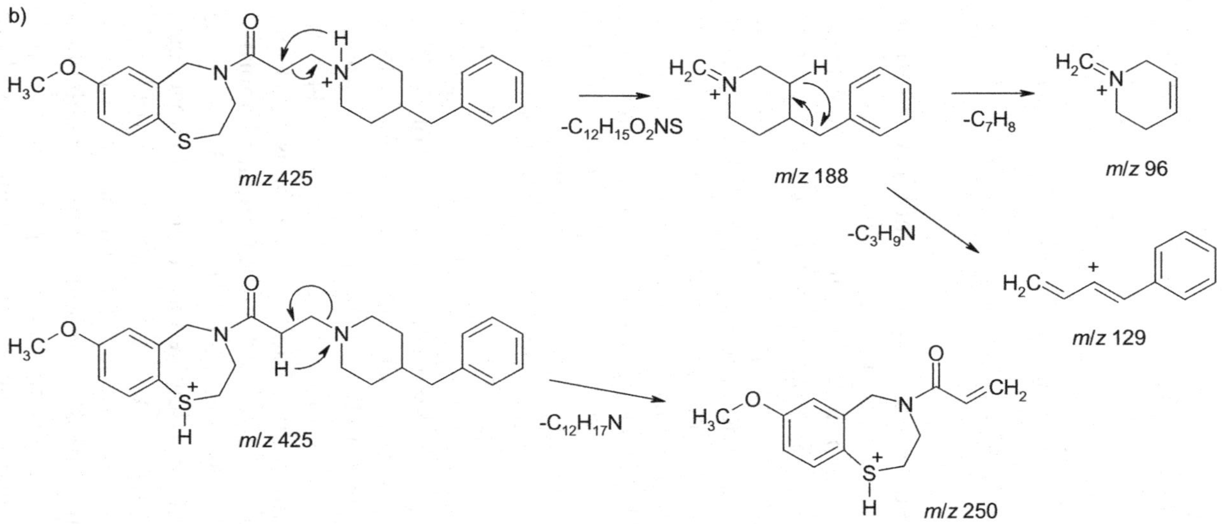

b)

H_3C O N O H N + S

m/z 425

$-C_{12}H_{15}O_2NS$

H_2C N + H

m/z 188

$-C_7H_8$

H_2C N +

m/z 96

$-C_3H_9N$

H_2C +

m/z 129

H_3C O N O N H S + H

m/z 425

$-C_{12}H_{17}N$

H_3C O N O CH_2 S + H

m/z 250

Scheme 4.10: Proposed dissociation routes of (a) S-107, and (b) JTV-519.[94]

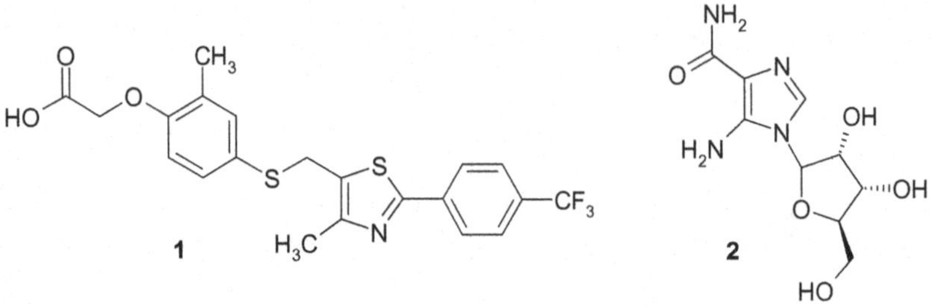

Figure 4.18: Structures of GW1516 (**1**) and 5-amino-4-imidazolecarboxamide ribonucleoside (AICAR, **2**)

resulted in an induction of oxidative genes,[98] and a modified substrate preference of skeletal muscles that caused a shift from carbohydrate to lipid consumption.[99] The gene signatures of the ligand- (i.e., GW1516) and exercise-induced muscle transcriptome were found to be approximately 50% overlapping, which demonstrated performance-mimicking properties of GW1516, especially since a total of 32% of the common target genes were attributed to the positive regulation of the aerobic capacity.[98] The combination of a pharmacological stimulation of the PPARδ receptor using GW1516 and exercise yielded significantly enhanced physical performance (longer distance and period of running), which was increased by approximately 70%. Exercise was shown to especially activate the adenosine monophosphate (AMP)-activated protein kinase (AMPK), which promotes particularly mitochondrial biogenesis, skeletal muscle gene expression, and oxidative metabolism.[100] The use of a natural and cell-permeable activator of AMPK, 5-amino-4-imidazolecarboxamide ribonucleoside (AICAR, Fig. 4.18, **2**) at 500 mg/kg/day effectively activated the AMPK signaling pathway and caused an improved endurance of untrained mice by 23–44% by upregulating 32 genes associated with oxidative metabolism without exercise.[98] The combination of GW1516 and AICAR induced approximately 40% of the genetic effects observed in the GW1516/exercise regimen, including genes linked to oxidative metabolism, angiogenesis, and glucose sparing. Consequently, new pharmacological routes to utilize the plasticity of muscle fibers are available that may allow reprogramming muscle cells to fatigue-resistant type-I fibers with enhanced mitochondrial content, which could significantly improve athletic performance. Misuse of either drug candidate or the combination of both by athletes is conceivable despite known health issues related to a long-term use of AICAR in particular[101] and police finding support the presumption that illicit applications have happened in the past.[102] Both compounds, GW1516 and AICAR, which have undergone advanced clinical trials, have been added to the section "Gene Doping" of the Prohibited List of the World Anti-Doping Agency that became effective in January 2009.[103]

CID of the protonated molecule of GW1516 ($[M+H]^+ = 454$) yielded only few abundant or characteristic product ions at *m/z* 396, 288, 257, 256, and 188 (Fig. 4.19a). The loss of $C_2H_2O_2$ (–58 Da) that yields the

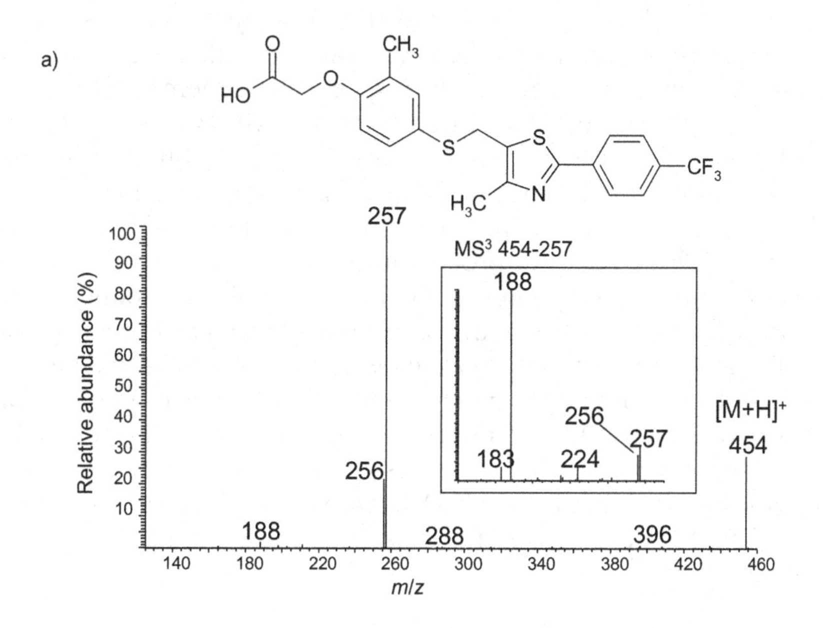

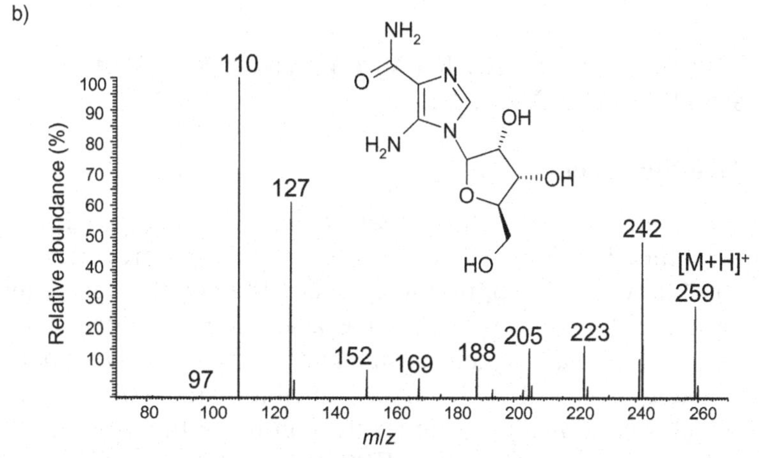

Figure 4.19: ESI product ion mass spectra of the protonated molecules $[M+H]^+$ of (a) GW1516 (mol wt = 453), and (b) 5-amino-4-imidazolecarboxamide ribonucleoside (AICAR, mol wt = 258) recorded on an LTQ-Orbitrap system.

ion at *m/z* 396 was attributed to the elimination of oxiran-2-one or glyoxal, which originates from the acetic acid moiety of **1**. Simultaneously, *o*-tolyloxy-acetic acid (–166 Da) is released from the protonated precursor ion to generate the product ion at *m/z* 288 (Scheme 4.11a).[80,132] In contrast to these low abundance product ions, an intense radical cation is obtained at *m/z* 257 that results from a homolytic cleavage of the S–C bond and the loss of a (4-mercapto-2-methyl-phenoxy)-acetic acid radical (–197 Da). Subsequently, *m/z* 257, which was assigned to the protonated 4,5-dimethyl-2-(4-trifluoromethyl-phenyl)-thiazole radical, releases either a hydrogen or a trifluoromethyl radical to yield the product ions at *m/z* 256 or 188, respectively (Scheme 4.11a).

The particular structural features of AICAR (Fig. 4.18, **2**) that includes an aminoimidazole moiety as well as a ribofuranose residue are reflected in its mass spectrometric dissociation behavior as shown with its product ion mass spectrum (obtained after positive ESI and CID; Fig. 4.19b). Numerous product ions originating from consecutive losses of water (–18 Da) and ammonia (–17 Da) were found at *m/z* 242, 241, 223, 205, and 188, and most abundant fragment ions resulting from the aminoimidazole-carboxamide nucleus were observed at *m/z* 127 and 110. Assuming an initial protonation of the primary amino function, the neutral loss of the ribose residue (2-hydroxymethyl-2,3-dihydro-furan-3,4-diol, –132 Da) gives rise to the protonated 5-amino-imidazole-4-carboxamide (*m/z* 127) that subsequently releases ammonia (–17 Da) to yield the cation of imidazole-4-carboxamide with *m/z* 110 (Scheme 4.11b).[132]

4.9 HYPOXIA-INDUCIBLE FACTOR (HIF)-STABILIZERS AND SIRTUIN ACTIVATORS

4.9.1 HIF-Stabilizers

The most common approach to treat anemia and associated diseases has been the administration of recombinantly produced EPO (rhEPO);[104] however, the option to correct EPO deficiencies by means of increasing EPO gene expression has also been evaluated since the mechanisms of cellular oxygen sensing and the hypoxia signal pathway were elucidated.[105–107] The hypoxia-inducible (transcription) factor (HIF), which was shown to be under the control of the arterial oxygen tension, plays a major role in the EPO gene expression. HIF is composed of a heterodimer that consists of the two subunits HIF-1α and HIF-1β, both of which are continuously produced and required to

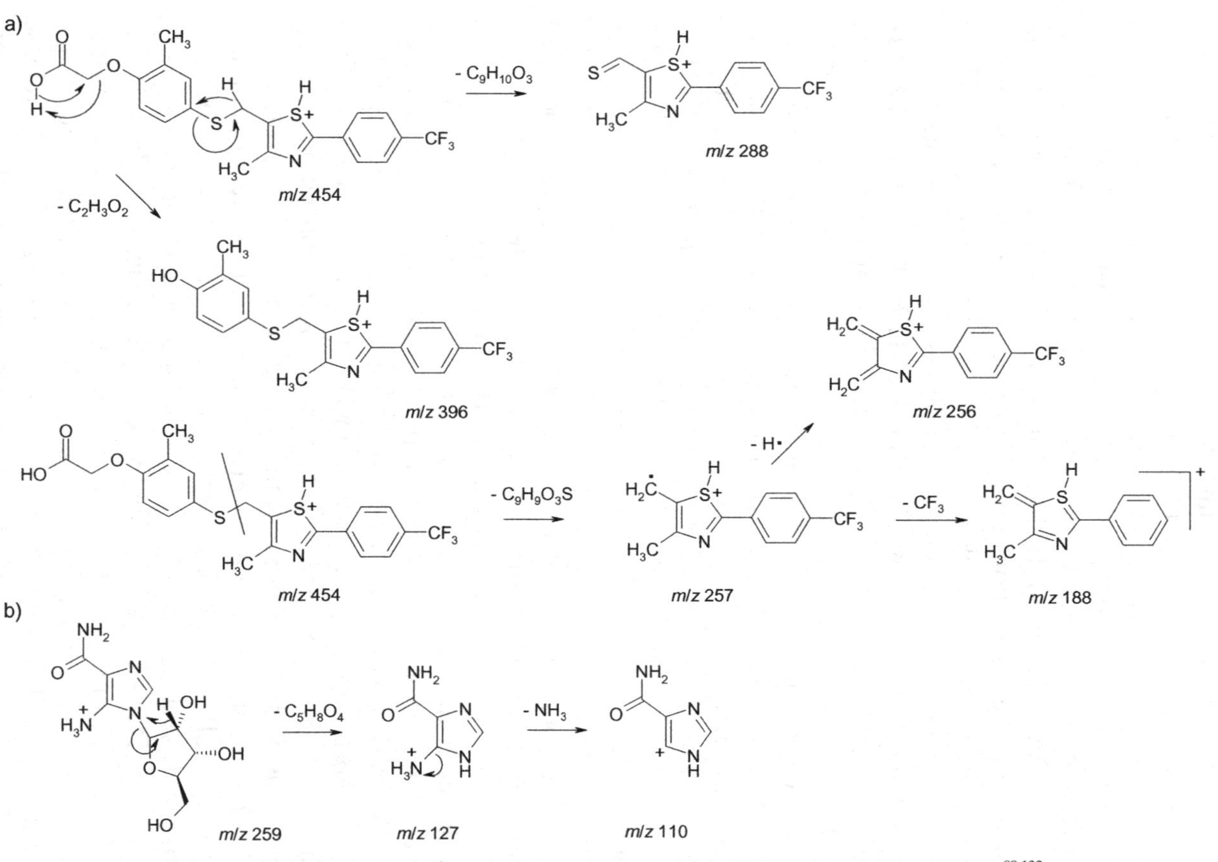

Scheme 4.11: Suggested dissociation pathways of (a) GW1516, and (b) AICAR.[80,132]

generate an active transcription complex in conjunction with several coactivators.[108] While HIF-1β is protected from proteolytic enzymes of the cytoplasm due to its sequestration within the nucleus, HIF-1α is exposed to cytoplasmic proline hydroxylases, which promote the hydroxylation of specific proline residues located in the oxygen-dependent degradation domain of HIF-1α under normoxic conditions. Being hydroxylated, HIF-1α is tagged by the von Hippel-Lindau gene product (pVHL), polyubiquinated, and finally degraded through the proteasomal pathway, i.e., the HIF-1α subunit is not stable under normoxia, which limits the stimulation of the EPO gene expression. Under hypoxic conditions, the oxygen levels required for the hydroxylation of HIF-1α are not reached, and the protein complex is able to enter the nucleus to form the heterodimer with the nuclear HIF-1β.[109] On the basis of these findings, orally available prolyl hydroxylase inhibitors were developed to mimic a reduced oxygen tension and stimulate the EPO gene expression. An advanced drug candidate is FG-2216[110,111] that reduced prolyl hydroxylase activity in non-human primate models (oral application of 60 mg of FG-2216 per kg bodyweight) and allowed to maintain hemoglobin levels comparable to those achieved using rhEPO in patients with chronic kidney disease.[109] The structure of FG-2216 has not been disclosed yet but was reported to be an isoquinoline derivative similar to the model compounds **1** and **2** (Fig. 4.20).[112] A potential for misuse in sports is obvious since the production of erythrocytes is stimulated and might contribute to enhanced athlete endurance.

A great variety of prolyl hydroxylase inhibitors (Fig. 4.20, **1–6**) with particular focus on the stabilization of HIF-1α have been clinically investigated. Those bearing an isoquinoline core structure such as FG-2216 have demonstrated advantageous medicinal properties, and two model substances (Fig. 4.20, **1** and **2**) were studied with regard to their mass spectrometric behavior.[113,114] Efficient positive ionization is accomplished using ESI, which presumably results in an initial protonation of *N*-2 as estimated from the proton affinity of native isoquinoline (951.7 kJ/mol).[115] Product ion mass spectra obtained after collision-induced dissociation included common eliminations e.g., the losses of water (–18 Da) and carbon monoxide (–28 Da, Fig. 4.21). In addition, unusual and, thus, specific dissociation routes accompanied by gas-phase reactions with oxygen or water were observed that resulted in a nominal loss of 11 Da, which was suggested to originate from the release of HCN (–27 Da) and a concurrent addition of oxygen (+16 Da) or the combination of an elimination of methyleneamine (–29 Da) and the addition of water (+18 Da). The introduction of deuterium atoms at the glycine residue of **2** provided information that support the latter

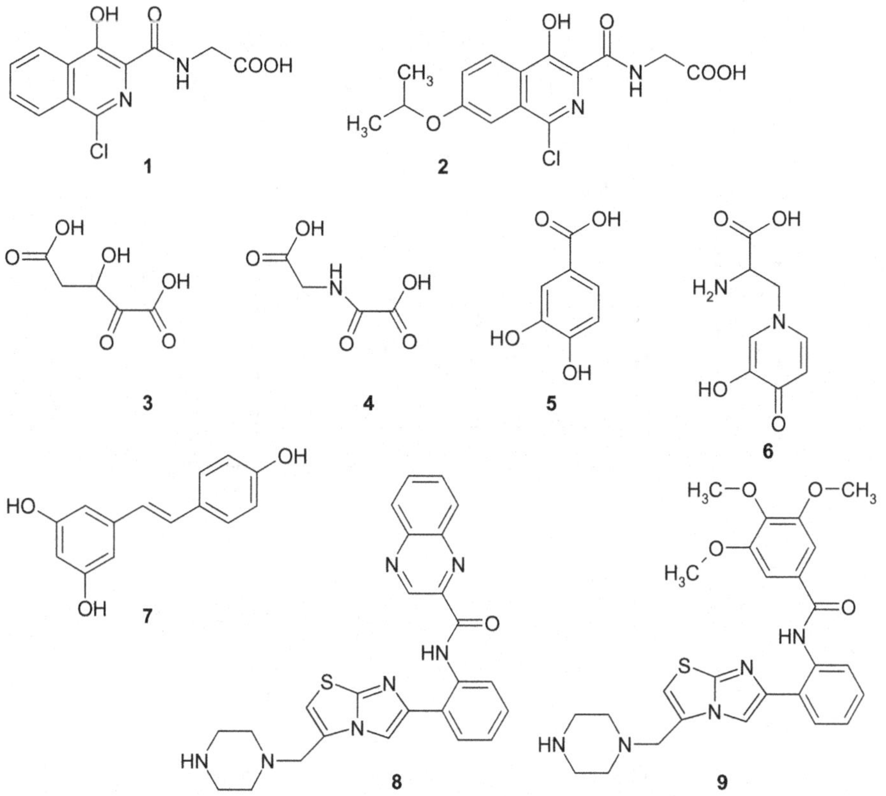

Figure 4.20: Structures of selected HIF-stabilizers (**1–6**) and sirtuin activators (**7–9**): isoquinoline model substance (**1**), isoquinoline model substance (**2**), 2-oxoglutarate (**3**), oxalyl glycine (**4**), 3,4-dihydroxybenzoic acid (**5**), L-mimosine (**6**), resveratrol (**7**), SRT-1720 (**8**), and SRT-1460 (**9**).

pathway as illustrated for compound **1** in Scheme 4.12a.[113,114,120] Evidence for the gas-phase composition of the product ions at *m/z* 224 and 282 derived from losses of 11 Da from *m/z* 235 and 293, respectively, was obtained by chemical synthesis of the putative structures and subsequent comparison of product ion mass spectra. The gas-phase addition reaction does not necessitate ion trapping but occurs also during in-space dissociation, and the oxygen or water does not originate from the solvent as D_2O and $H_2{}^{18}O$ were employed. However, the protonated 1-chloro-4-hydroxy-isoquinoline-3-carboxylic acid (*m/z* 224) and its isopropoxy derivative (*m/z* 282) as obtained from **1** and **2**, respectively (Fig. 4.21), apparently comprise a stabilized core structure that allows an efficient re-generation under ion storage conditions, which enabled "ping-pong" experiments between *m/z* 206 and 224 up to MS^7.[113]

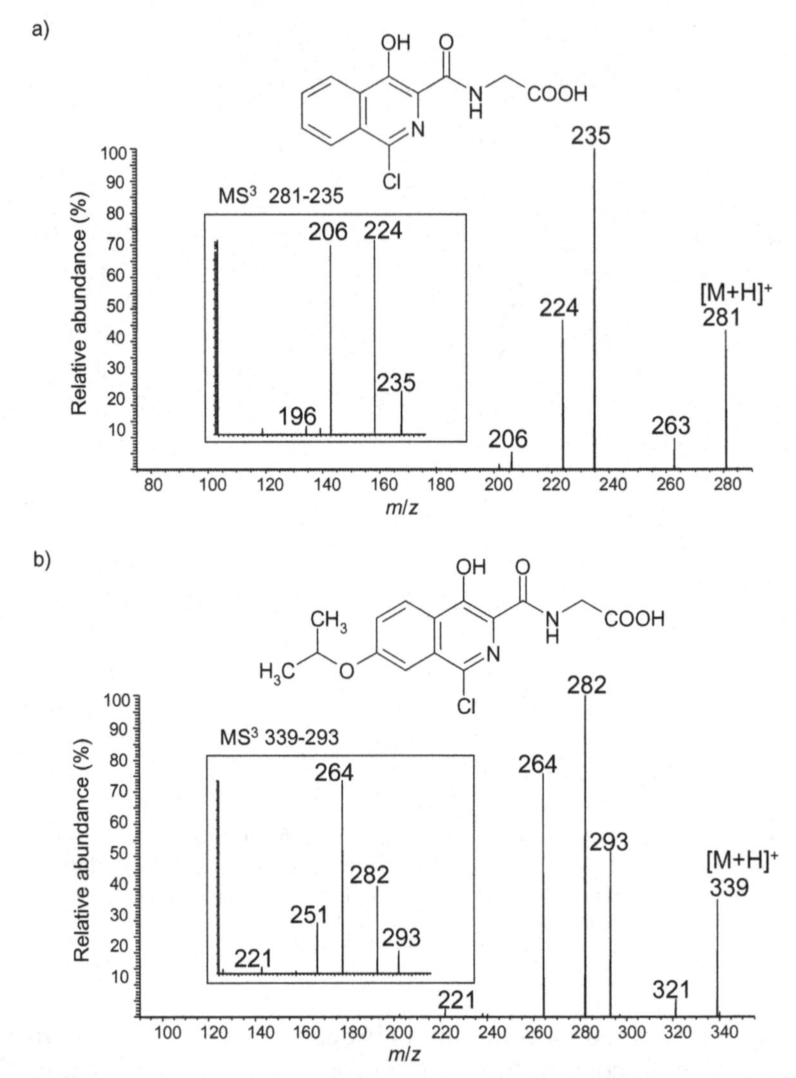

Figure 4.21: ESI product ion mass spectra of the protonated molecules $[M+H]^+$ of (a) isoquinoline-derived HIF stabilizer **1** (mol wt = 280) and (b) isoquinoline-derived HIF stabilizer **2** (mol wt = 338), recorded on an LTQ-Orbitrap system.

4.9.2 Sirtuin Activators

Mammalian sirtuins (SIRTs, silent information regulator transcript) were identified as conserved structural homologues to the yeast silent information regulator protein (Sir2), which has been shown to mediate longevity under calorie restriction conditions.[116] The increased lifespan associated with calorie restriction in general has been attributed to reduced rates of cancer, diabetes, inflammation, and cardiovascular diseases, and SIRT1 (as one of the currently seven identified members of the mammalian sirtuin family) was shown to play a key role in these processes.[117] SIRT1 has been studied in great detail and functions as a nicotinamide adenine dinucleotide $(NAD)^+$-dependent protein deacetylase with various different therapeutic indications. Among the numerous potential effects of SIRT1 are reduced neurodegeneration, elevated mitochondriogenesis, increased fat metabolism and insulin sensitivity, etc. which resulted in a broad interest in small molecule activators of SIRT1 as therapeutics for the treatment of diseases of aging.[118–120] Several natural SIRT1 activators were identified in the past including predominantly polyphenolic plant metabolites such as chalcones (e.g., butein), flavones (e.g., quercetin), and stilbenes (e.g. resveratrol, Fig. 4.20, **7**), with the latter being the most potent compound that is marketed currently as active pharmaceutical ingredient as well as nutritional supplement. In addition to these natural products, entirely synthetic SIRT1 activators have been studied including SRT-1720 and SRT-1460 (Fig. 4.20, **8** and **9**, respectively), which are structurally unrelated to any of the previously identified SIRT-activating substances but orally available and significantly more potent than resveratrol.[120] Numerous substrates and interactors of SIRT1 were determined including the peroxisome proliferator-activated receptor-γ coactivator 1α (PGC1α), which is activated by deacetylation via SIRT1 and essential for mitochondrial biogenesis. This and other (partially hypothesized) beneficial effects of SIRT1 activation such as cardioprotection against oxidative stress and increased fatty acid oxidation have made sirtuins an attractive class of new therapeutics but also a new option of performance manipulation in sport.

4.9.2.1 Resveratrol The mass spectrometric behavior of resveratrol (3,5,4′-trihydroxystilbene, Fig. 4.20, **7**) was studied using negative ESI, high resolution/high accuracy mass spectrometry and deuterium labeling experiments to elucidate the generation of abundant product ions such as *m/z* 185, 183, 159, 157, and 143 (Fig. 4.22a).[121] Deprotonation was suggested to occur at the 4′-located hydroxyl function to produce the precursor ion at *m/z* 227, which eliminated ketene (−42 Da) by a charge-driven or charge-remote pathway to yield the product ion at *m/z* 185. Further dehydrogenation led to the ion at *m/z* 183, and losses of carbon monoxide (−28 Da) or ketene (−42 Da) gave rise to the product ions at *m/z* 157 and 143, respectively (Scheme 4.12b). Comprehensive rearrangements were proposed to result in bicyclic or tricyclic product ions of *m/z* 185, 183, 157, and 143,[121] but the analysis of the recently commercialized four-fold deuterated resveratrol (2′,3′,5′,6′-$^2H_3$-resveratrol) demonstrated the exclusion of the

a)

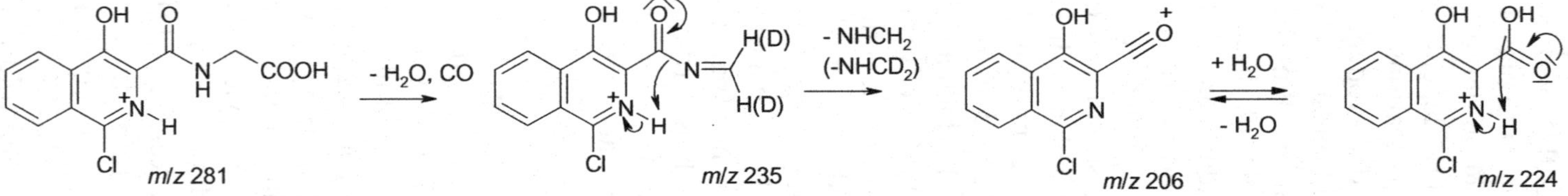
m/z 281
- H2O, CO
m/z 235
- NHCH2
(-NHCD2)
m/z 206
+ H2O
- H2O
m/z 224

b)

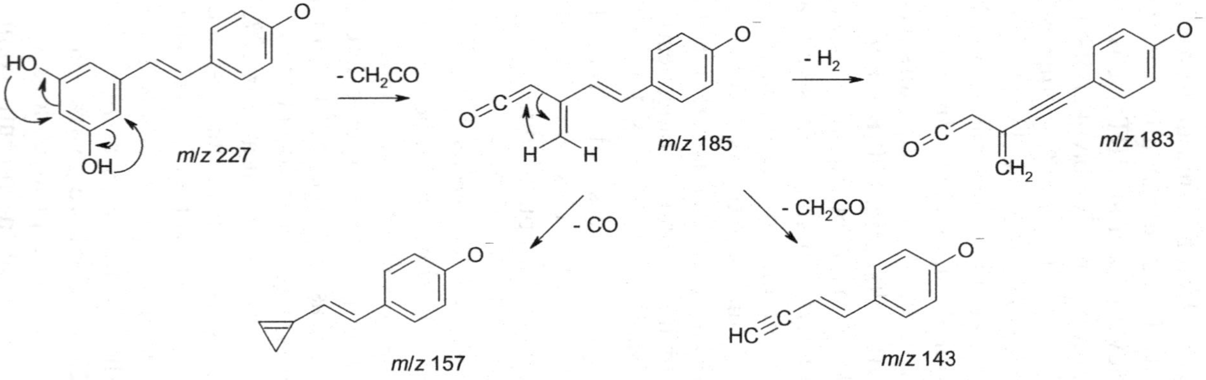
m/z 227
- CH2CO
m/z 185
- H2
m/z 183
- CO
- CH2CO
m/z 157
m/z 143

c)

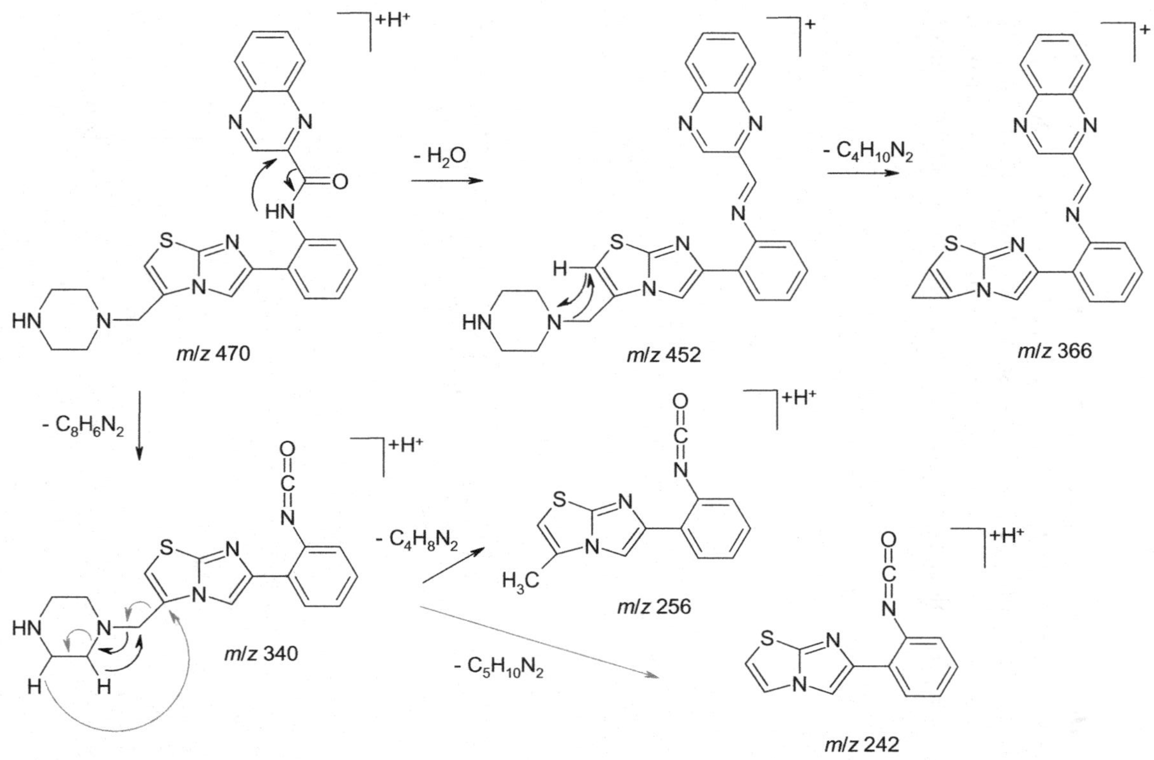

- H_2O
- $C_4H_{10}N_2$
- $C_8H_6N_2$
- $C_4H_8N_2$
- $C_5H_{10}N_2$

m/z 470, *m/z* 452, *m/z* 366, *m/z* 340, *m/z* 256, *m/z* 242

Scheme 4.12: Suggested dissociation pathways of (a) isoquinoline-derived HIF-stabilizer **1**, and the sirtuin activators resveratrol (b) and SRT-1720 (c).[113,114,120]

introduced labels from the product ion formations, which suggests that alternative routes are possibly present, potentially based on a deprotonated 5-(4-hydroxy-phenyl)-3-methylene-penta-1,4-dien-1-one intermediate.

4.9.2.2 SRT-1720 A typical representative of fully synthetic SIRT1 activators is SRT-1720 (Fig. 4.20, **8**), which yields a protonated molecule using positive ESI that efficiently dissociates upon CID (Fig. 4.22b). Various ionization sites are provided including the quinoxaline, imidazo[2,1-b]thiazole, and the piperazine moieties, which complicated

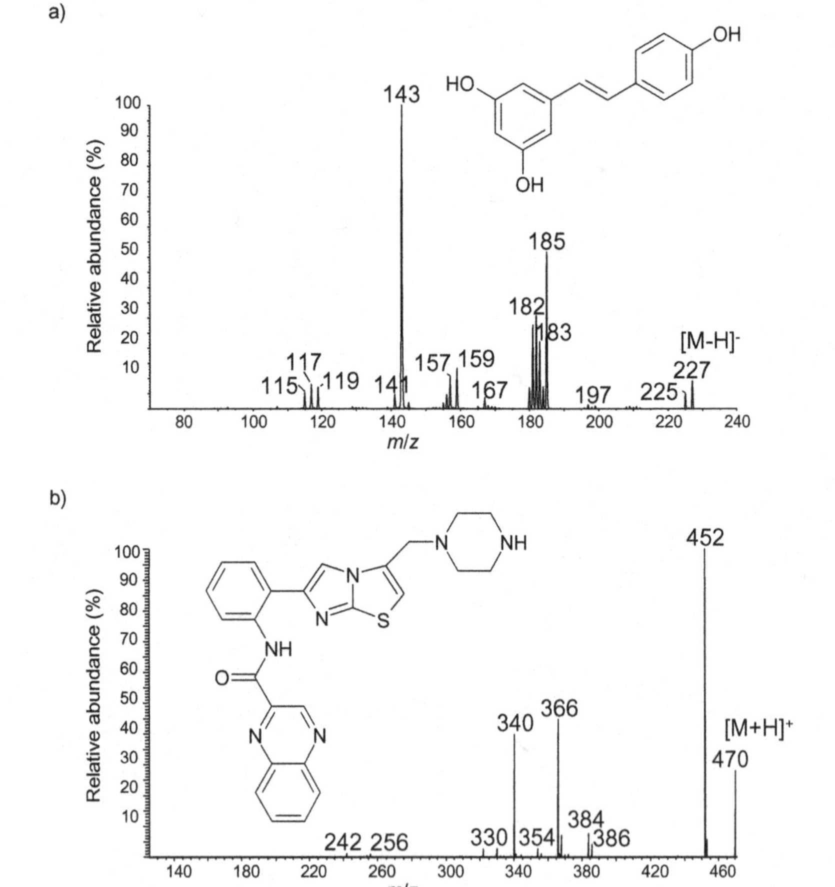

Figure 4.22: ESI product ion mass spectra of (a) the deprotonated molecule $[M-H]^-$ of resveratrol (mol wt = 228), recorded on an API2000 QTrap system and (b) the protonated molecule $[M+H]^+$ of the sirtuin activator SRT-1720 (mol wt = 469), recorded on an LTQ-Orbitrap system.

the location of the introduced proton; however, by means of H/D-exchange experiments, its immediate elimination with a water molecule (−18 Da) to yield the product ion at *m/z* 452 was demonstrated. The ion at *m/z* 452 subsequently released piperazine (−86 Da) to give rise to *m/z* 366. Alternatively, the protonated molecule of SRT-1720 eliminated quinoxaline (−130 Da) to produce the ion at *m/z* 340, which further dissociated into *m/z* 256 and 242, presumably by losses of 1,2,3,6-tetrahydropyrazine (−84 Da) and 1-methyl-1,2,3,4-tetrahydropyrazine (−98 Da), respectively (Scheme 4.12c).

4.10 β-RECEPTOR BLOCKING AGENTS

Due to the need of extensive derivatization for GC-MS analysis (see Chapter 3), β-receptor blocking agents such as atenolol, bupranolol, metoprolol, and propranolol (Fig. 4.23, **1–4**) have been early subjects of LC-MS/MS analysis. The presence of a secondary amino function in all β-blockers provides sufficient proton affinity for positive ESI, and core structure-specific as well as individual product ions are derived from target analytes using low energy CID as summarized in Table 4.7.

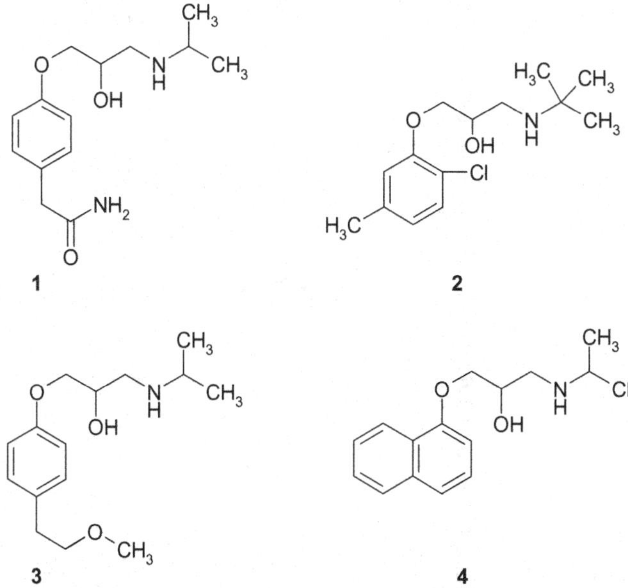

1

2

3

4

Figure 4.23: Structures of selected β-blockers: atenolol (**1**), bupranolol (**2**), metoprolol (**3**), and propranolol (**4**).

TABLE 4.7: Characteristic Product Ions of Selected β-blockers Generated by Positive ESI and CID

Compound	Category	Mol Wt (Da)	Precursor Ion (m/z)	Product Ions (m/z)									
				$[M+H]^+$-56	$[M+H]^+$-77	Individual			Common				
Acebutolol	a	336	337	—	260	218	180	148	116	74	72	—	56
Alprenolol	a	249	250	—	173	145	131	91	116	74	72	—	56
Atenolol	a	266	267	—	190	145	107	91	116	74	72	—	56
Befunolol	a	291	292	—	215	250	203	177	—	74	72	—	56
Betaxolol	d	307	308	—	—	121	91	55	116	74	72	—	56
Bisoprolol	d	325	326	—	—	133	107	89	116	74	72	—	56
Bunitrolol	b	248	249	193	—	120	102	—	—	—	—	57	—
Bupranolol	b	271	272	216	—	155	125	91	—	74	—	57	56
Butofilolol	b	311	312	256	—	221	209	109	—	74	—	57	56
Carazolol	a	298	299	—	222	194	184	139	116	74	72	—	56
Carteolol	b	282	283	237	—	202	164	122	—	74	—	57	56
Carvedilol	d	406	407	—	—	224	222	100	—	—	—	—	56
Celiprolol	b	379	380	324	—	307	251	100	—	74	—	57	56
Cloranolol	b	291	292	236	—	175	145	109	—	74	—	57	56
Esmolol	a	295	296	—	219	145	133	91	116	74	72	—	56
Indenolol	a	247	248	—	171	145	128	98	116	74	72	—	56

Labetalol	d	328	329	—	—	294	162	91	—	—	—	—	—
Levobunolol	b	291	292	236	—	201	145	91	—	74	—	57	56
Mepindolol	a	262	263	—	186	160	148	130	116	74	72	—	56
Metipranolol	a	309	310	—	233	191	165	135	116	74	72	—	56
Metoprolol	a	267	268	—	191	159	133	77	116	74	72	—	56
Moprolol	a	239	240	—	163	121	98	77	116	74	72	—	56
Nadolol	b	309	310	254	—	201	145	115	—	74	—	57	56
Nebivolol	d	405	406	—	—	151	123	103	—	—	—	—	—
Nifenalol	c	224	225	—	—	165	119	118	—	—	—	—	—
Oxprenolol	a	265	266	—	—	225	98	—	116	74	72	—	56
Penbutolol	b	291	292	236	—	168	133	105	—	74	—	57	56
Pindolol	a	248	249	—	172	146	144	134	116	74	72	—	56
Propranolol	a	259	260	—	183	157	155	129	116	74	72	—	56
Sotalol	c	272	273	—	—	213	133	106	—	—	—	—	—
Talinolol	b	363	364	308	—	226	209	100	—	74	—	57	56
Timolol	b	316	317	261	—	188	144	113	—	74	—	57	56
Toliprolol	a	223	224	—	147	212	119	91	—	74	72	—	56

a: isopropylamino-oxypropan-2-ol side chain, b: *tert.*-butylamino-oxypropan-2-ol side chain, c: phenylethanolamine derivative, d: other.

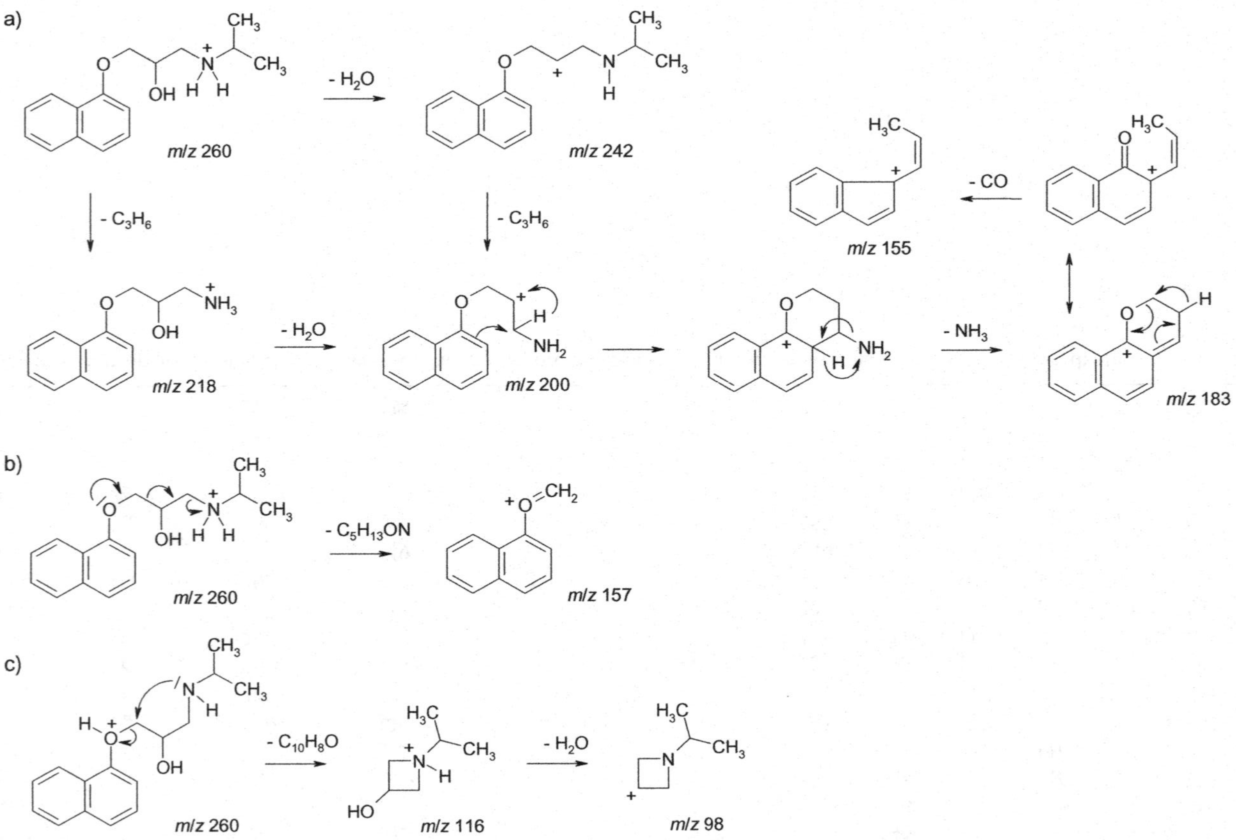

Scheme 4.13: Suggested dissociation pathways of propranolol.[9,122]

A common feature of β-receptor blocking drugs bearing an oxypropanolamine side chain terminating in an isopropyl function is the occurrence of abundant ions at *m/z* $[M+H]^+$-77, which is suggested to result from the combined elimination of water (–18 Da) and isopropylamine (–59 Da), and *m/z* 116, the proposed generation of which is illustrated with propranolol in Scheme 4.13c. In contrast, those β-blockers comprising a *tert.*-butyl group instead of the isopropyl residue (e.g., nadolol), yielded abundant ions that indicate the preferred elimination of isobutene (–56 Da, Table 4.7, Fig. 4.24b). In

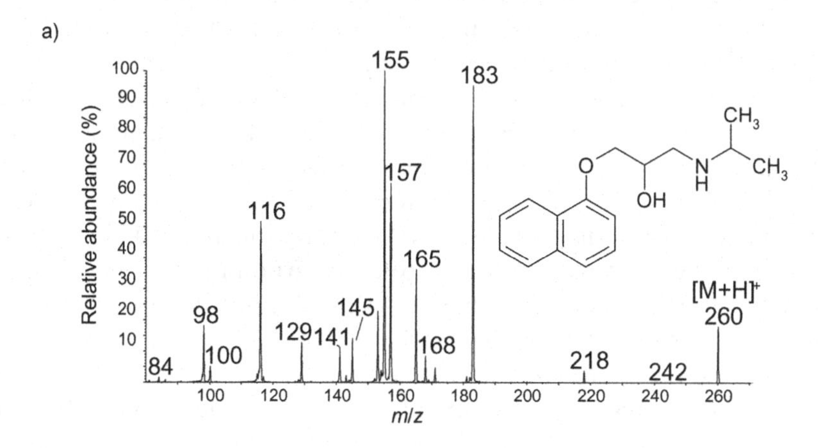

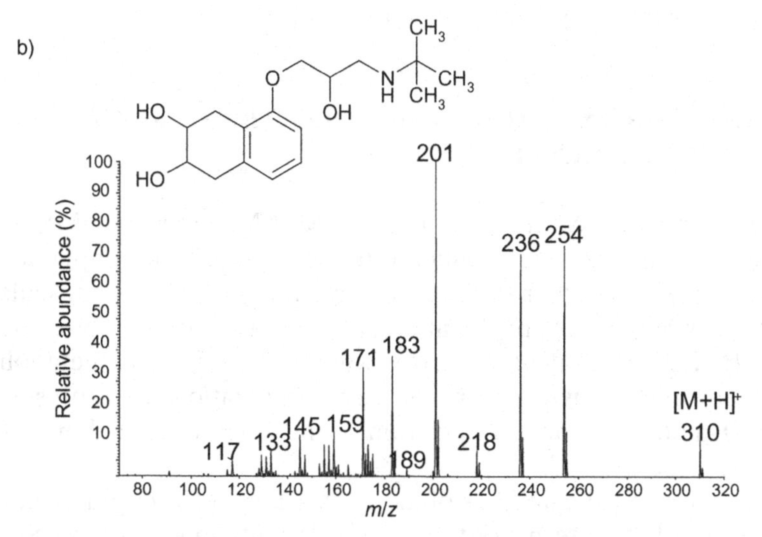

Figure 4.24: ESI product ion mass spectra of the protonated molecule $[M+H]^+$ of (a) propranolol (mol wt = 259), and (b) nadolol (mol wt = 309), recorded on an API2000 QTrap system.

addition, numerous individual but characteristic product ions are found in ESI-MS/MS spectra of β-blockers,[9] which were studied in great detail, e.g., for propranolol (Fig. 4.24a) using selectively deuterium- and ^{18}O-labeled analytes.[122] One of the most abundant product ions derived from the protonated precursor ion at *m/z* 260 was observed at *m/z* 183, which was suggested to result from consecutive eliminations of water (−18 Da), propene (−42 Da), and ammonia (−17 Da) to yield a tricyclic product ion (*m/z* 183). The introduction of five deuterium atoms in the propanolamine moiety as well as at the 2′- and 4′-position of the naphthyl residue supported the proposed mechanism as all labels of the oxypropanolamine side chain were retained during the dissociation procedure while one deuterium atom (presumably the one located at position 2′) was eliminated from the naphthyl core (Scheme 4.13a).[9,122] The subsequent release of carbon monoxide yielded the intense product ion at *m/z* 155, which was assigned to a 1-alkylated indene structure. The elimination of isopropylamine (−58 Da) and acetaldehyde (or ethenol, −44 Da) from the protonated molecule at *m/z* 260 possibly forms the methylene-naphthalen-1-yl-oxonium ion at *m/z* 157 (Scheme 4.13b), which was shown to further release carbon monoxide (−28 Da) to produce the ion at *m/z* 129 (Fig. 4.24a).[122] The product ion characterizing the isopropylamino-propanoyl residue was found at *m/z* 116, which was suggested to constitute protonated 1-isopropyl-azetidin-3-ol (Scheme 4.13c)[9] that further eliminates water (−18 Da) to yield the ion at *m/z* 98. Also here, stable isotope labeling supported the assumed dissociation routes illustrated in Scheme 4.13c.

4.11 GLUCURONIC ACID AND SULFATE CONJUGATES OF TARGET ANALYTES

One of the major advantages of LC-ESI-MS(/MS) over GC-MS is the capability to analyze intact glucuronic acid or sulfate conjugates of target compounds and/or their phase-I-metabolites. This is particularly useful, e.g., when sample preparation steps required to hydrolyze phase-II-metabolites are difficult to verify, when intact phase-II-metabolites are to determine, or when more than one conjugation site is present in a phase-I-metabolite and the position of the glucuronic acid or sulfate moiety is to be located.

There are several examples of analytes relevant for doping controls that were studied in terms of their phase-II-metabolism using ESI-MS (/MS) such as anabolic steroids, [2,5,28,123–128] narcotics,[129] stimulants,[130] etc. These studies outlined the fact that the CID behavior of the conjugates is closely related to that of the aglycons and asulfates, which facilitates the interpretation of respective product ion mass spectra as illustrated by means of morphine and morphine-3-glucuronide as well as *O*-desmethyl-methoxyphenamine and the corresponding glucuronic acid conjugate (Fig. 4.25). Commonly, the conjugates dissociate under elimination of the glucuronic acid or sulfate moiety, followed by fragmentation pathways that are in close accordance to the dissociation routes found with the native drugs or respective phase-I-metabolites.

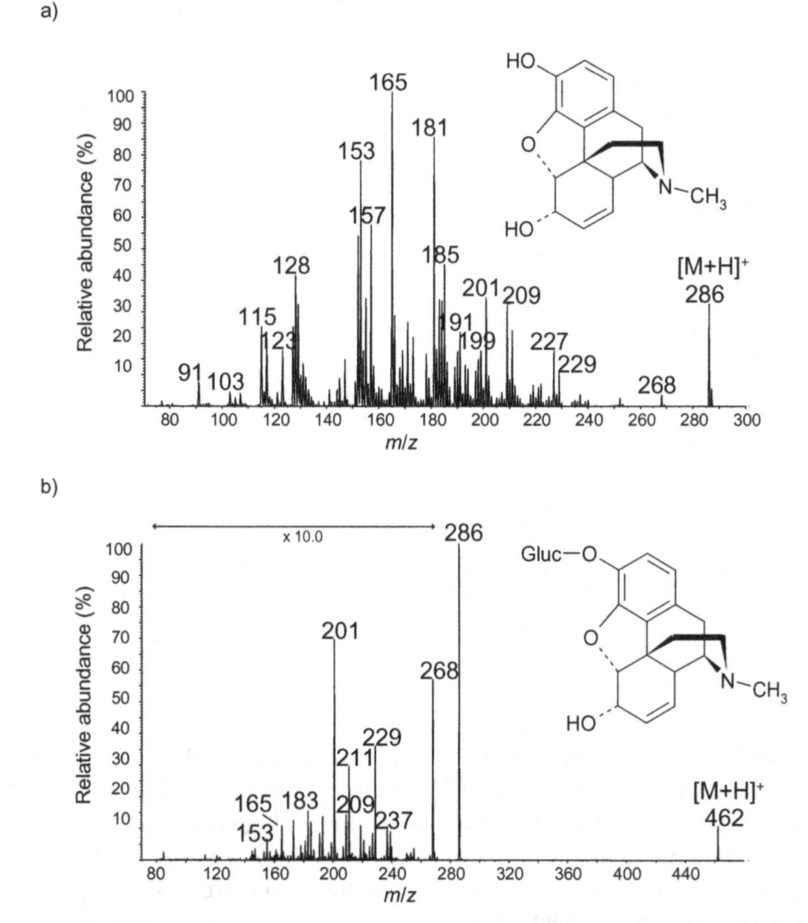

Figure 4.25: ESI product ion mass spectra of the protonated molecule $[M+H]^+$ of (a) morphine (mol wt = 285), (b) morphine-3-O-glucuronide (mol wt = 461), (c) O-desmethylmethoxyphenamine (mol wt = 166), and (d) O-desmethylmethoxyphenamine glucuronide, recorded on an API2000 QTrap system.

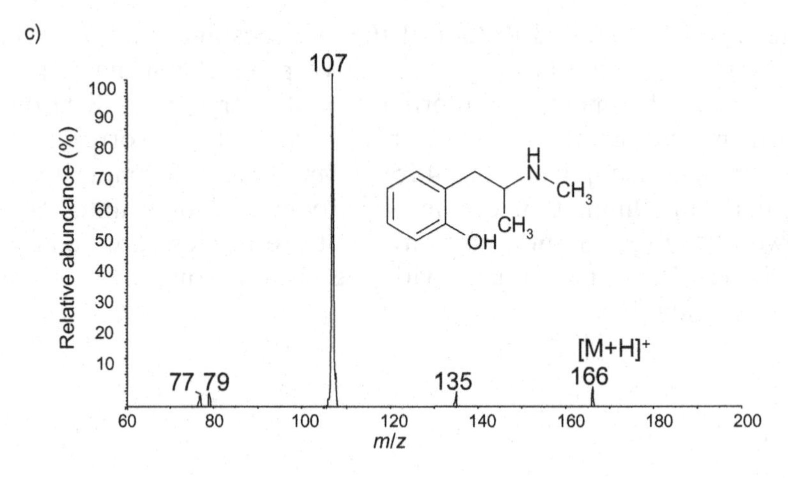

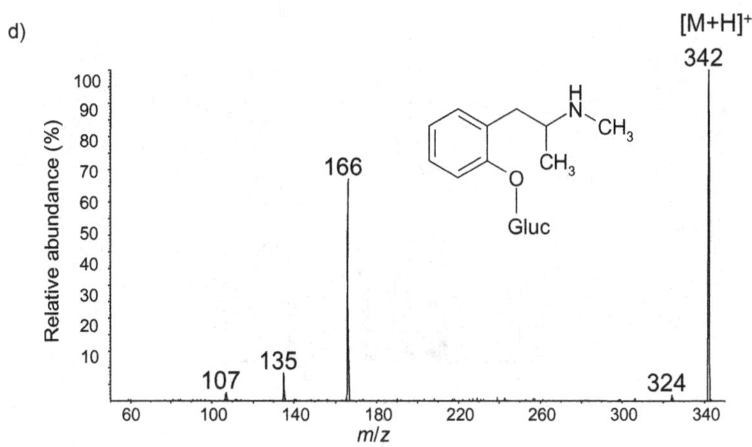

Figure 4.25: (*Continued*)

REFERENCES

1. Thevis, M., and Schänzer, W. (2007) Current role of LC-MS(/MS) in doping control. *Analytical and Bioanalytical Chemistry*, **388**, 1351–1358.
2. Bean, K.A., and Henion, J.D. (1997) Direct determination of anabolic steroid conjugates in human urine by combined high-performance liquid chromatography and tandem mass spectrometry. *Journal of Chromatography B*, **690**, 65–75.
3. Mück, W.M., and Henion, J.D. (1990) High-performance liquid chromatography/tandem mass spectrometry: Its use for the identification of stanozolol and its major metabolites in human and equine urine. *Biomedical and Environmental Mass Spectrometry*, **19**, 37–51.

4. Bowers, L.D., and Fregien, K. (1993) HPLC/MS confirmation of peptide hormones in urine: An evaluation of limit of detection, in *Recent Advances in Doping Analysis* (eds M. Donike, H. Geyer, A. Gotzmann, U. Mareck-Engelke, and S. Rauth), Sport&Buch Strauß, Cologne, pp. 175–184.
5. Bowers, L.D., and Sanaullah. (1996) Direct measurement of steroid sulfate and glucuronide conjugates with high-performance liquid chromatography-mass spectrometry. *Journal of Chromatography B*, **687**, 61–68.
6. Shackleton, C.H.L., Chuang, H., Kim, J., *et al.* (1997) Electrospray mass spectrometry of testosterone esters: Potential for use in doping control. *Steroids*, **62**, 523–529.
7. Sanz-Nebot, V., Toro, I., Berges, R., *et al.* (2001) Determination and characterization of diuretics in human urine by liquid chromatography coupled to pneumatically assisted electrospray ionization mass spectrometry. *Journal of Mass Spectrometry*, **36**, 652–657.
8. Thieme, D., Grosse, J., Lang, R., *et al.* (2001) Screening, confirmation and quantification of diuretics in urine for doping control analysis by high-performance liquid chromatography-atmospheric pressure ionisation tandem mass spectrometry. *Journal of Chromatography B*, **757**, 49–57.
9. Thevis, M., Opfermann, G., and Schänzer, W. (2001) High speed determination of beta-receptor blocking agents in human urine by liquid chromatography/tandem mass spectrometry. *Biomedical Chromatography*, **15**, 393–402.
10. Thevis, M., Opfermann, G., and Schänzer W. (2004) *N*-methyl-*N*-trimeth ylsilyltrifluoroacetamide promoted synthesis and mass spectrometric characterization of deuterated ephedrines. *European Journal of Mass Spectrometry*,**10**, 673–681.
11. Thevis, M., Sigmund, G., Thomas, A., *et al.* (2009) Doping control analysis of metamfepramone and two major metabolites using liquid chromatography: Tandem mass spectrometry. *European Journal of Mass Spectrometry*, **15**, 507–515.
12. Yan, J., Liu, Z., Yan, C., *et al.* (2006) Analysis of strychnos alkaloids using electrospray ionization Fourier transform ion cyclotron resonance multi-stage tandem mass spectrometry. *Rapid Communications in Mass Spectrometry*, **20**, 1335–1344.
13. Choi, Y.H., Sohn, Y-M., Kim, C.Y., *et al.* (2004) Analysis of strychnine from detoxified Strychnos nux-vomica [corrected] seeds using liquid chromatography-electrospray mass spectrometry. *Journal of Ethno-Pharmacology*, **93**, 109–112.
14. Fandino, A.S., Karas, M., Toennes, S.W., and Kauert, G. (2002) Identification of anhydroecgonine methyl ester N-oxide, a new metabolite of anhydroecgonine methyl ester, using electrospray mass spectrometry. *Journal of Mass Spectrometry*, **37**, 525–532.
15. Wang, P., and Bartlett, M.G. (1998) Collision-induced dissociation mass spectra of cocaine, and its metabolites and pyrolysis products. *Journal of Mass Spectrometry*, **33**, 961–967.

16. McLuckey, S.A., Goeringer, D.E., and Glish, G.L. (1992) Collisional activation with random noise in ion trap mass spectrometry. *Analytical Chemistry*, **64**, 1455–1460.
17. Raith, K., Neubert, R., Poeaknapo, C., *et al.* (2003) Electrospray tandem mass spectrometric investigations of morphinans. *Journal of the American Society for Mass Spectrometry*, **14**, 1262–1269.
18. Poeaknapo, C., Fisinger, U., Zenk, M.H., and Schmidt, J. (2004) Evaluation of the mass spectrometric fragmentation of codeine and morphine after 13C-isotope biosynthetic labeling. *Phytochemistry*, **65**, 1413–1420.
19. Zhang, Z., Yan, B., Liu, K., *et al.* (2008) Fragmentation pathways of heroin-related alkaloids revealed by ion trap and quadrupole time-of-flight tandem mass spectrometry. *Rapid Communications in Mass Spectrometry*, **22**, 2851–2862.
20. Song, F., Meng, C., and Liu, S. (1999) Quantitative analysis of pethidine using liquid secondary ion and tandem mass spectrometry. *Rapid Communications in Mass Spectrometry*, **13**, 478–480.
21. Thevis, M., and Schänzer, W. (2007) Mass spectrometry in sports drug testing: Structure characterization and analytical assays. *Mass Spectrometry Reviews*, **26**, 79–107.
22. Thevis, M., and Schänzer, W. (2005) Mass spectrometric analysis of androstan-17beta-ol-3-one and androstadiene-17beta-ol-3-one isomers. *Journal of the American Society for Mass Spectrometry*, **16**, 1660–1669.
23. Budzikiewicz, H., and Djerassi, C. (1962) Mass spectrometry in structural and stereochemical problems I. Steroid ketones. *Journal of the American Chemical Society*, **84**, 1430–1439.
24. Shapiro, R.H., Williams, D.H., Budzikiewicz, H., and Djerassi, C. (1964) Mass spectrometry in structural and stereochemical problems. LIII. Fragmentation and hydrogen transfer reactions of a typical 3-keto steroid, 5α-androstan-3-one. *Journal of the American Chemical Society*, **86**, 2837–2845.
25. Gurst, J.E., and Djerassi, C. (1964) Mass spectrometry in structural and stereochemical problems. LIX. Mechanism of the formal loss of acetone from 2-oxo-5α-steroids. *Journal of the American Chemical Society*, **86**, 5542–5550.
26. Ma, Y-C., and Kim, H-Y. (1997) Determination of steroids by liquid chromatography/mass spectrometry. *Journal of the American Society for Mass Spectrometry*, **8**, 1010–1020.
27. Williams, T.M., Kind, A.J., Houghton, E., and Hill, D.W. (1999) Electrospray collision-induced dissociation of testosterone and testosterone hydroxy analogs. *Journal of Mass Spectrometry*, **34**, 206–216.
28. Thevis, M. (2001) *Synthese und Charakterisierung von Glucuronidkonjugaten anabol-androgener Steroidhormone mittels Gaschromatographie/Massenspektrometrie, Flüssigkeitschromatographie/Massenspektrometrie und kernmagnetischer Resonanzspektroskopie*. Shaker Verlag, Aachen.

29. Pozo, O.J., Van Eenoo, P., Deventer, K., *et al.* (2008) Collision-induced dissociation of 3-keto anabolic steroids and related compounds after electrospray ionization. Considerations for structural elucidation. *Rapid Communications in Mass Spectrometry*, **22**, 4009–4024.

30. Thevis, M., Geyer, H., Mareck, U., and Schänzer, W. (2005) Screening for unknown synthetic steroids in human urine by liquid chromatography-tandem mass spectrometry. *Journal of Mass Spectrometry*, **40**, 955–962.

31. Wendler, N.L. (1963) Rearrangements in steroids, in *Molecular Rearrangements Part 1* (ed P. de Mayo), Wiley-Interscience, New York, pp. 1020–1034.

32. Segaloff, A., and Gabbard, R.B. (1964) Anti-androgenic activity of 17,17-dimethyl-18-norandrost-13-enes. *Steroids*, **4**, 433–443.

33. Cohen, A., Cook, J.W., and Hewett, C.L. (1935) The synthesis of compounds related to sterols, bile acids, and oestrus-producing hormones. Part VI. Experimental evidence of the complete structure of Oestrin, Equilin, and Equilenin. *Journal of the Chemical Society*, 445–455.

34. Johns, W.F. (1961) Retropinacol rearrangement of estradiol 3-methyl ether. *Journal of Organic Chemistry*, **26**, 4583–4591.

35. Shapiro, R.H., and Djerassi, C. (1964) Mass spectrometry in structural and stereochemical problems. L. Fragmentation and hydrogen migration reactions of α,β-unsaturated 3-keto steroids. *Journal of the American Chemical Society*, **86**, 2825–2832.

36. Schänzer, W., Geyer, H., Fusshöller, G., *et al.* (2006) Mass spectrometric identification and characterization of a new long-term metabolite of metandienone in human urine. *Rapid Communications in Mass Spectrometry*, **20**, 2252–2258.

37. Fusshöller, G., Mareck, U., Schmechel, A., and Schänzer, W. (2007) Long-term detection of metandienone abuse by means of the new metabolite 17β-hydroxymethyl-17α-methyl-18-norandrost-1,4,13-trien-3-one, in *Recent Advances in Doping Analysis* (eds W. Schänzer, H. Geyer, A. Gotzmann, and U. Mareck), Sport&Buch Strauss, Cologne, pp. 393–396.

38. Budzikiewicz, H., Djerassi, C., and Williams, D.H. (1964) *Structure Elucidation of Natural Products by Mass Spectrometry Volume II: Steroids, Terpenoids, Sugars, and Miscellaneous Classes*. Holden-Day, Inc., San Francisco.

39. Catlin, D.H., Sekera, M.H., Ahrens, B.D., *et al.* (2004) Tetrahydrogestrinone: Discovery, synthesis, and detection in urine. *Rapid Communications in Mass Spectrometry*, **18**, 1245–1249.

40. Karni, M., and Mandelbaum, A. (1980) The "Even-Electron-Rule." *Organic Mass Spectrometry*, **15**, 53–64.

41. Thevis, M., Guddat, S., and Schänzer, W. (2009) Doping control analysis of trenbolone and related compounds using liquid chromatography-tandem mass spectrometry. *Steroids*, **74**, 315–321.

42. Thevis, M., Makarov, A.A., Horning, S., and Schänzer, W. (2005) Mass spectrometry of stanozolol and its analogues using electrospray ionization and collision-induced dissociation with quadrupole-linear ion trap and linear ion trap-orbitrap hybrid mass analyzers. *Rapid Communications in Mass Spectrometry*, **19**, 3369–3378.
43. McKinney, A.R., Suann, C.J., Dunstan, A.J., *et al.* (2004) Detection of stanozolol and its metabolites in equine urine by liquid chromatography-electrospray ionization ion trap mass spectrometry. *Journal of Chromatography B*, **811**, 75–83.
44. Hsu, F.F., and Turk, J. (2000) Charge-remote and charge-driven fragmentation processes in diacyl glycerophosphoethanolamine upon low-energy collisional activation: A mechanistic proposal. *Journal of the American Society for Mass Spectrometry*, **11**, 892–899.
45. Cadilla, R., and Turnbull, P. (2006) Selective androgen receptor modulators in drug discovery: Medicinal chemistry and therapeutic potential. *Current Topics in Medicinal Chemistry*, **6**, 245–270.
46. Chen, F., Rodan, G.A., and Schmidt, A. (2002) Development of selective androgen receptor modulators and their therapeutic applications. *National Journal of Andrology*, **8**, 162–168.
47. Chen, J., Hwang, D.J., Bohl, C.E., *et al.* (2005) A selective androgen receptor modulator for hormonal male contraception. *Journal of Pharmacology and Experimental Therapeutics*, **312**, 546–553.
48. Chen, J., Kim, J., and Dalton, J.T. (2005) Discovery and therapeutic promise of selective androgen receptor modulators. *Molecular Interventions*, **5**, 173–188.
49. Gao, W., and Dalton, J.T. (2007) Expanding the therapeutic use of androgens via selective androgen receptor modulators (SARMs). *Drug Discovery Today*, **12**, 241–248.
50. Kilbourne, E.J., Moore, W.J., Freedman, L.P., and Nagpal, S. (2007) Selective androgen receptor modulators for frailty and osteoporosis. *Current Opinion in Investigational Drugs*, **8**, 821–829.
51. Negro-Vilar, A. (1999) Selective androgen receptor modulators (SARMs): A novel approach to androgen therapy for the new millennium. *Journal of Clinical Endocrinology and Metabolism*, **84**, 3459–3462.
52. Thevis, M., and Schänzer, W. (2008) Mass spectrometry of selective androgen receptor modulators. *Journal of Mass Spectrometry*, **43**, 865–876.
53. GTx, Inc. (2006) Ostarine achieved the primary endpoint of increasing lean body mass and a secondary endpoint of improving functional performance. Available at http://www.salesandmarketingnetwork.com/news_release.php?ID=2015328. Accessed 08-12-2006.
54. Gao, W., Kim, J., and Dalton, J.T. (2006) Pharmacokinetics and pharmacodynamics of nonsteroidal androgen receptor ligands. *Pharmaceutical Research*, **23**, 1641–1658.

55. Wu, Z., Gao, W., Phelps, M.A., *et al.* (2004) Favorable effects of weak acids on negative-ion electrospray ionization mass spectrometry. *Analytical Chemistry*, **76**, 839–847.
56. Gao, W., Wu, Z., Bohl, C.E., *et al.* (2006) Characterization of the in vitro metabolism of selective androgen receptor modulator using human, rat, and dog liver enzyme preparations. *Drug Metabolism and Disposition*, **34**, 243–253.
57. Thevis, M., Kamber, M., and Schänzer, W. (2006) Screening for metabolically stable aryl-propionamide-derived selective androgen receptor modulators for doping control purposes. *Rapid Communications in Mass Spectrometry*, **20**, 870–876.
58. Wu, D., Wu, Z., Yang, J., *et al.* (2006) Pharmacokinetics and metabolism of a selective androgen receptor modulator (SARM) in rats—Implication of molecular properties and intensive metabolic profile to investigate ideal pharmacokinetic characteristics of a propanamide in preclinical study. *Drug Metabolism and Disposition*, **34**, 483–494.
59. Kuuranne, T., Leinonen, A., Schänzer, W., *et al.* (2008) Aryl-propionamide-derived selective androgen receptor modulators: LC-MS/MS characterization of the in vitro synthesized metabolites for doping control purposes. *Drug Metabolism and Disposition*, **36**, 571–581.
60. Ostrowski, J., Kuhns, J.E., Lupisella, J.A., *et al.* (2006) Pharmacological and x-ray structural characterization of a novel selective androgen receptor modulator: potent hyperanabolic stimulation of skeletal muscle with hypostimulation of prostate in rats. *Endocrinology*, **48**, 4–12.
61. Sun, C., Robl, J.A., Wang, T.C., *et al.* (2006) Discovery of potent, orally-active, and muscle-selective androgen receptor modulators based on an N-aryl-hydroxybicyclohydantoin scaffold. *Journal of Medicinal Chemistry*, **49**, 7596–7599.
62. Hamann, L.G., Manfredi, M.C., Sun, C., *et al.* (2007) Tandem optimization of target activity and elimination of mutagenic potential in a potent series of N-aryl bicyclic hydantoin-based selective androgen receptor modulators. *Bioorganic & Medicinal Chemistry Letters*, **17**, 1860–1864.
63. Shen, J.X., and Brodbelt, J. (1996) Reactions of hydantoin and succinimide anticonvulsants with dimethyl ether ions in a quadrupole ion trap mass spectrometer. *Journal of Mass Spectrometry*, **31**, 1389–1398.
64. Wang, F., Ma, S., Zhang, D., and Cooks, R.G. (1998) Proton affinity and gas-phase basicity of urea. *Journal of Physical Chemistry*, **102**, 2988–2994.
65. Thevis, M., Kohler, M., Schlörer, N., *et al.* (2008) Mass spectrometry of hydantoin-derived selective androgen receptor modulators. *Journal of Mass Spectrometry*, **43**, 639–650.
66. Wysocki, V.H., Tsaprailis, G., Smith, L.L., and Breci, L.A. (2000) Mobile and localized protons: A framework for understanding peptide dissociation. *Journal of Mass Spectrometry*, **35**, 1399–1406.

67. Paizs, B., and Suhai, S. (2005) Fragmentation pathways of protonated peptides. *Mass Spectrometry Reviews*, **24**, 508–548.
68. Edwards, J.P., West, S.J., Pooley, C.L., *et al.* New nonsteroidal androgen receptor modulators based on 4-(trifluoromethyl)-2(1H)-pyrrolidino[3,2-g] quinolinone. *Bioorganic & Medicinal Chemistry Letters*, **8**, 745–750.
69. Edwards, J.P., Higuchi, R.I., Winn, D.T., *et al.* (1999) Nonsteroidal androgen receptor agonists based on 4-(trifluoromethyl)-2H-pyrano[3,2-g] quinolin-2-one. *Bioorganic & Medicinal Chemistry Letters*, **9**, 1003–1008.
70. Hamann, L.G., Mani, N.S., Davis, R.L., *et al.* (1999) Discovery of a potent, orally active, nonsteroidal androgen receptor agonist: 4-ethyl-1,2,3,4-tetrahydro-6- (trifluoromethyl)-8-pyridono[5,6-g]- quinoline (LG121071). *Journal of Medicinal Chemistry*, **42**, 210–212.
71. van Oeveren, A., Motamedi, M., Mani, N.S., *et al.* (2006) Discovery of 6-N,N-bis(2,2,2-trifluoroethyl)amino- 4-trifluoromethylquinolin-2(1H)-one as a novel selective androgen receptor modulator. *Journal of Medicinal Chemistry*, **49**, 6143–6146.
72. van Oeveren, A., Motamedi, M., Martinborough, E., *et al.* (2007) Novel selective androgen receptor modulators: SAR studies on 6-bisalkylamino-2-quinolinones. *Bioorganic & Medicinal Chemistry Letters*, **17**, 1527–1531.
73. van Oeveren, A., Pio, B.A., Tegley, C.M., *et al.* (2007) Discovery of an androgen receptor modulator pharmacophore based on 2-quinolinones. *Bioorganic & Medicinal Chemistry Letters*, **17**, 1523–1526.
74. Wang, F., Liu, X.Q., Li, H., *et al.* (2006) Structure of the ligand-binding domain (LBD) of human androgen receptor in complex with a selective modulator LGD2226. *Acta Crystallographica Section F: Structural Biology and Crystallization Communications*, **62**, 1067–1071.
75. Rosen, J., and Negro-Vilar, A. (2002) Novel, non-steroidal, selective androgen receptor modulators (SARMs) with anabolic activity in bone and muscle and improved safety profile. *Journal of Musculoskeletal & Neuronal Interactions*, **2**, 222–224.
76. Levsen, K., Schiebel, H.M., Terlouw, J.K., *et al.* (2007) Even-electron ions: A systematic study of the neutral species lost in the dissociation of quasi-molecular ions. *Journal of Mass Spectrometry*, **42**, 1024–1044.
77. Thevis, M., Kohler, M., Maurer, J., *et al.* (2007) Screening for 2-quinolinone-derived selective androgen receptor agonists in doping control analysis. *Rapid Communications in Mass Spectrometry*, **21**, 3477–3486.
78. Hanada, K., Furuya, K., Yamamoto, N., *et al.* (2003) Bone anabolic effects of S-40503, a novel nonsteroidal selective androgen receptor modulator (SARM), in rat models of osteoporosis. *Biological and Pharmaceutical Bulletin*, **26**, 1563–1569.
79. Hanada, K., Furuya, K., Inoguchi, K., *et al.* (2007) Tetrahydroquinoline derivatives (EP 1221 439 B1). Kaken Pharmaceutical Co., Ltd., Tokyo.
80. Thevis, M., Beuck, S., Thomas, A., *et al.* (2009) Doping control analysis of emerging drugs in human plasma: Identification of GW501516, S-107,

JTV-519, and S-40503. *Rapid Communications in Mass Spectrometry*, **23**, 1139–1146.

81. Thevis, M., Kohler, M., Thomas, A., *et al.* (2008) Doping control analysis of tricyclic tetrahydroquinoline-derived selective androgen receptor modulators using liquid chromatography/electrospray ionization tandem mass spectrometry. *Rapid Communications in Mass Spectrometry*, **22**, 2471–2478.
82. Murena, F., and Gioia, F. (1998) Catalytic hydroprocessing of chlorobenzene-pyridine mixtures. *Journal of Hazardous Materials*, **60**, 271–285.
83. Eckert-Maksic, M., Hodoscek, M., Kovacek, D., *et al.* (1997) Theoretical model calculations of the absolute proton affinities of benzonitrile, nitroso- and nitrobenzene. *Journal of Molecular Structure (Theochem)*, **417**, 131–143.
84. Thevis, M., Schmickler, H., and Schänzer, W. (2003) Effect of the location of hydrogen abstraction on the fragmentation of diuretics in negative electrospray ionization mass spectrometry. *Journal of the American Society for Mass Spectrometry*, **14**, 658–670.
85. Thevis, M., Schmickler, H., and Schänzer, W. (2002) Mass spectrometric behavior of thiazide-based diuretics after electrospray ionization and collision-induced dissociation. *Analytical Chemistry*, **74**, 3802–3808.
86. Garcia, P., Popot, M.A., Fournier, F., *et al.* (2002) Gas-phase behavior of negative ions produced from thiazidic diuretics under electrospray conditions. *Journal of Mass Spectrometry*, **37**, 940–953.
87. Deventer, K., Delbeke, F.T., Roels, K., and Van Eenoo, P. (2002) Screening for 18 diuretics and probenecid in doping analysis by liquid chromatography-tandem mass spectrometry. *Biomedical Chromatography*, **16**, 529–535.
88. Garcia, P., Popot, M-A., Bonnaire, Y., and Tabet, J.C. (1996) Identification of thiazidic diuretics in equine urine by using electrospray tandem mass spectrometry. *Proceedings of the 11th International Conference of Racing Analysts and Veterinarians*, 498–502.
89. Gross, M.L. (1992) Charge-remote fragmentations: Method, mechanism and applications. *International Journal of Mass Spectrometry and Ion Processes*, **118/119**, 137–165.
90. Biancotto, G., Angeletti, R., Piro, R.D., *et al.* (1997) Ion trap high-performance liquid chromatography/multiple mass spectrometry in the determination of beta-agonists in bovine urines. *Journal of Mass Spectrometry*, **32**, 781–784.
91. Cai, J., and Henion, J. (1997) Quantitative multi-residue determination of beta-agonists in bovine urine using on-line immunoaffinity extraction-coupled column packed capillary liquid chromatography-tandem mass spectrometry. *Journal of Chromatography B, Biomedical Sciences and Applications*, **691**, 357–370.

92. Thevis, M., Opfermann, G., and Schänzer, W. (2003) Liquid chromatography/electrospray ionization tandem mass spectrometric screening and confirmation methods for β_2-agonists in human or equine urine. *Journal of Mass Spectrometry*, **38**, 1197–1206.
93. Thevis, M., Opfermann, G., Krug, O., and Schänzer, W. (2004) Electrospray ionization mass spectrometric characterization and quantitation of xanthine derivatives using isotopically labelled analogues: An application for equine doping control analysis. *Rapid Communications in Mass Spectrometry*, **18**, 1553–1560.
94. Thevis, M., Beuck, S., Thomas, A., *et al.* (2009) Screening for the calstabin-ryanodine-receptor complex stabilizers JTV-519 and S-107 in doping control analysis. *Drug Testing and Analysis*, **1**, 32–42.
95. Wang, Y.X., Lee, C.H., Tiep, S., *et al.* (2003) Peroxisome-proliferator-activated receptor delta activates fat metabolism to prevent obesity. *Cell*, **113**, 159–170.
96. Krämer, D.K., Al-Khalili, L., Guigas, B., *et al.* (2007) Role of AMP kinase and PPARdelta in the regulation of lipid and glucose metabolism in human skeletal muscle. *Journal of Biological Chemistry*, **282**, 19313–19320.
97. Wang, Y.X., Zhang, C.L., Yu, R.T., *et al.* (2004) Regulation of muscle fiber type and running endurance by PPARdelta. *PLoS Biology*, **2**, e294.
98. Narkar, V.A., Downes, M., Yu, R.T., *et al.* (2008) AMPK and PPARdelta agonists are exercise mimetics. *Cell*, **134**, 405–415.
99. Brunmair, B., Staniek, K., Dorig, J., *et al.* Activation of PPAR-delta in isolated rat skeletal muscle switches fuel preference from glucose to fatty acids. *Diabetologia*, **49**, 2713–2722.
100. Reznick, R.M., and Shulman, G.I. (2006) The role of AMP-activated protein kinase in mitochondrial biogenesis. *The Journal of Physiology*, **574**, 33–39.
101. Goodyear, L.J. (2008) The exercise pill—Too good to be true? *New England Journal of Medicine*, **359**, 1842–1844.
102. Benkimoun, P. (2009) Police find unlicensed drugs after trawling bins. *British Medical Journal*, **339**, b4201.
103. World Anti-Doping Agency (2009) *The 2009 Prohibited List*. Available at http://www.wada-ama.org/rtecontent/document/2009_Prohibited_List_ENG_Final_20_Sept_08.pdf. Accessed 02-01-2009.
104. Jelkmann, W. (2007) Erythropoietin after a century of research: Younger than ever. *European Journal of Haematology*, **78**, 183–205.
105. Safran, M., and Kaelin, W.G., Jr. (2003) HIF hydroxylation and the mammalian oxygen-sensing pathway. *Journal of Clinical Investigation*, **111**, 779–783.
106. Metzen, E., and Ratcliffe, P. (2004) HIF hydroxylation and cellular oxygen sensing. *Biological Chemistry*, **385**, 223–230.

107. Bruick, R. (2003) Oxygen sensing in the hypoxic response pathway: Regulation of the hypoxia-inducible transcription factor. *Genes & Development*, **17**, 2614–2623.
108. Jelkmann, W. (2004) Molecular biology of erythropoietin. *Internal Medicine*, **43**, 649–659.
109. Mikhail, A., Covic, A., and Goldsmith, D. (2008) Stimulating erythropoiesis: Future perspectives. *Kidney & Blood Pressure Research*, **31**, 234–246.
110. Wang, Q., Gou, G., Guenzler, V., *et al.* (2004) Stimulation of erythropoiesis and treatment of anemia in rodents by oral administration of FG-2216, a novel HIF-prolyl hydroxylase inhibitor. *Journal of the American Society of Nephrology: JASN*, **15**, 773A.
111. Hsieh, M.M., Linde, N.S., Wynter, A., *et al.* (2007) HIF prolyl hydroxylase inhibition results in endogenous erythropoietin induction, erythrocytosis, and modest fetal hemoglobin expression in rhesus macaques. *Blood*, **110**, 2140–2147.
112. Fibrogen (2005) EP 1 538 160 A1, European Patent Office.
113. Thevis, M., Kohler, M., Schlörer, N., and Schänzer, W. (2008) Gas phase reaction of substituted isoquinolines to carboxylic acids in ion trap and triple quadrupole mass spectrometers after electrospray ionization and collision-induced dissociation. *Journal of the American Society for Mass Spectrometry*, **19**, 151–158.
114. Beuck, S., Schwabe, T., Grimme, S., *et al.* (2009) Unusual mass spectrometric dissociation pathway of protonated isoquinoline-3-carboxamides due to multiple reversible water adduct formation in the gas phase. *Journal of the American Society for Mass Spectrometry*, **20**, 2034–2048.
115. Lide, D.R. (2008) *Handbook of Chemistry and Physics*.CRC Press, Boca Raton.
116. Howitz, K.T., Bitterman, K.J., Cohen, H.Y., *et al.* (2003) Small molecule activators of sirtuins extend Saccharomyces cerevisiae lifespan. *Nature*, **425**, 191–196.
117. Baur, J.A., Pearson, K.J., Price, N.L., *et al.* (2006) Resveratrol improves health and survival of mice on a high-calorie diet. *Nature*, **444**, 337–342.
118. Lavu, S., Boss, O., Elliott, P.J., and Lambert, P.D. (2008) Sirtuins—Novel therapeutic targets to treat age-associated diseases. Nature reviews. *Drug Discovery*, **7**, 841–853.
119. Elliott, P.J., and Jirousek, M. (2008) Sirtuins: Novel targets for metabolic disease. *Current Opinion in Investigational Drugs*, **9**, 371–378.
120. Milne, J.C., and Denu, J.M. (2008) The Sirtuin family: Therapeutic targets to treat diseases of aging. *Current Opinion in Chemical Biology*, **12**, 11–17.
121. Stella, L., De Rosso, M., Panighel, A., *et al.* (2008) Collisionally induced fragmentation of [M-H](-) species of resveratrol and piceatannol investigated by deuterium labelling and accurate mass measurements. *Rapid Communications in Mass Spectrometry*, **22**, 3867–3872.

122. Upthagrove, A.L., Hackett, M., and Nelson, W.L. (1999) Fragmentation pathways of selectively labeled uropranolol using electrospray ionization on an ion trap mass spectrometer and comparison with ions formed by electron impact. *Rapid Communications in Mass Spectrometry*, **13**, 534–541.

123. Kuuranne, T., Pystynen, K.H., Thevis, M., Leinonen, A., *et al.* (2008) Screening of in vitro synthesised metabolites of 4,9,11-trien-3-one steroids by liquid chromatography mass spectrometry. *European Journal of Mass Spectrometry*, **14**, 181–189.

124. Kuuranne, T., Aitio, O., Vahermo, M., *et al.* (2002) Enzyme-assisted synthesis and structure characterization of glucuronide conjugates of methyltestosterone (17a-methylandrost-4-en-17b-ol-3-one) and nandrolone (estr-4-en-17b-ol-3-one) metabolites. *Bioconjugate Chemistry*, **13**, 194–199.

125. Kuuranne, T., Kurkela, M., Thevis, M., *et al.* Glucuronidation of anabolic androgenic steroids by recombinant human UDP-glucuronosyltransferases. *Drug Metabolism and Disposition*, **31**, 1117–1124.

126. Kuuranne, T., Vahermo, M., Leinonen, A., and Kostiainen, R. (2000) Electrospray and atmospheric pressure chemical ionization tandem mass spectrometric behavior of eight anabolic steroid glucuronides. *Journal for the American Society for Mass Spectrometry*, **11**, 722–730.

127. Hintikka, L., Kuuranne, T., Aitio, O., *et al.* (2008) Enzyme-assisted synthesis and structure characterization of glucuronide conjugates of eleven anabolic steroid metabolites. *Steroids*, **73**, 257–265.

128. Pozo, O.J., Van Eenoo, P., Van Thuyne, W., *et al.* (2008) Direct quantification of steroid glucuronides in human urine by liquid chromatography-electrospray tandem mass spectrometry. *Journal of Chromatography A*, **1183**, 108–118.

129. Murphy, C.M., and Huestis, M.A. (2005) LC-ESI-MS/MS analysis for the quantification of morphine, codeine, morphine-3-beta-D-glucuronide, morphine-6-beta-D-glucuronide, and codeine-6-beta-D-glucuronide in human urine. *Journal of Mass Spectrometry*, **40**, 1412–1416.

130. Thevis, M., Sigmund, G., Koch, A., *et al.* (2008) Doping control analysis of methoxyphenamine using liquid chromatography-tandem mass spectrometry. *European Journal of Mass Spectrometry*, **14**, 145–152.

131. Thevis, M., Opfermann, G., Bommerich, U., and Schänzer, W. (2005) Characterization of chemically modified steroids for doping control purposes by electrospray ionization tandem mass spectrometry. *Journal of Mass Spectrometry*, **40**, 494–502.

132. Thevis, M., Thomas, A., Kohler, M., *et al.* (2009) Emerging drugs: Mechanism of action, mass spectrometry and doping control analysis. *Journal of Mass Spectrometry*, **44**, 442–460.

5 Structure Characterization of High Molecular Weight Target Analytes: Electrospray Ionization

Many peptide hormones and proteins have been prohibited in sports since 1989 even in the absence of detection methods that enable the identification of administration to athletes. The advent of matrix-assisted laser desorption ionization (MALDI) or ESI combined with MS(/MS) and, thus, the option to efficiently ionize high—molecular weight compounds and elucidate structural features provided important information for the characterization and development of analytical assays to uncover the misuse of a challenging subset of banned substances.

5.1 HUMAN CHORIONIC GONADOTROPHIN (hCG)

One of the first peptide hormones included in routine doping control procedures (based on immunochemical methods) was human chorionic gonadotrophin (hCG),[1–5] a dimeric glycoprotein that belongs to the same glycoprotein hormone family as luteinizing hormone (LH), follicle stimulating hormone (FSH), and thyroid stimulating hormone (TSH). hCG was identified in urine as the "anterior pituitary hormone" in 1927 by Aschheim and Zondek,[6–9] being an indicator of pregnancy, which has been used for past[10,11] and present pregnancy tests since. The assay established by Aschheim and Zondek was based on the subcutaneous injection of approximately 100 μL of urine (collected from potentially pregnant women) into infertile mice. In case of pregnancy and, thus, the presence of hCG, morphological and functional alterations of the murine ovary were observed including ovulation, follicular maturation, luteinization, etc. Only 2% of false test results

Mass Spectrometry in Sports Drug Testing: Characterization of Prohibited Substances and Doping Control Analytical Assays, By Mario Thevis
Copyright © 2010 John Wiley & Sons, Inc.

were obtained; however, mice needed to be sacrificed for each analysis. An alternative approach was established by Galli-Mainini in 1947, who suggested injecting 10 mL of urine of presumably pregnant females subcutaneously into the male toad "Bufo arenarum Hensel," which caused the batrachia to spawn or induced spermatorrhea and allowed for the detection of sperms in frog urine in case of a test person's gravidity.[12] A series of other batrachia was tested in the following years yielding false positive test results in 0.4–3% of all measured specimens, which were further extended to serum samples.[13] In contrast to the Aschheim-Zondek assay, the frogs commonly recovered from the testing procedure. The method was frequently employed until the 1960s, when it was substituted by more sensitive and entirely immunological procedures.[14,15]

hCG consists of non-covalently linked α- and β-subunits and yields an average molecular weight of 37,500 Da resulting from a 14 kDa α- and 23.5 kDa β-subunit, both of which contain two *N*-linked carbohydrate moieties.[16] While all α-subunits of hCG, LH, FSH, and TSH are identical, the β-subunit of hCG bears four additional *O*-linked and sialic acid end-capped oligosaccharide residues within the C-terminal peptide region, which represents an extension of 24 amino acids compared to the LH β-subunit (Fig. 5.1). The prolongation of the protein and concurrent additional glycosylation results in a significantly increased half-life of hCG, making it a naturally occurring long-acting analogue to LH.[17] Clinically, hCG administration is indicated in cases of infertility associated with hypogonadism resulting from dysfunctions in the hypothalamic-pituitary axis and, thus, gonadotrophin deficiency.

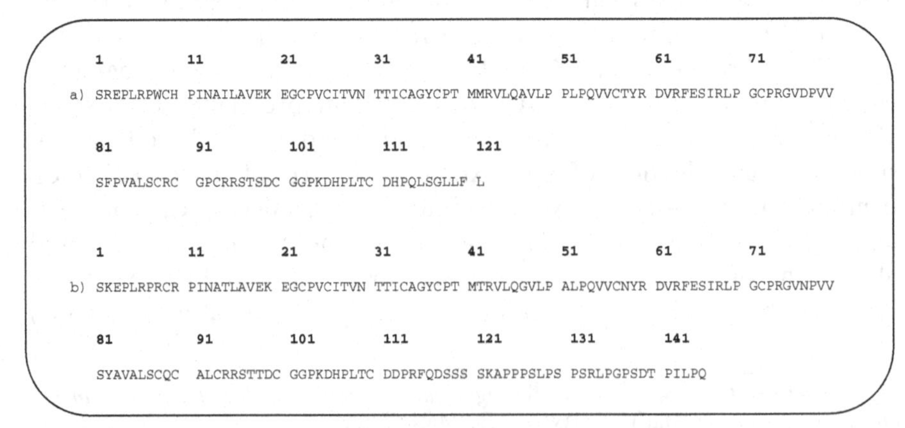

Figure 5.1: Primary structures of the β-subunits of LH (a) and hCG (b).

In terms of doping, hCG can be abused to artificially increase plasma testosterone levels. The assumed reasons for this include (a) the stimulation of the testicular testosterone production to counteract the anabolic steroid-induced reduced testis size, and (b) the attempt to mask anabolic steroid misuse being detected by altered urinary steroid profiles or suspicious testosterone /epitestosterone ratios (T/E, see Chapter 6). Consequently, hCG has been prohibited in sports for male athletes since 1987, and the first arbitrary threshold levels of 25 international units (IU) per L of urine were corrected to 5 IU/L.[3,4] In 2005, hCG was also banned for female athletes, but due to the facts that performance-enhancing effects are rather unlikely for women and the routine screening for pregnancy represented a significant invasion of female athletes' privacy, the ban has been maintained for men only since 2006. Tests for hCG are commonly conducted by means of immunoassays, but characterization of the analyte[18–20] and first analytical assays were also established using MS-based approaches.[21–23] Due to the considerable heterogeneity of the glycoprotein, bottom-up methods were preferred over top-down analyses, and peptides derived from tryptic digest as well as carbohydrate moieties released by peptide-*N*4-(acetyl-β-glucosaminyl)-asparagine amidase (PNGase F) were used to elucidate the target analyte's structure in detail. As expected and aimed, the amino acid composition and primary structure of human and synthetic hCG are identical; however, since humans lack the enzyme CMP-*N*-acetyl neuraminic acid hydroxylase, which is required to convert *N*-acetyl neuraminic acid (Neu5Ac) into *N*-glycolyl neuraminic acid (Neu5Gc), the latter is found in carbohydrate residues of pharmaceutical preparations of hCG produced in Chinese hamster ovary (CHO) cells up to 2% of the entire sialic acid content but not in human urinary hCG; hence, a structural difference is given that might allow to distinguish between the recombinant and natural glycoprotein using mass spectrometry.[20]

5.2 ERYTHROPOIETINS (EPO)

As early as 1977, Miyake and associates purified milligram amounts of urinary erythropoietin (uEPO) from 2550 L of urine that was collected from patients suffering from aplastic anemia.[24] The glycoprotein of mainly renal origin with an approximate molecular weight of 34 kDa, comprises 165 amino acid residues, three tetra-antennary *N*-linked (Asn 24, 38 and 83), and one *O*-linked (Ser 126) carbohydrate side chain as well as two disulfide bonds (Cys 7-Cys 161 and Cys 29-Cys 33)

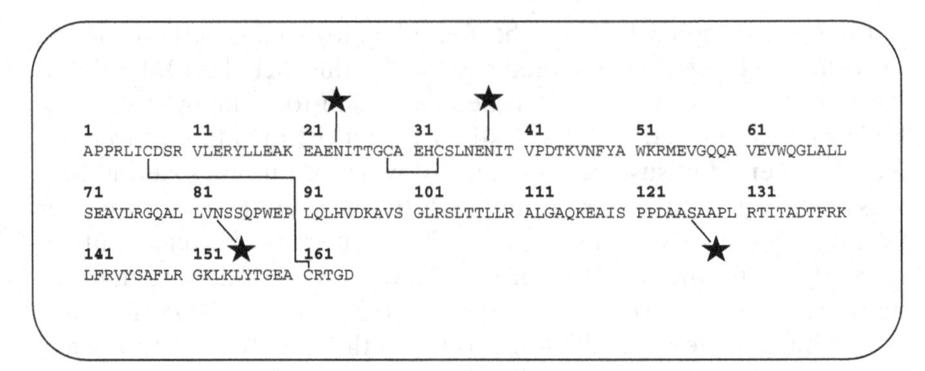

Figure 5.2: Primary structure of human erythropoietin with an average molecular weight of 34 kDa. Locations of glycan side chains are indicated with stars (★).

(Fig. 5.2). It is an essential growth factor for erythrocytic progenitors in the bone marrow by inhibiting apoptosis and, thus, maintaining the growth of red blood cells. The peptide core of EPO is crucial for receptor-binding while the importance of the carbohydrate structure, which accounts for approximately 40% of the molecular weight, is necessary to allow *in vivo* survival of the hormone.[25] Comprehensive reviews on the molecular biology of EPO and erythropoiesis-stimulating agents were published by several groups.[26–30] Since EPO has become available by biotechnological production in 1989, a variety of clinically approved recombinant human EPO preparations as well as derivatives such as CERA (continuous erythropoiesis receptor activator) have been commercialized (Table 5.1), and numerous additional biosimilars were released since the patent for erythropoietin alpha expired in Europe in 2004.[31] Moreover, an entirely synthetic erythropoietin derivative has recently been described named SEP (synthetic erythropoiesis protein).[32] SEP was assembled by sequential chemo-selective ligation reactions from individual peptide segments yielding a protein core composed of 166 amino acid residues. The backbone differs from human EPO at 7 positions. Three asparagines (residues 24, 38, and 83) as well as one serine (residue 126) that are usually connected to carbohydrate moieties of EPO were substituted by lysine residues, two of which (24 and 126) are linked to precision polymers that terminate in negative charge-control units. Glutamates at position 89 and 117 were exchanged by cysteine residues due to ligation reactions during synthesis, but sulfhydryl functions were derivatized to mimic missing glutamates. Finally, a C-terminal arginine is present that is removed from circulating and urinary EPO.

TABLE 5.1: Details on Selected Erythropoietin Products[a]

Product	Manufacturer	Cell Line	Erythropoietin Subtype			
			α	β	δ	ω
Epogen	Amgen	CHO	x			
Epopen	Esteve	CHO	x			
Epoxitin	Janssen-Cilag	CHO	x			
Eprex	Janssen	CHO	x			
Erypo	Janssen-Cilag	CHO	x			
Espo	Kirin	CHO	x			
Globuren	Cilag	CHO	x			
Procrit	Amgen	CHO	x			
Epoch	Chugai	CHO		x		
Epogin	Chugai	CHO		x		
Eritrogen	Roche	CHO		x		
Erantin	Boehringer Mannheim	CHO		x		
Erythrostim	Microgen	CHO		x		
Marogen	Chugai	CHO		x		
Neorecormon	Roche	CHO		x		
Recormon	Boehringer Mannheim	CHO		x		
Dynepo	Aventis Pharma	human			x	
Epomax	Elanex	BHK				x
Hemax	Elanex	BHK				x
Repotin	Bodene	BHK				—
Aranesp[b]	Amgen	CHO	x			
Mircera[c]	Roche			x		
SEP (synthetic erythropoiesis protein)[d]	Gryphon Therapeutics					

CHO = Chinese hamster kidney; BHK = baby hamster kidney.
[a] Refs. 190 and 241.
[b] Five amino acid substitutions (Ala30Asn, His32Thr, Pro87Val, Trp88Asn, Pro90Thr), hyperglycosylation by two additional carbohydrate side chains.
[c] Erythropoietin β, covalently linked to a 30 kDa methoxy polyethylene glycol (PEG) nucleus.
[d] Synthetic peptide backbone, Asn 24, 38 and 83 and Ser 126 are substituted by Lys; positions 24 and 126 bear precision polymers with terminal charge-control units; Glu 89 and 117 are exchanged by derivatized Cys.

Since 1990, EPO has been prohibited by the IOC and WADA, and a study published by Videman in 2000 stressed the rationale;[33] a considerable increase of total hemoglobin mass in elite cross country skiers has been measured since recombinant EPO has become commercially available. Hence, the search for an adequate assay to determine the misuse of EPO in sports has been an important item of anti-doping research. With regard to mass spectrometry, research activities have been limited primarily due to the complexity and nearly identical composition of uEPO and its recombinant analogues. While the primary and secondary structures of human urinary EPO and erythropoietins alpha, beta, delta, omega, and zeta are identical,[34,35] minor differences in glycan structures including different sialylation levels and modifications such as sulfonation have been determined.[36–39] In contrast to these, darbepoetin alpha (Table 5.1) differs considerably from human EPO due to the substitution of 5 amino acid residues (Ala30Asn, His32Thr, Pro87Val, Trp88Asn, and Pro90Thr), which allow for additional oligosaccharide attachment at asparagines. Causing a prolonged biological activity of the drug.[40] Numerous studies have described the possibility to characterize recombinant as well as natural EPO by means of capillary electrophoresis,[41] isoelectric focusing (IEF),[42–44] two-dimensional gel electrophoresis,[45] MALDI-MS(/MS) and ESI-MS(/MS),[46–48] as well as fast atom bombardment MS and GC-MS,[49,50] and fundamental principles have been applied to allow the determination of recombinant EPO abuse in sports; however, except for darbepoetin alpha,[51] no MS-based approach is currently applicable to human doping controls (see also Chapter 7). Nevertheless, the minor but still significantly different glycosylation signatures particularly concerning Neu5Gc (see section 5.1) might provide a tool for future mass spectrometric assays to uncover EPO misuse.

5.3 SYNACTHEN

Human adrenocorticotrophin (ACTH), a peptide hormone consisting of 39 amino acid residues (Fig. 5.3a) was identified as an important factor of the hypothalamus-pituitary-adrenal system in the 1950s,[52] and numerous studies have been conducted ever since to elucidate its structure,[53–55] plasma concentrations,[56] structure-activity relationships,[57] degradation products,[58] and synthetic analogues.[59–61] ACTH was shown to

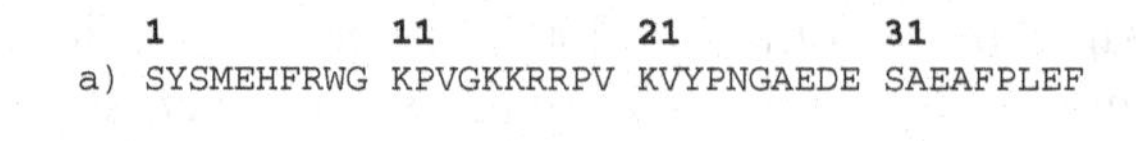

Figure 5.3: Primary structures of (a) adrenocorticotrophic hormone (mol $wt_{monoisotopic}$ = 4538.3 Da), and (b) Synacthen (mol $wt_{monoisotopic}$ = 2931.6 Da).

derive from, among others, pro-opiomelanocortin (POMC) by tissue-selective proteolytic processing in the anterior pituitary.[62] Released from corticotroph cells into circulation, it plays an important physiological role by stimulating cortisol production in the *zona fasciculata* of the adrenal cortex. ACTH binds with high affinity to receptors located at the surface of adrenal cortical cells and exclusively regulates the biosynthesis of the main endogenous glucocorticoid cortisol.[63] The primary structure of ACTH (Fig. 5.3a) has been highly conserved in a variety of species, and studies on the structure-activity relationships proved that only the first 24 amino acid residues are required for full bioactivity of ACTH.[64] Fragments composed of fewer than 20 amino acids require substitutions to allow for bioactivity,[60] and the function of residues 25–39 is attributed primarily to the stabilization of circulating ACTH.[65] Due to the importance of ACTH for the endogenous cortisol production, a fully bioactive synthetic analogue termed Synacthen was introduced to the pharmaceutical market in 1961[59,61] consisting of the first 24 amino acid residues of ACTH (Fig. 5.3b). Medical applications including tests regarding adrenal function and dysfunction[66–68] as well as its administration in cases of glucocorticosteroid long-term therapy[69] using depot formulations[70] were investigated. The first mentioned purposes have become the major application of the synthetic ACTH analogue as significant increases of serum cortisol levels in a dose-dependent manner have been reported after administration of Synacthen.[71,72]

In sports, the use of corticosteroids and respective releasing hormones has been restricted by regulative authorities such as WADA,[73] and Synacthen in particular has been studied intensively by means of mass spectrometry since its use was reported and admitted by cyclists in the course of recent doping scandals.[74,75] The dissociation pattern of Synacthen yields a series of triply and quadruply charged y-ions (y_{18}-y_{22}) and additional informative product ions resulting from a typical proline-directed fragmentation (Fig. 5.4a).[76] In particular b_{23}^{4+} resulting from the loss of the C-terminal proline residue and y_{13}^{2+} demonstrate the influence of proline positions on product ion spectra compositions. The predominantly generated y-ions are suggested to originate from a stabilization of the introduced charges of the precursor ion within the region of the amino acid residues 8–24, which represent the basic core of Synacthen. CID then causes an N-terminal dissociation of the multiply protonated species yielding mainly y-ions (Fig. 5.4a); however, the one of the most diagnostic product ions obtained from Synacthen was observed at *m/z* 223.1, which is suggested to account for the a_2-ion as its elemental composition was determined as $C_{11}H_{15}O_3N_2$.[77] In addition,

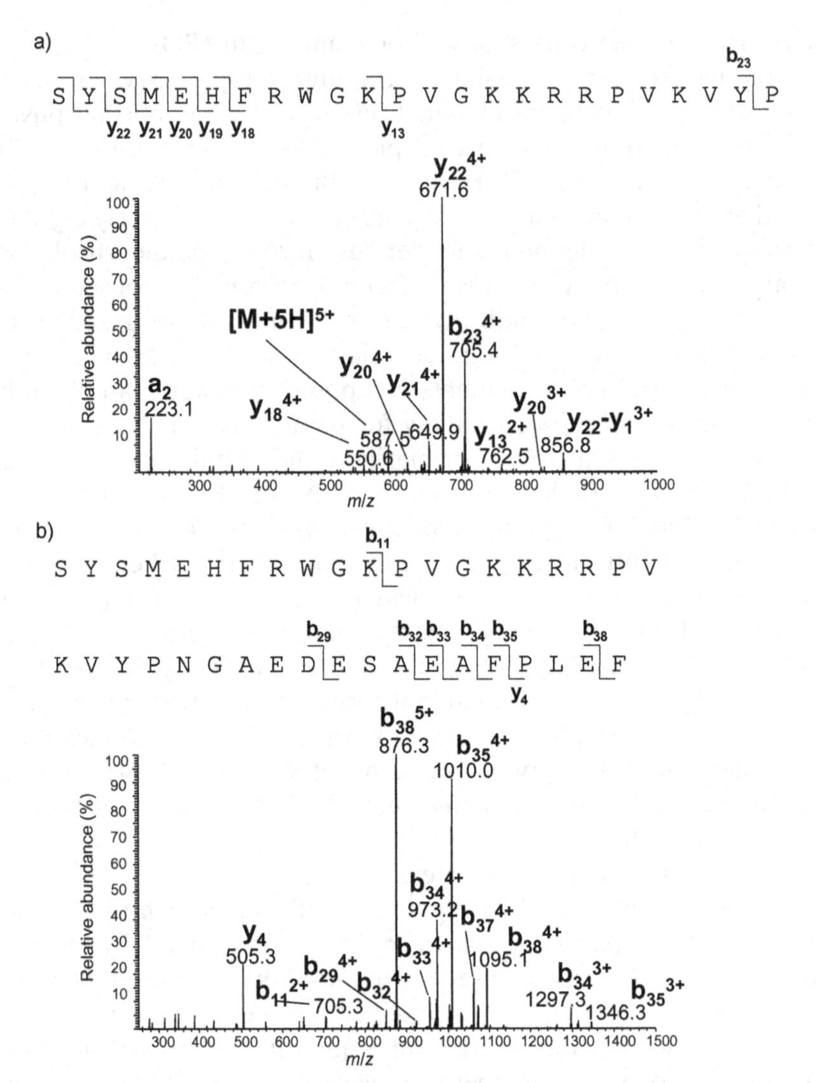

Figure 5.4: ESI product ion mass spectra of the five-fold charged precursor ions $[M+5H]^{5+}$ of (a) *m/z* 587 of Synacthen, and (b) *m/z* 909 of ACTH, measured on an LTQ-Orbitrap system.

the corresponding y_{22}^{4+}-ion is found at *m/z* 671.6 representing the base peak of the product ion spectrum.

In contrast to Synacthen, ACTH yielded a significantly different product ion mass spectrum that is predominantly composed by b-ions (Fig. 5.4b). ACTH bears an extended and acidic C-terminus compared to Synacthen, which was suggested to provoke the location of charges introduced during the ionization process in the center of the peptide, and four- and five-fold charged b-ion series mainly originating from the C-terminal region of ACTH are generated.

5.4 INSULINS

Insulin and its synthetic analogues have been considered relevant for doping controls because insulin has demonstrated an enormous functional versatility since its existence was hypothesized in 1913 by Sir Edward Sharpey-Schäfer.[78] As early as 1889, von Mering and Minkowski demonstrated that removal of the pancreas is followed by severe hyperglycemia leading to fatal diabetes,[79] which outlined that a hormone being responsible for the regulation of carbohydrate metabolism is produced and secreted from the pancreas. Subsequent investigations primarily conducted by Frederick G. Banting, Charles H. Best, James B. Collip, and John J.R. Macleod, who shared the Nobel Prize in Physiology or Medicine awarded in 1923, led to the discovery and identification of the active component, insulin, which originates from the islet of Langerhans (referred to as the β-cells of the pancreas). A comprehensive review on the history, the discovery of insulin and the corresponding controversy, was recently published.[80]

Human insulin is a small protein with an average molecular weight of 5807 Da. It is composed of two peptides (A- and B-chain) that are cross-linked by two disulfide bonds as illustrated in Figure 5.5a. Its precursor proinsulin eliminates the connecting peptide (C-peptide) that links the C-terminus of the insulin B-chain to the N-terminus of the insulin A-chain via two additional amino acid residues each, to produce insulin, which is stored with equimolar amounts of C-peptide in mature granules of the β-cells.[81] Insulin is released into the blood stream upon stimulation by glucose[82] and other nutrients such as protein hydrolysates,[83–85] selected amino acids (leucine, glutamic acid)[86] and their derivatives, particularly designed substrates (e.g., esters of succinic acid),[87–90] as well as hormones and neurotransmitters.[81] The key role of insulin is the regulation of carbohydrate metabolism, and several additional anabolic and anti-catabolic properties were attributed to this versatile peptide hormone.[91,92] *Diabetes mellitus* is considered an epidemic of the 21st century due to an enormous increase of incidences up to an estimated number of 246 million patients worldwide.[93,94] Consequently, numerous synthetic insulin preparations have been made available by pharmaceutical companies, and most drugs designed for the treatment of insulin-dependent *diabetes mellitus*

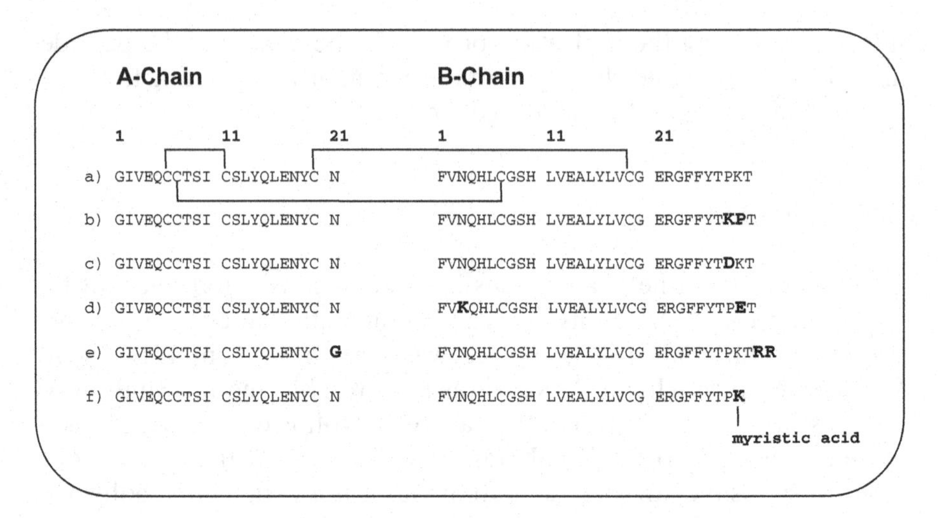

Figure 5.5: Primary structures of (a) human insulin (mol $wt_{monoisotopic}$ = 5803.6 Da), (b) Humalog LisPro (mol $wt_{monoisotopic}$ = 5803.6 Da), (c) Novolog Aspart (mol $wt_{monoisotopic}$ = 5821.6 Da), (d) Glulisine Apidra (mol $wt_{monoisotopic}$ = 5818.6 Da), (e) Lantus Glargine (mol $wt_{monoisotopic}$ = 6058.8 Da), and (f) Detemir (mol $wt_{monoisotopic}$ = 5913.8 Da). Disulfide bonds are indicated in case of human insulin (a) only but are present in all synthetic derivatives.

(IDDM) are now composed of modified insulins that possess either rapid- or long-acting properties. Human insulin tends to self-associate and generate non-covalent hexamers,[95] a fact that considerably contributes to the unfavorable injection-to-onset profiles of several recombinant human insulin formulations.[96] After subcutaneous injection, the insulin aggregates are required to dissociate into bioavailable monomers, which results in so-called lag-phase periods of 45–120 minutes; hence, rapid-acting insulins such as Humalog LisPro, Novolog Aspart, and Glulisine Apidra (Figure 5.5b–d) and long-acting insulins (e.g. Lantus Glargine, and Detemir, Fig. 5.5e–f) were introduced to allow either bioavailability within 10–15 minutes after administration[97–102] or constant basal insulin plasma levels,[103–105] respectively. The rapid- or long-acting insulins comprise slightly different primary structures compared to human insulin with, e.g., Humalog containing proline and lysine residues at B_{28} and B_{29} (Fig. 5.5b), aspartic acid at position B_{28} in case of Novolog (Fig. 5.5c), lysine and glutamic acid residues instead of asparagine and lysine in the B-chain of Glulisine Apidra (Fig. 5.5d), two additional arginines at the C-terminus of the B-chain and an A-chain C-terminal glycine instead of an asparagine residue in case of Lantus Glargine (Fig. 5.5e), or Detemir that represents the truncated des-B30 insulin, the C-terminal lysine of which is acylated with myristic acid (Fig. 5.5f).

Qualitative and quantitative analysis of insulin and its related compounds in body fluids is of paramount importance for various fields of clinical, forensic, and doping control analytical chemistry. Patients suffering from different types of diabetes under individual conditions require accurate analytical results for an adequate medical treatment. In addition, physicians need to identify potential situations of Munchhausen (by proxy), which commonly represent surreptitious applications of insulin by individuals or a third party to pretend a severe disease.[106] Moreover, forensic scientists identified the power of insulin as a lethal weapon in numerous cases of unlawful killing since 1958.[107–109] Subsequently, a great number of attempted or successful homicides and suicides by administration of the pancreatic hormone was found.[110,111] In contrast to medical or forensic reasons, athletes have been tempted to abuse insulins in order to artificially improve performance. Assumed beneficial effects are an improved recovery and loading of glycogen stores, but also anabolic as well as anti-catabolic properties of hyperinsulinemic clamps.[91,112,113] Rapid-acting insulins have been classified as particularly "tempting" because of their improved controllability compared to conventional human insulin formulations;[114–118] however, indications that long-acting insulins have been misused in elite sport were also obtained recently.[119] Different types of immunoassays, such as radioimmunoassay (RIA), enzyme-linked immunosorbent assay (ELISA), or microparticle enzyme immunoassay (MEIA) are available for identification and quantitation of human and animal insulin[120] since the first RIA assay was introduced by Yalow and Berson in 1959.[121] However, several shortcomings were observed, primarily concerning the accuracy and, thus, the cross-reactivity of immunoassays to precursors (such as proinsulin or split proinsulin), degradation products, or analogues to human insulin. Improved immunoassays have mainly eliminated these disadvantages,[98,122,123] but the need for unambiguous results as provided by mass spectrometry has yielded several new applications published within the last decade.

Intact insulins are efficiently ionized using positive ESI to yield multiply charged molecules. The accurate determination of a target analyte's molecular weight is a key factor providing utmost confidence in analytical results; hence, high resolution/high accuracy mass spectrometry has become a valuable tool to identify insulins relevant for doping controls, but is not a requirement.[124] Except for Humalog LisPro (Fig. 5.5b), all currently available synthetic insulins differ from human insulin by more than 15 Da in molecular weight (Fig. 5.5), which allows a facile

a)

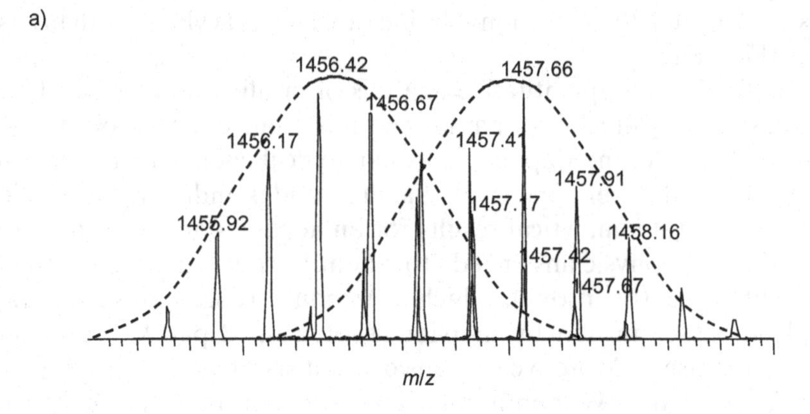

b)

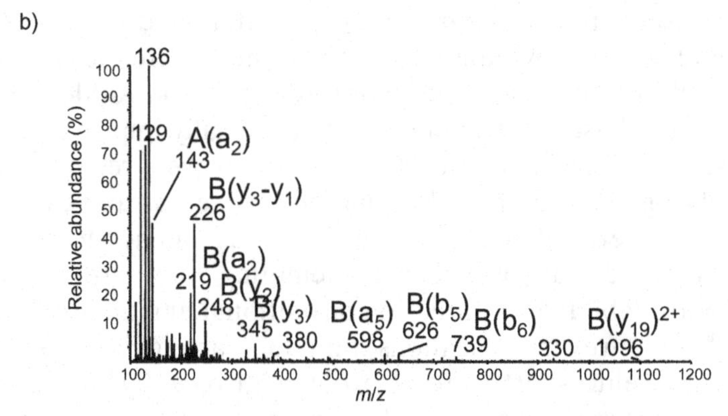

Figure 5.6: (a) Four-fold charged molecules of Glulisine Apidra (left) and Novolog Aspart (right), recorded on an LTQ-Orbitap using a resolving power of 100.000. Although two species are visible, the deconvolution of a mixture of these compounds is difficult. (b) ESI product ion mass spectrum of the five-fold charged precursor ion $[M+5H]^+$ *m/z* 1162 of human insulin, measured on a QTrap 4000 system.

distinction of endogenous insulin from synthetic counterparts; however, due to multiple charges, minor differences between synthetic drugs such as Novolog Aspart and Glulisine Apidra complicate the unequivocal determination of molecular weights in case of mixtures even at a resolving power of 100,000, as illustrated in Figure 5.6a. Hence, low- or high-resolution product-ion mass spectra are necessary to provide comprehensive information to allow the identification of the drug targets (Fig. 5.7a–c). Due to the interconnection of two peptides (A- and B-chain) as well as the additional disulfide bond that links the residues Cys_7 and Cys_{11} of the A-chain, ionized insulins possess a considerable stability under low-energy CID conditions. Ion trap MS/MS capabilities have demonstrated numerous favorable identification properties for peptides and proteins, but have also been limited by the so-called "low-mass cut-off" that results from the instability of small product ions that travel in the axial direction toward the ion trap end caps. This phenomenon was well-described by stability diagrams based on the Mathieu equation,[125] and attempts were made in the past to extend the stability domain for ion trap tandem mass spectrometry.[126] Due to the importance of product ions that represent the C-terminus of the B-chain (i.e., y_3 or y_2) to characterize the primary modification site of synthetic insulins, tandem mass analyzers were used such as quadrupole/linear ion trap,[127–129] quadrupole/time-of-flight,[130] or, more recently, hybrid instruments composed by linear ion trap/orbitrap[131] that use higher energy curved linear trap (C-trap) dissociation (HCD).[132] Informative product ion mass spectra were obtained for instance by means of a QTrap mass spectrometer as illustrated in Figures 5.6b and 5.7. Precursor ions $[M+5H]^{5+}$ were entirely dissociated to yield amino acid sequence tags and immonium ions that allow the identification of respective analytes. Human insulin (Fig. 5.6b) gives rise to intense product ions at *m/z* 345, 327, 226, 219, and 143 that represent $(B)y_3$, $(B)y_3$-H_2O, $(B)y_3$-y_1, $(B)a_2$, and $(A)a_2$, respectively, and immonium ions were observed at *m/z* 136 (Tyr), 129 (Arg), and 120 (Phe). Except for Humalog LisPro, all synthetic insulins differ from human insulin by molecular weight, and provide a set of diagnostic product ions that allow for their unambiguous identification using tandem mass spectrometry. Due to identical elemental compositions and closely related primary sequences of Humalog LisPro and human insulin (Fig. 5.5), a characteristic criterion (i.e., product ion) was required to differentiate these two analytes, and proline-directed dissociation yielded an abundant and informative product ion for Humalog LisPro at *m/z* 217 (y_2), which is not generated from human insulin (Figures 5.6b and 5.7a). Thus, the occurrence of the singly charged ion at *m/z* 217 in product ion spectra of multiply charged precursor ions of human insulin and Humalog LisPro (e.g., *m/z* 1162.4) or the four-fold charged counterpart to *m/z* 217 at *m/z* 1398 (comprising the intact A-chain and B_{1-28}) indicates the presence of the rapid-acting synthetic analogue.[133]

5.5 HEMOGLOBIN-BASED OXYGEN CARRIERS (HBOCs)

The shortage of blood products, the problems arising from possible bacterial, viral and prion transmissions, the significant costs for storage, and problems with type- and cross-matching has led to numerous

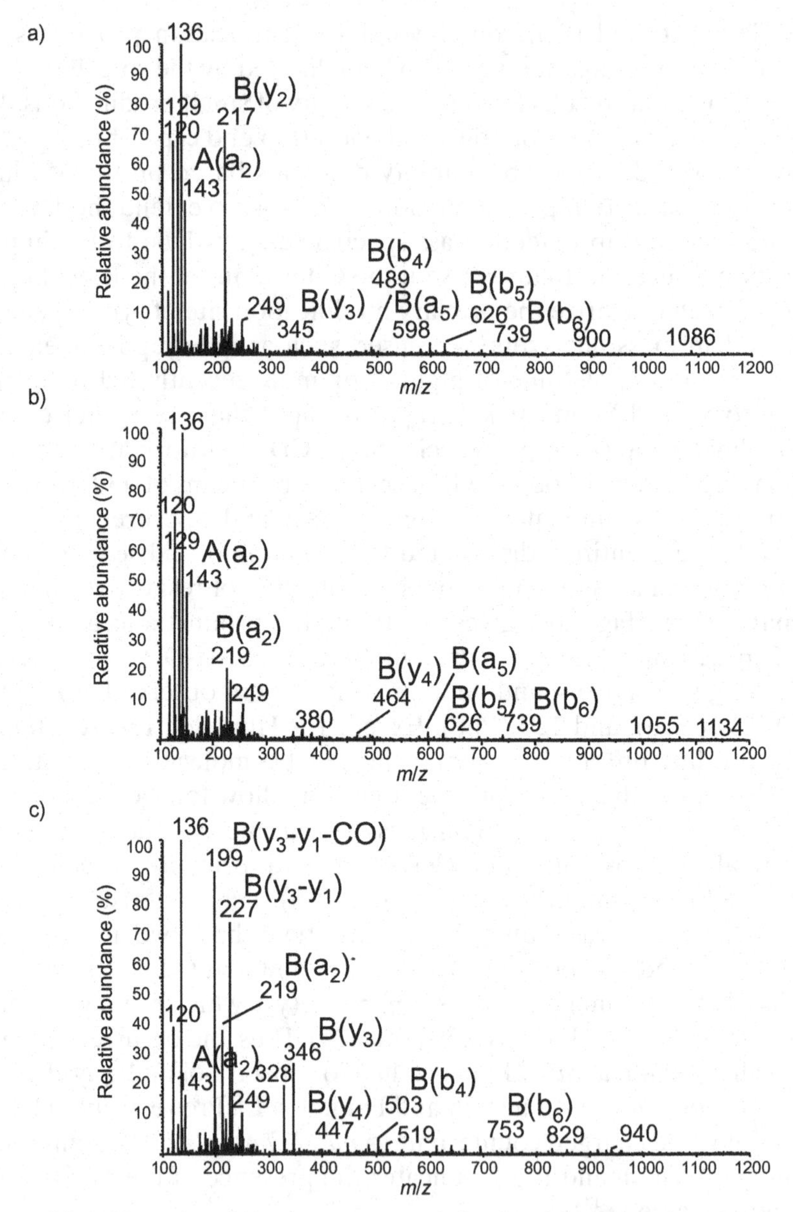

Figure 5.7: ESI product ion mass spectra of the five-fold charged precursor ions $[M+5H]^{5+}$ of (a) *m/z* 1162 of Humalog LisPro, (b) *m/z* 1166 of Novolog Aspart, and (c) *m/z* 1165 of Glulisine Apidra measured on a QTrap 4000 system.

studies to find suitable artificial blood substitutes. A promising approach has been hemoglobin-based oxygen carriers (HBOCs),[134] the clinical investigations of which started as early as the late 1800s,[135] but successful developments with clinical approval for human and/or veterinary use were accomplished only a few years ago.[136] Native cell-free hemoglobin (Hb) lacks the capability of reversible oxygen binding and release (owing to the absence of its polyanionic effector 2,3-diphosphoglycerate)[137] and its tetrameric structure dissociates into α,β-dimers causing severe nephrotoxicity.[138] Numerous studies have been performed to address the advantages and drawbacks of various chemically modified Hb. Intra- and/or inter-molecular cross-linking of Hb has proven to maintain the stability of hemoglobin molecules and to improve persistence as well as oxygen off-loading by means of a decreased oxygen affinity. In addition, Hb itself does not comprise AB0 antigens and allows long-term storage. Hence, Hb from different sources such as outdated human blood donations, bovine blood as well as recombinant preparation were utilized to prepare synthetic blood substitutes that possess all clinically important properties and demonstrate advantages over blood bottles such as virtually unlimited availability, purity, and independence in terms of blood types. However, to date fewer than ten products have been under clinical consideration (Table 5.2), a fact that stresses the challenging requirements artificial blood substitutes have to fulfill.[136,139,140]

Although scientific data about the effect of HBOCs on endurance and athletic performance is very limited and skepticism regarding a positive influence is prevalent,[141,142] artificial oxygen carriers are considered relevant for doping controls. Several studies were conducted to characterize bovine Hb-based oxygen carriers by means of mass spectrometry.[143–146] Using acidic hydrolysis, peptides were obtained, which dissociated under ESI-CID conditions by eliminating 159 Da in a triple quadrupole mass spectrometer. The species with a mass of 159 Da eliminated from precursor ions at *m/z* 299 and 399 was not further characterized but seemed to be generated from cross-linked bovine Hb only. In another assay, enzymatic digestion with endoproteinase Glu-C outlined the fact that the cross-linking reagent employed in the production process of selected HBOCs caused modifications at the N-terminus of bovine hemoglobin beta-chains, providing target molecules for an HBOC-specific screening. The peptide MLTAEE was shown to carry a modified amino function of the N-terminal methionine, which gives rise to precursor ions at *m/z* 759 and 761 to generate characteristic product ions at *m/z* 170 and 172, respectively. Trout and co-workers[143,147] as well as Thevis and associates[146,148] used trypsin digestion of HBOCs

TABLE 5.2: Details on Selected Hemoglobin-based Oxygen Carriers[a]

Product	Source of Hemoglobin	Chemical Modification	Average M_r	Company	Status / Clinical Trials
Hemopure	Bovine	Inter-/intramolecularly cross-linked by glutaraldehyde	250 kDa	Biopure, Cambridge, MA, USA	Approved for human use in South Africa (2001), filed for biologic license application by FDA (USA) / Phase III
Oxyglobin	Bovine	Inter-/intramolecularly cross-linked by glutaraldehyde	200 kDa	Biopure, Cambridge, MA, USA	Approved for dogs (1999)
Hemolink	Human	Inter-/intramolecularly cross-linked by O-raffinose	64–500 kDa	Hemosol, Mississauga, ON, Canada	Phase II/III, clinical study suspended
PolyHeme	Human	Inter-/intramolecularly cross-linked by glutaraldehyde	256 kDa	Northfield, Evanston, IL, USA	Phase III
Hemospan	Human	Thiolation and conjugation to maleimide polyethylene glycol	95 kDa	Sangart Corporation, San Diego, CA, USA	Phase III
HemAssist	Human	Intramolecularly cross-linked by diaspirin	64 kDa	Baxter, Deerfield, IL, USA	Phase III , clinical studies discontinued
Optro	Recombinant (E. coli)	Lys108 (beta) mutation (Hb Presbyterian), intramolecularly cross-linked by diaspirin	64 kDa	Somatogen, Boulder, CA, USA	Phase II/III, clinical study discontinued
PHP	Human	Pyridoxalated and polyoxyethylene conjugated	187 kDa	Apex, Plainfield, IL, USA	Phase III
PEG-Hb	Bovine	Polyethylene glycol conjugated (pegylated)	not reported	Enzon, Piscataway, NJ, USA	Phase II, clinical study discontinued

[a] Refs. 147, 190, 242, and 243.

TABLE 5.3: Selected Predicted Peptide Masses $[M+H]^+$ of Alpha- and Beta-chains of Human and Bovine Hemoglobin after Trypsin Digestion[a]

Peptide	Origin	Mass (Da)	Position	Amino Acid Sequence
Alpha-chains				
aT12	b	2969.6	100–127	LLSHSLLVTLASHLPSDFTP AVHASLDK
aT12	h	2967.6	100–127	LLSHCLLVTLAAHLPAEFTP AVHASLDK
aT9	b	2367.2	69–90	AVEHLDDLPGALSELSDLHA HK
aT6	b+h	1833.9	41–56	TYFPHFDLSHGSAQVK
aT4	b+h	1529.7	17–31	VGAHAGEYGAEALER
aT13	b	1279.7	128–139	FLANVSTVLTSK
aT13	h	1252.7	128–139	FLASVSTVLTSK
aT5	b+h	1071.6	32–40	MFLSFPTTK
aT11	b+h	818.4	93–99	VDPVNFK
aT1	h	729.4	1–7	VLSPADK
aT1	b	703.4	1–7	VLSAADK
aT3	b+h	532.3	12–16	AAWGK
Beta-chains				
bT6	b	2090.0	40–58	FFESFGDLSTADAVMNNPK
bT6	h	2058.9	41–59	FFESFGDLSTPDAVMGNPK
bT10	h	1669.9	67–82	VLGAFSDGLAHLDNLK
bT4	h	1314.7	18–30	VNVDEVGGEALGR
bT5	b+h	1274.7	31–39/40	LLVVYPWTQR
bT4	b	1101.6	19–29	VDEVGGEALGR
bT10	b	1097.5	66–75	VLDSFSNGMK
bT1	h	952.5	1–8	VHLTPEEK
bT2	b	950.5	8–16	AAVTAFWGK
bT2	h	932.5	9–17	SAVTALWGK
bT1	b	821.4	1–7	MLTAEEK

[a]Ref. 244.
b = bovine hemoglobin; h = human hemoglobin.

and ESI-CID to characterize artificial oxygen carriers exploiting the detectable chemical differences in primary structures of bovine and human hemoglobins as well as cross-links introduced into the hemoglobin molecules. Sequence homology of human and bovine hemoglobin is only 85% giving rise to bovine-hemoglobin specific peptides upon trypsin degradation (Table 5.3), and a typical product ion mass spectrum of a peptide used as diagnostic marker for bovine Hb is presented in Figure 5.8. The presence of peptide(s) uniquely generated from bovine Hb in athletes' doping control samples would be proof of drug misuse and doping. In addition, the modification of lysines in case of the bovine Hb-derived compounds Hemopure and Oxyglobin provide substantiating evidence for cross-linked hemoglobin as trypsin

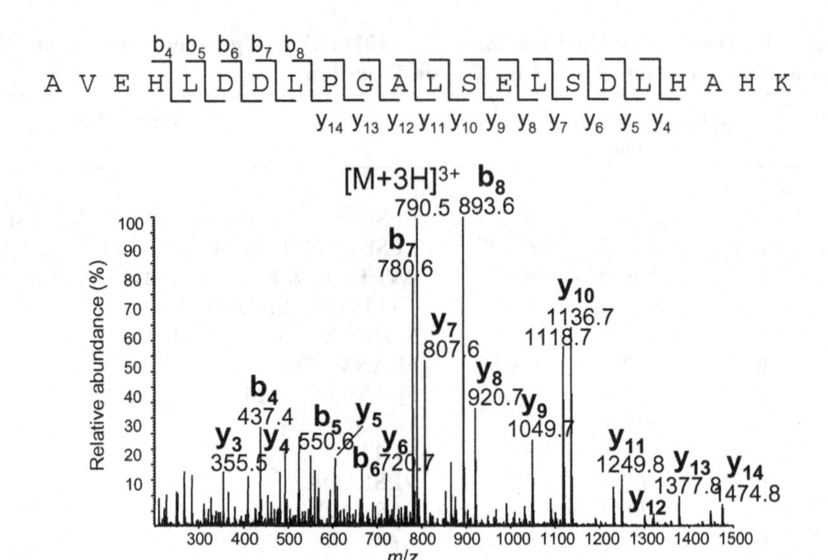

Figure 5.8: ESI product ion mass spectrum of the aT9 peptide characterizing the α-chain of bovine hemoglobin, measured on a QTrap 4000 system.

does not cleave Hb at cross-linked positions. This results in significantly different ratios of trypsin-derived peptides from bovine Hb and Hemopure.[143]

5.6 HUMAN GROWTH HORMONE (hGH)

Human growth hormone (hGH, Fig. 5.9a) has been subject of numerous studies since it was identified as the growth-promoting principle of the human pituitary gland in 1956.[149] Isolated and prepared from hypophysis, its structural heterogeneity became evident although firstly, and mainly spuriously, attributed to laboratory artifacts or impurities;[150] however, in-depth investigations allowed the elucidation of numerous hGH variants (Fig. 5.9b, Table 5.4), which were found to originate from an alternative splice site at the mRNA level,[151,152] from post-translational modifications such as N-terminal acylation, deamidation, or glycosylation,[153,154] as well as from oligomerization.[155–157] In addition, artificial products such as scissions within the primary structure supposedly resulting from harsh extraction conditions or long-term storage have been identified.[150,158] The heterogeneous family of growth hormones is of great interest in various scientific fields including clinical diagnostics as well as doping control analysis. Several assays enabling

a)

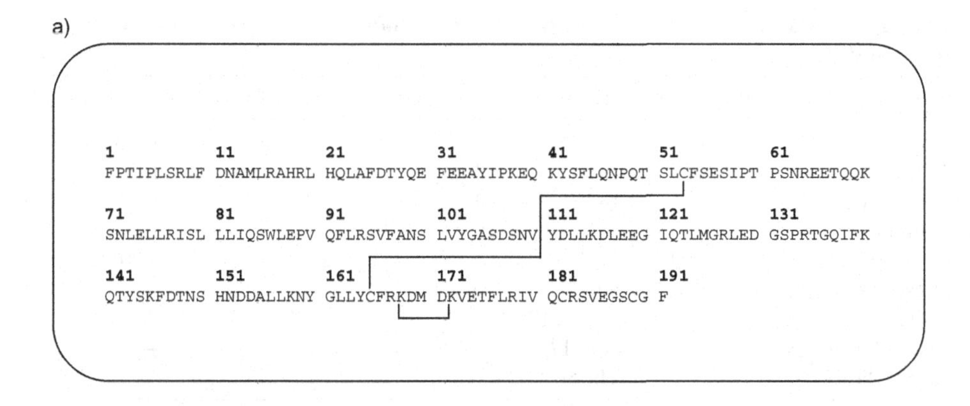

b)

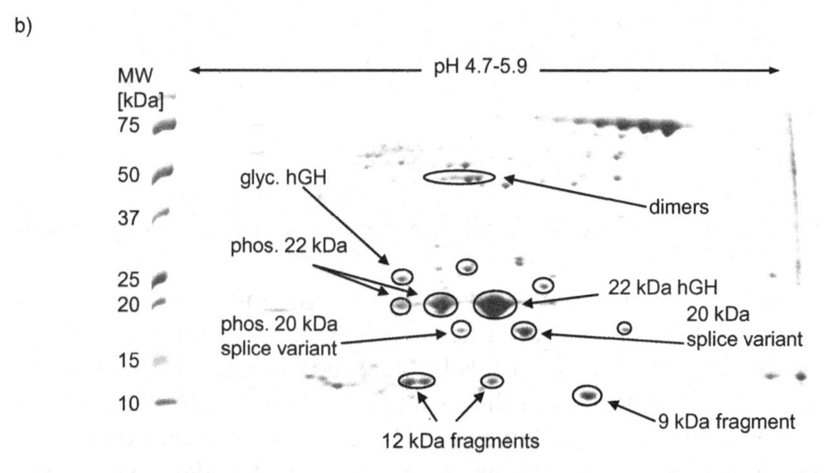

Figure 5.9: (a) Primary structure of human growth hormone (mol $wt_{monoisotopic}$ = 22115.1 Da) and (b) 2D-gel electrophoresis of proteins extracted from a pituitary. Spots containing growth hormone variants and fragments (circled) include the major isoforms with 22 and 20 kDa as well as respective phosphorylated species, glycosylated hGH, dimers, and various fragments of 9 and 12 kDa size.

the determination of hGH levels as well as the potential misuse of recombinant preparations have been published primarily utilizing sophisticated immunoassays.[156,159–166] In order to optimize immunoassays in terms of cross reactivity as well as specificity, the characterization of antigens and their natural and artificial variants is of paramount importance and has been subject of considerable research in the past, employing preferably proteomics strategies. The subjects of previous investigations have included natural monomers (e.g., 22 kDa and 20 kDa variants),[167,168] covalent as well as non-covalent oligomers,[156,157,169–173]

TABLE 5.4: Major Isoforms and Post-translational Modifications of Human Growth Hormone[a]

	Precursor	Isoform 1	Isoform 2	Isoform 3	Isoform 4
Mol wt (Da)[b]	24847.3	22129.1	20274.0	17843.3	17083.4
Amino acids	217	191	176	153	145
Eliminated amino acids[c]	—	1–26	1–26, 58–72	1–26, 111–148	1–26, 117–162
PI (theor.)	5.16	5.27	5.28	6.35	5.90
Phosphorylation site[c]	—	Ser 77, 132, 176	Ser 176	—	—
Deamidation[c]	—	Asn 178	—	—	—
Glycosylation[c,d]	—	Thr 86	—	—	—

[a] Refs. 154 and 190.
[b] Molecular masses are calculated with reduced cysteine residues.
[c] Amino acid residue numbering includes the signal peptide 1–26.
[d] Glycosylation identified as HexHexNAc*2 NeuAc yielding a monoisotopic mass of 23062.5 Da.
— not reported.

and chemical, proteolytic, or metabolic degradation products of growth hormones.[150,172,174]

In terms of doping control analysis, growth hormone has gained much attention for more than a decade. Numerous studies have been conducted in order to verify or falsify the hypothesis of a performance-enhancing effect of hGH, and the results have been controversial.[175–181] While hGH administration demonstrably reverses symptoms of respective deficiency disorders such as increased body fat, decreased bone density, and impaired muscle strength, the application of hGH to normal subjects even in supraphysiological dosages did not provide evidence for improved exercise tolerance or endurance,[182,183] and a net anabolic effect on whole body protein metabolism has been concluded.[184] Nevertheless, "street talk" and numerous confessions of elite athletes have substantiated the suspicion that hGH is frequently misused in sports and necessitates efficient doping controls.

hGH was studied extensively using top-down as well as bottom-up sequencing approaches, yielding characteristic mass spectra and unequivocal information on its composition and modification. Using low- and high-resolution/high–accuracy mass spectrometry,[76,154,185,186] exact molecular weights of major isoforms were obtained, and complementary information derived from comprehensive amino acid sequence tags from various different positions of the molecules mostly confirmed the early postulations regarding particular structural features of hGH and its natural variants. Electrospray ionization of hGH yields multiply charged molecules generating a charge envelope commonly ranging from 12+ to 22+ and CID of e.g., the 18-fold charged species gives rise to sequence tags originating from different regions of the protein. As illustrated in Figure 5.10a, consecutively multiply charged b-ions provide detailed, though limited, information on the composition of the

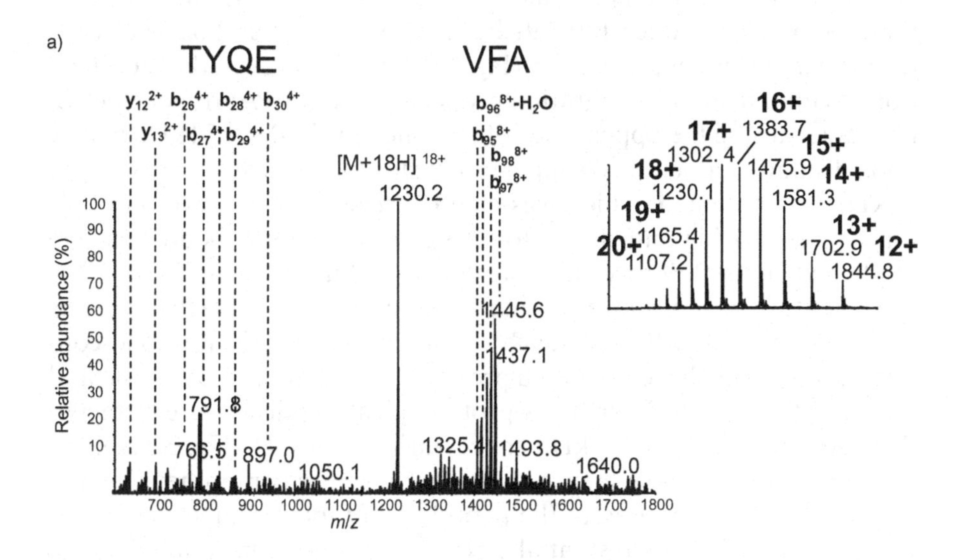

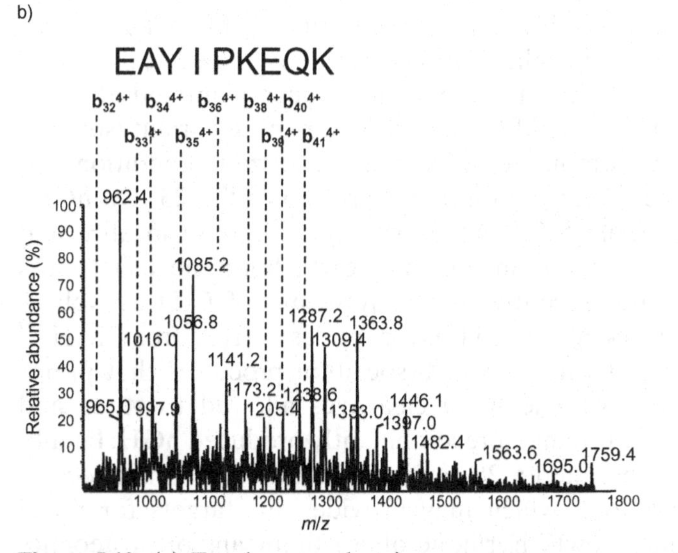

Figure 5.10: (a) Top-down product ion mass spectrum generated from the 18-fold charged molecule of the 22 kDa isoform of hGH; and (b) top-down product ion mass spectrum generated from the 14-fold charged molecule (m/z = 1148.3) of the 16 kDa artefact of hGH (both measured on a 4000 QTrap system).

target analyte, which could serve identification purposes if combined with accurate molecular weight determination.

Moreover, also artifacts resulting from (inadequate) long-term storage were elucidated, such as the hGH “variant” migrating in SDS-PAGE gel electrophoresis according to a 24 kDa-sized isoform. A degenerative cleavage of the peptide chain between phenylalanine 139 and lysine 140 yielded additional N- and C-termini without liberating the amino acid residues 140–191 because of a disulfide bond between the cysteine residues 53 and 165 (Fig. 5.9a);[187,188] however, the modification resulted in altered migration properties in SDS-PAGE gel electrophoresis, causing the appearance of the “nicked” hGH at 24 kDa which, upon disulfide reduction, disappeared. Instead, a band at approximately 16 kDa was obtained, which was characterized by top-down and bottom-up sequencing and shown to result from a peptide bond cleavage at phenylalanine 139 to yield a protein of 16061.5 Da (Fig. 5.10b).[189,190]

Separating proteins of a pituitary extract by 2D-gel electrophoresis (Fig. 5.9b) revealed the presence of numerous growth hormone-containing spots, which were studied preferably using trypsin digestion and LC-MS/MS analysis. Numerous proteotypical peptides were detected that identified, e.g., the 20 kDa splice variant of hGH (Fig. 5.11a), the mass and amino acid sequence tags of which were comprehensive and unambiguously attributed to the peptide T4 (Table 5.5). Desiderio and co-workers provided substantial evidence for post-translational modifications and isoforms of pituitary growth hormones by means of MS-based techniques.[172,186] Phosphorylations were determined at the serine residues 51, 106, and 150, deamidations have been identified at glutamine 137 and asparagine 152,[191] and four major splice variant isoforms of pituitary growth hormone have been characterized in addition to a glycosylation of the 22 kDa isoform (referred to as the 23 kDa hGH) as summarized in Table 5.4.[154] An approach on how to localize and characterize disulfide bonds in human growth hormone using mass spectrometry was also recently shown by means of CID as well as electron-transfer dissociation (ETD) of peptides derived from hGH.[192] The complementary nature of the dissociation processes that prefer either backbone or disulfide bond cleavages allowed to study and determine the correct folding of recombinantly produced hGH. Finally, recent studies demonstrated a 2% amino acid modification in recombinant GH preparations, which may provide new targets for novel assays to determine growth hormone abuse by means of proteomics strategies.[193] Major modifications as determined by mass spectrometry were deamidations of asparagines 149 and 152, and, interestingly, exchanges of methionines 14, 125, and 170 by isoleucines as shown with the example of the peptide T2 in Figure 5.11b. The latter phenomenon

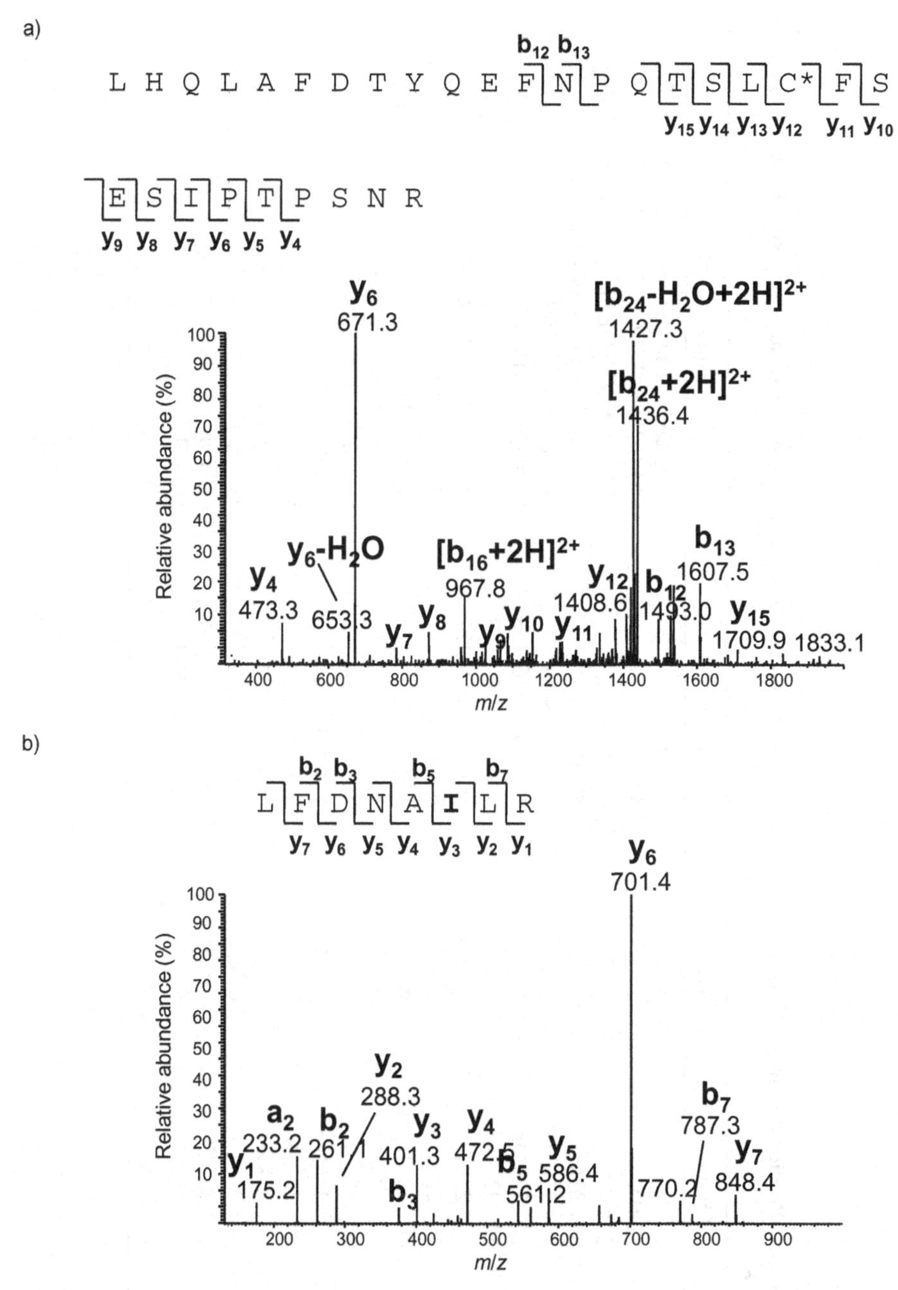

Figure 5.11: ESI product ion mass spectra of (a) the three-fold charged precursor ion $[M+3H]^{3+}$ of *m/z* 1182 of T4 derived from the 20 kDa splice variant of hGH (*the cysteine was derivatized to its acrylamide adduct), and (b) the two-fold charged precursor ion $[M+2H]^{2+}$ of *m/z* 481.1 of the modified peptide T2 comprising an isoleucine (bold) instead of a methionine at position 14 of hGH (corresponding to position 6 of T2), both measured on an LTQ system.

TABLE 5.5: Selected Predicted Peptide Masses $[M+H]^+$ of the 22 kDa and 20 kDa Splice Variants of Human Growth Hormone after Trypsin Digestion

Peptide	Origin	Mass (Da)	Position	Amino Acid Sequence
T4	20 kDa	3470.6	20–49	LHQLAFDTYQEFNPQTSLCFSESIPTPSNR
T6	22 kDa	2616.2	42–64	YSFLQNPQTSLCFSESIPTPSNR
T4	22 kDa	2342.1	20–38	LHQLAFDTYQEFEEAYIPK
T8 / T10	20 / 22 kDa	2262.1	80–100 / 95–115	SVFANSLVYGASDSNVYDLLK
T7 / T9	20 / 22 kDa	2055.2	63–79 / 78–94	ISLLLIQSWLEPVQFLR
T13 / T15	20 / 22 kDa	1489.7	131–141 / 146–158	FDTNSHNDDALLK
T14 / T16	20 / 22 kDa	1148.6	144–152 / 159–167	NYGLLYCFR
T9 / T11	20 / 22 kDa	1361.7	101–112 / 116–127	DLEEGIQTLMGR
T2	20 / 22 kDa	979.5	9–16	LFDNAMLR
T1	20 / 22 kDa	930.5	1–8	FPTIPLSR
T6 / T8	20 / 22 kDa	844.5	56–62 / 71–77	SNLELLR
T19 / T21	20 / 22 kDa	785.3	169–176 / 184–191	SVEGSCGF
T10 / T12	20 / 22 kDa	773.4	113–119 / 128–134	LEDGSPR
T17 / T19	20 / 22 kDa	764.4	158–163 / 173–178	VETFLR
T5 / T7	20 / 22 kDa	762.4	50–55 / 65–70	EETQQK
T11 / T13	20 / 22 kDa	693.3	120–125 / 135–140	TGQIFK
T12 / T14	20 / 22 kDa	626.3	126–130 / 141–145	QTYSK
T18 / T20	20 / 22 kDa	618.3	164–168 / 179–183	IVQCR
T16 / T18	20 / 22 kDa	508.2	154–157 / 169–172	DMDK

is presumably owing to false translation of the rare codon AGG in *E. coli* and may allow the detection of the investigated rGH Genotropin by providing specific target peptides, but the major issue of short half-life and low plasma as well as urine concentrations will remain.[176,194]

5.7 SERMORELIN (GEREF)

As GH deficiency is a serious health issue and numerous different symptoms such as growth retardation, somatotroph irresponsiveness and frailty occur with decreased GH serum levels, several strategies to correct these conditions were evaluated, and the search for drugs allowing influence on the GH/insulin-like growth factor-1 (IGF-1) axis has started as early as 1977 with the synthesis of GH releasing peptides,[195] a selection of which is listed in Table 5.6. Based on the knowledge of the direct relation of growth, aging, and anabolism with the GH/IGF-1 axis,[196] numerous studies were conducted on the synthesis of compounds able to trigger the release of GH in animals and humans. Momany and Bowers[197–202] prepared peptide analogues of Leu- and Met-enkephalins that were capable of stimulating isolated rat brain cells to secrete GH. These syntheses yielded, among others, the GH releasing peptide 6 (GHRP-6, Table 5.6) that has become a reference compound for comparison with new drug candidates aiming growth hormone release. The human hypothalamic growth hormone releasing hormone (GHRH), which consists of 44 amino acid residues and a C-terminal carboxamide moiety (Fig. 5.12a), was characterized from human pancreatic tumors, and the GH releasing activity was identified to reside within the first 27 amino acids. Consequently, numerous synthetic analogues with and

TABLE 5.6: Primary Structures of Peptides with Growth Hormone Releasing Activity

Compound	Sequence
GHRP-1	Ala-His-D-βNal-Ala-Trp-D-Phe-Lys-NH_2
GHRP-2	D-Ala-D-βNal-Ala-Trp-D-Phe-Lys-NH_2
GHRP-4	D-Trp-Ala-Trp-D-Phe-NH_2
GHRP-5	Tyr-D-Trp-Ala-Trp-D-Phe-NH_2
GHRP-6	His-D-Trp-Ala-Trp-D-Phe-Lys-NH_2
Hexarelin	His-D-Mrp-Ala-Trp-D-Phe-Lys-NH_2
Ipamorelin	Aib-His-D-2-Nal-D-Phe-Lys-NH_2
Alexamorelin	Ala-His-D-Mrp-Ala-Trp-D-Phe-Lys-NH_2

Non-standard abbreviations: Aib = aminoisobutyric acid; Nal = naphthylalanine; Mrp 2-methyltryptophane.

```
   1          11         21         31         41
a) YADAIFTNSY RKVLGQLSAR KLLQDIMSRQ QGESNQERGA RARL

b) YADAIFTNSY RKVLGQLSAR KLLQDIMSR-NH2
```

Figure 5.12: Primary structures of a) GHRH (mol $wt_{monoisotopic}$ = 5037.6 Da), and (b) Geref (GHRH 1-29-NH_2, Sermorelin, mol $wt_{monoisotopic}$ = 3355.8 Da).

without amino acid substitution were prepared, some of which were found to provide so-called superactivity as compared to human GHRH. However, the superior efficacy as determined in *in vitro* and animal studies failed to demonstrate the potency in humans, and only one product, namely GHRH 1-29-NH_2 (Geref, Fig. 5.12b), has been marketed for the treatment of GH-deficient patients. Ever since, several studies have proved the beneficial effects of GHRH 1-29-NH_2 on skin and lean body mass, particularly in men using a recommended daily subcutaneous application of 30 μg/kg. In contrast to human GH replacement therapies, side effects such as hypertension and water/salt retention and elevated lipoprotein levels were not observed, indicating a promising clinical utility of synthetic GHRH; however, therapeutic applications have become rare. Beneficial effects of therapies based on an increased growth hormone release are expected in the treatment of GH deficiencies in children and adults assuming a dysfunctional hypothalamo-pituitary unit. The treatment of catabolic states and age-related frailty is another promising application. Rejuvenation of the GH/IGF-1 axis by growth hormone secretagogues (GHS) has been described in which 70–90-year-old subjects produced serum IGF-1 levels and GH pulse amplitudes that are typically observed in healthy persons younger than 30 years. Improvements of body composition, strength, and bone density were reported, and increased levels of IGF-1 were sustained for more than 12 months.[203,204] Additional possible benefits include positive effects on metabolism (resulting in an increased lean body mass and reduced fat deposition), on skeletal muscle tissue, on the aging brain, as well as the aging immune system. These facts might lead to a misuse of these compounds by athletes aiming to artificially improve their physical constitution and resulting athletic performance, a suspicion that has been substantiated with findings of Geref, which was recently confiscated in house searches of subjects suspicious for doping offences.

The peptide Geref, which contains 29 amino acids with a C-terminal amidation, dissociates considerably after positive ESI and subsequent

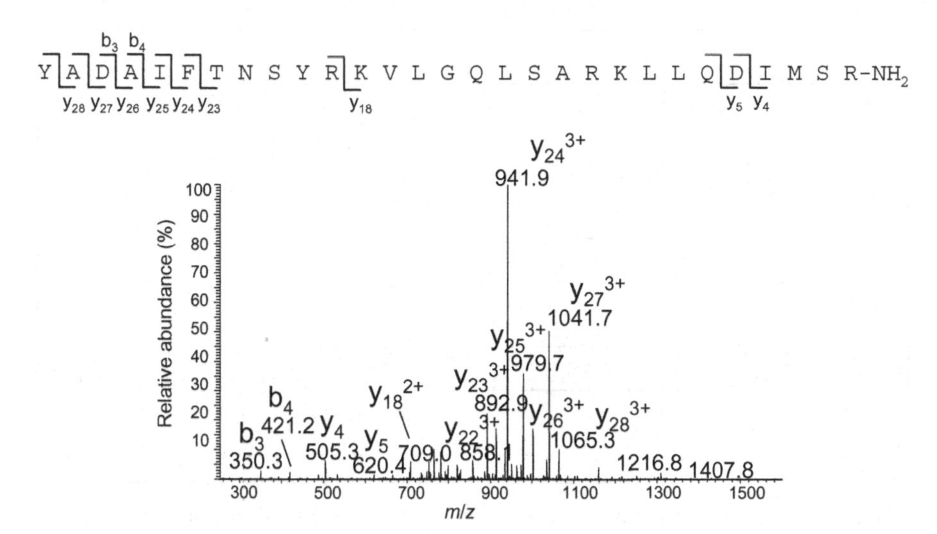

Figure 5.13: ESI product ion mass spectrum of the three-fold charged precursor ions $[M+3H]^{3+}$ of *m/z* 1118 of Geref, measured on an LTQ-Orbitrap system.

CID, yielding comprehensive y-ion-based amino acid sequence tags (Fig. 5.13). Although Geref represents a truncated version of the endogenously produced GHRH, its natural occurrence has not been reported; consequently, the analyte as such can be used as target compound for doping control purposes employing mass spectrometric methods in contrast to GH, the product released by GHRH or Geref.

5.8 INSULIN-LIKE GROWTH FACTOR-1 (IGF-1)

The relevance of hGH and its releasing hormone(s) in sports drug testing has further necessitated the consideration of another growth promoting compound, namely insulin-like growth factor-1 (IGF-1). The class of insulin-like growth factors was isolated and characterized in 1976 by Rinderknecht and Humbel,[205–207] yielding a predominating form termed IGF-1 that consists of 70 amino acid residues and bears three disulfide bonds (Fig. 5.14a).[208,209] The majority of IGF-1 is produced in the liver, and its biological activity is modulated by six binding proteins (IGFBPs) concurrently elevating its plasma half-life.[210] IGF-1 mediates many of the effects of hGH in an autocrine/paracrine mode, e.g., anabolism in cartillage,[211] bone,[212] and skeletal muscle system.[213] Especially when injected or infused locally into target tissues, IGF-1 causes a significant increase in protein and DNA content[214] as well as

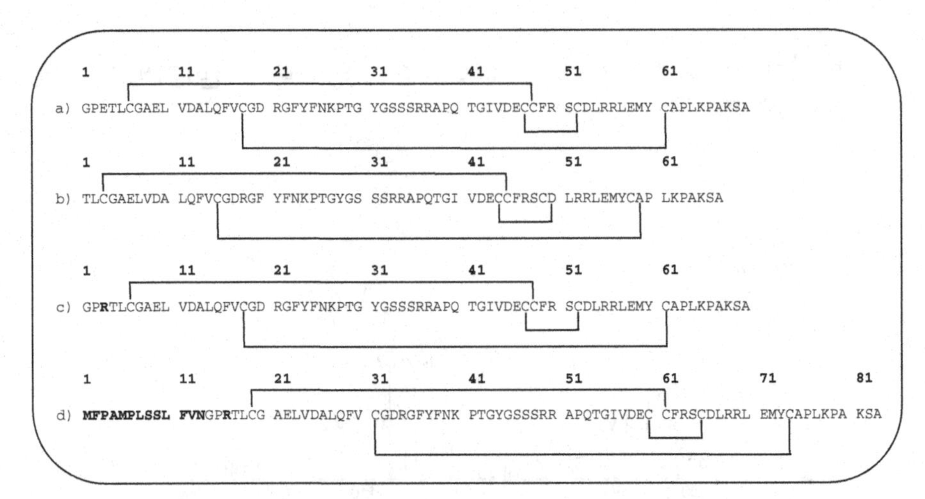

Figure 5.14: Primary structures of (a) human insulin-like growth factor 1 (IGF-1, mol $wt_{monoisotopic}$ = 7643.6 Da), (b) des(1-3)IGF-1 (mol $wt_{monoisotopic}$ = 7360.5 Da), (c) R^3-IGF-1 (mol $wt_{monoisotopic}$ = 7670.6 Da), and (d) long-R^3-IGF-1 (mol $wt_{monoisotopic}$ = 9105.4 Da). Modifications compared to IGF-1 are highlighted with bold letters.

longitudinal bone growth.[211] Connected with these performance-enhancing properties is the capability of IGF-1 to stimulate proliferation and differentiation of cells, making IGF-1 a potential risk factor for cancer.[215–218] Therapeutically, the administration of IGF-1 (commonly 8–12 mg/day) has been approved only for very few indications such as the long-term treatment of growth failure in children with severe primary IGF-1 deficiency;[219] however, several additional target diseases (e.g., myotonic muscular dystrophy and Lou Gehrig's disease)[220] are under investigation. Moreover, various research programs have been conducted aiming the elucidation of the relationship between the primary structure of the IGF-1 molecule and its biological effects.[210,221–226] Based on these structure-activity-relationship (SAR) studies, IGF-1 was modified to yield more potent analogues such as des(1-3)IGF-1 that lacks the first three amino acids of human IGF-1, R^3IGF-1 that is obtained from an amino acid exchange Glu3Arg, and long-R^3IGF-1, which bears an extension of 13 amino acids at the N-terminus in addition to the aforementioned Glu3Arg modification (Fig. 5.14b–d). These compounds demonstrated considerably reduced affinities to the plasma-binding proteins of IGF-1 and an increased bioavailability and potency at target tissues[221,223,227,228] in spite of a reduced receptor binding affinity, presumably because of a faster clearance into the target tissue compared with native IGF-1.[225,229]

The growth promoting properties in particular have made IGF-1 an important target for doping controls, and since 1999 the use of this peptide hormone is explicitly prohibited in sports. Consequently, several studies concerning the mass spectrometric characterization of IGF-1 and its synthetic analogues were conducted, which were subsequently employed in MS-based analytical assays to qualitatively and quantitatively measure these analytes in blood and urine specimens. Under ESI conditions, IGF-1 and derivatives yield considerably stable and multiply charged molecules that generate few but informative product ions upon CID (Figures 5.15 and 5.16). Those are predominantly composed

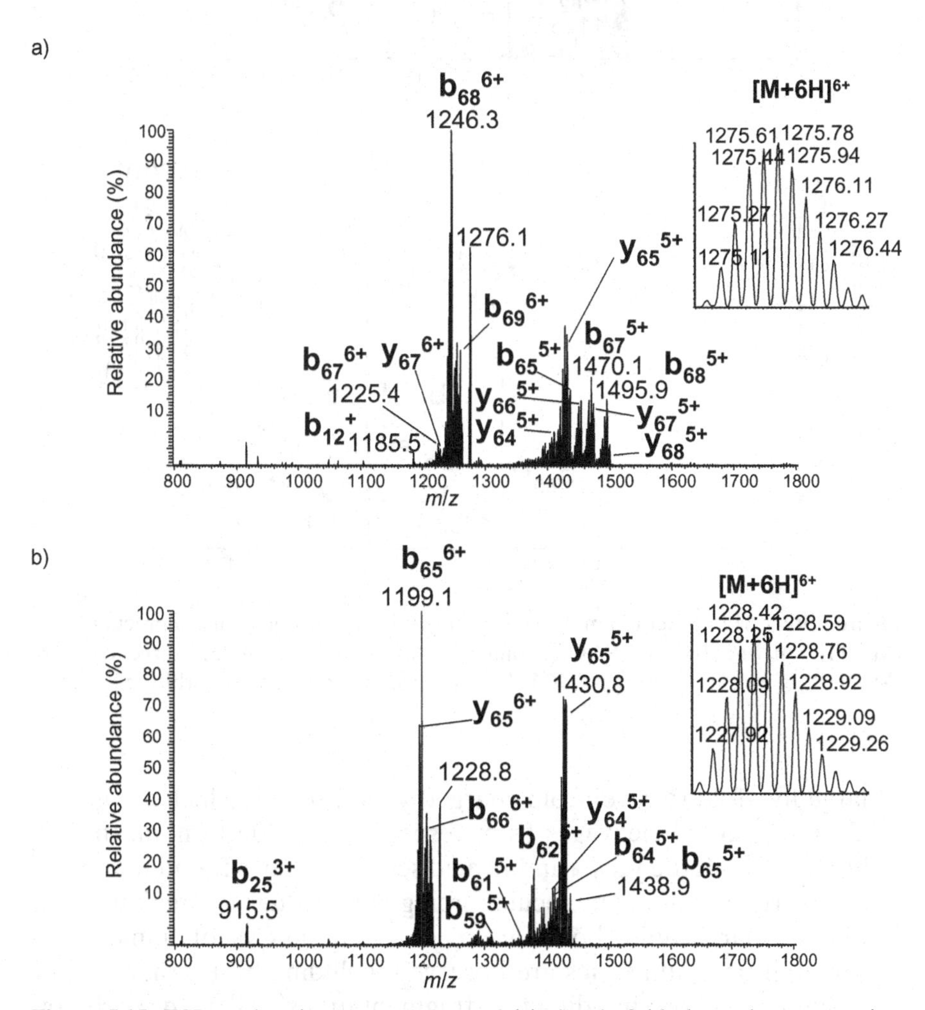

Figure 5.15: ESI product ion mass spectra of (a) the six-fold charged precursor ion $[M+6H]^{6+}$ of *m/z* 1275 of IGF-1, and (b) the six-fold charged precursor ion $[M+6H]^{6+}$ of *m/z* 1228 of des(1-3)IGF-1, both measured on an LTQ-Orbitrap system.

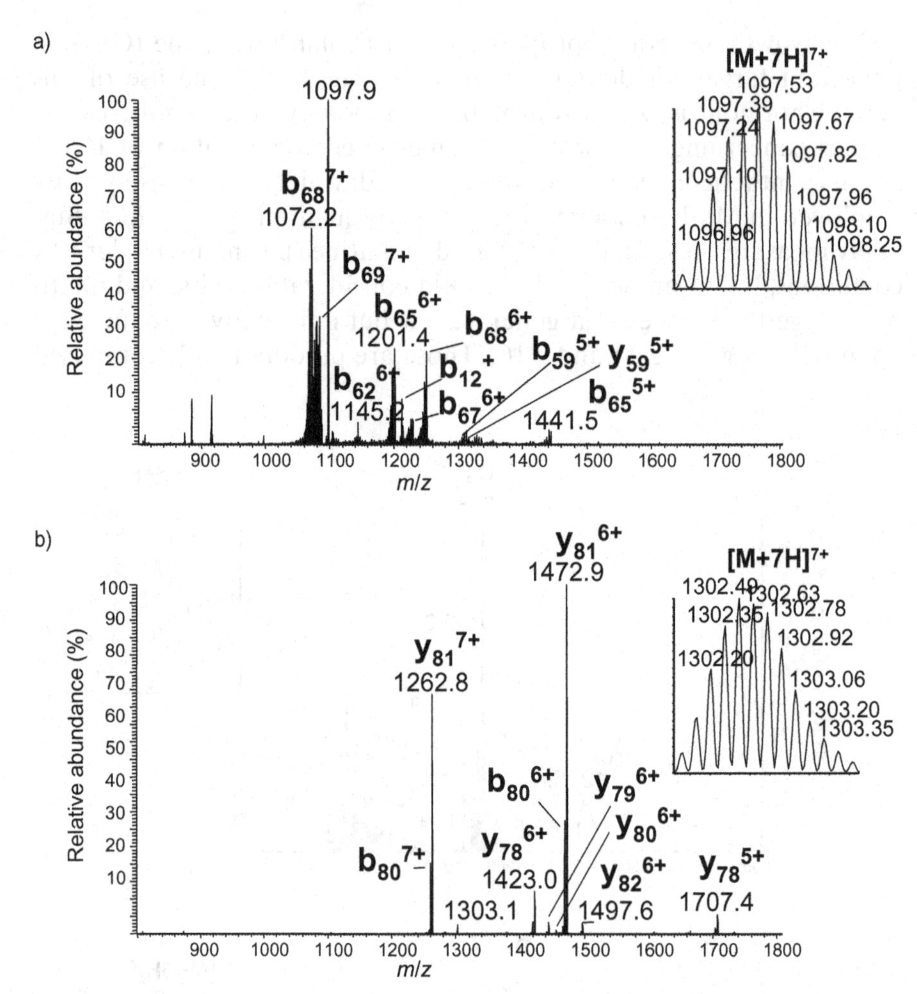

Figure 5.16: ESI product ion mass spectra of (a) the seven-fold charged precursor ion $[M+7H]^{7+}$ of *m/z* 1097 of R^3IGF-1, and (b) the seven-fold charged precursor ion $[M+7H]^{7+}$ of *m/z* 1302 of long-R^3IGF-1, both measured on an LTQ-Orbitrap system.

of large five-fold to seven-fold charged y- and b-ions, which are generated by sequential cleavages from respective peptide termini. In case of human IGF-1, the y_{64}-y_{68} amino acid sequence tag is found that highlights the rigidity of the molecule forcing the residues to break up until intramolecular disulfide bonds are reached (e.g., at Cys_6 of human IGF-1). In addition, b-ion series are observed including b_{65}-b_{69} that further indicate typical proline-directed fragmentations.[230] The dissociation behavior of des(1-3)IGF-1 is highly comparable to that of IGF-1 although the N-terminally located proline residue (Pro_2) is missing, and abundant b_{62}-b_{66} product ions originating from peptide bond cleavages that correspond to those yielding b_{65}-b_{69} in case of IGF-1 are detected. Moreover, the y_{65}-ion representing the peptide backbone after release of the N-terminal amino acid residues up to the disulfide-linked cysteine is found with great intensity (Fig. 5.15b). R^3IGF-1 (Fig. 5.16a) appears to produce fewer y-ions compared to IGF-1 but the principal dissociation pattern follows the same routes as IGF-1. In contrast, the CID of long-R^3IGF-1 (Fig. 5.16b) is significantly influenced by proline-directed fragmentation processes, which cause the formation of few but highly abundant product ions representing a series of y-ions encompassing the region around y_{78}–y_{82}, which includes two proline residues with preferred dissociation sites.

5.9 GONADORELIN (LHRH)

The physiological stimulus for the gonads to produce testosterone follows a complex cascade including gonadorelin (luteinizing hormone releasing hormone, LHRH) and various feedforward- and suppressing-effects due to the combined release of the LH and FSH from the pituitary gland in a pulsatile manner. LHRH is composed of ten amino acid residues (Fig. 5.17) and is produced in the hypothalamus.[231–233] Its isolation and characterization was accomplished in 1971 outlining a C-terminal amidation of glycine and cyclic-dehydratation of the N-terminal glutamic acid (pyroglutamic acid, Pyr) yielding the bioactive substance (Pyr-His-Trp-Ser-Tyr-Gly-Leu-Arg-Pro-Gly-NH_2).[234] LHRH is secreted in intervals of approximately 90–120 minutes, a fact that proved crucial for the gonadal steroid regulation, as only short-term stimulation leads to an increased release of LH and FSH.[231,232,235–238] In case of continuous hormone infusion, inhibitory effects on the gonadotrophin (i.e., LH and FSH) release were reported due to LHRH receptor down-regulation by persistent exposition.[236] Upon activation of these receptors, the peptide hormone is internalized into the cells and subsequently exposed to peptidases;[231,239] hence, the intact hormone is found only in the microcirculation between the hypothalamus and the pituitary gland.

LHRH and selected synthetic analogues were developed for the treatment of conditions that necessitate a hypogonadal state (endometriosis, prostate or breast cancer) or to promote the descent of testicles in children 12–24 months old; however, the pulsatile application of exogenous LHRH, for instance, by intranasal administration to healthy young men was shown to lead to a significant increase of the plasma

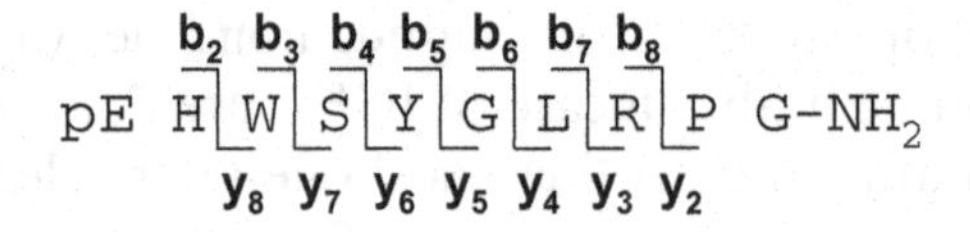

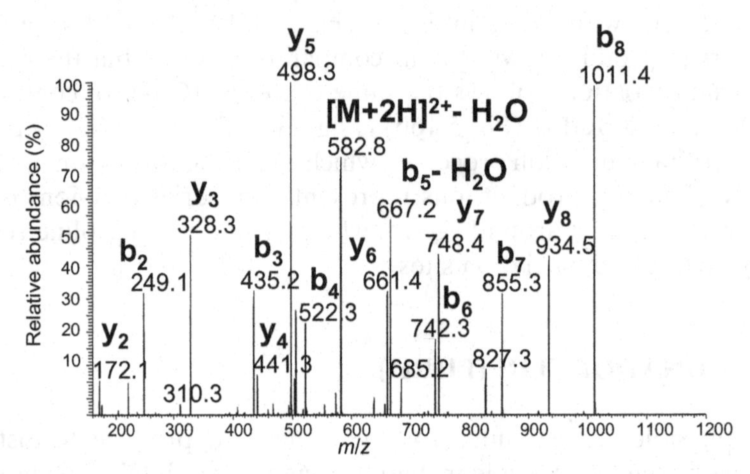

Figure 5.17: ESI product ion mass spectrum of the doubly charged precursor ion $[M+2H]^{2+}$ of *m/z* 592 of luteinizing hormone releasing hormone (LHRH, mol $wt_{monoisotopic}$ = 1181.6 Da), measured on an LTQ system.

LH and testosterone levels, a major aspect entailing relevancy for a potential misuse in elite sports as performance-enhancing or masking agent.[235,240] Consequently, the decapeptide LHRH was studied by mass spectrometry to allow its detection in plasma or urine doping control samples. ESI of LHRH yields a doubly charged precursor ion at *m/z* 592 that generates characteristic b- and y-ion series after collisional activation that allows the reconstruction of nearly the entire peptide composition (Fig. 5.17). A preferred proline-directed dissociation yielding the abundant ion b_8 is observed representing the most intense ion of the b_2-b_8 ion series, but also comprehensive y-ion sequence tags ranging from y_2 to y_8 are produced, which enable an unequivocal identification of the target analyte.

REFERENCES

1. Kicman, A.T., Brooks, R.V., and Cowan, D.A. (1991) Human chorionic gonadotrophin and sport. *British Journal of Sports Medicine*, **25**, 73–80.
2. Kicman, A.T., and Cowan, D.A. (1992) Peptide hormones and sport: Misuse and detection. *British Medical Bulletin*, **48**, 496–517.

3. Laidler, P., Cowan, D.A., Hider, R.C., and Kicman, A.T. (1994) New decision limits and quality-control material for detecting human chorionic gonadotropin misuse in sports. *Clinical Chemistry*, **40**, 1306–1311.
4. Delbeke, F.T., Van Eenoo, P., and De Backer, P. (1998) Detection of human chorionic gonadotrophin misuse in sports. *International Journal of Sports Medicine*, **19**, 287–290.
5. Stenman, U.H., Unkila-Kallio, L., Korhonen, J., and Alfthan, H. (1997) Immunoprocedures for detecting human chorionic gonadotropin: Clinical aspects and doping control. *Clinical Chemistry*, **43**: 1293–1298.
6. Aschheim, S., and Zondek, B. (1927) Hypophysenvorderlappenhormon und Ovarialhormon im Harn von Schwangeren. *Klinische Wochenschrift*, **6**, 1322.
7. Zondek B., and Aschheim, S. (1927) Das Hormon des Hypophysenvorderlappens. *Klinische Wochenschrift*, **6**, 248–252.
8. Aschheim, S., and Zondek, B. (1928) Schwangerschaftsdiagnose aus dem Harn (durch Hormonnachweis). *Klinische Wochenschrift*, **7**, 8–9.
9. Zondek, B., and Aschheim, S. (1928) Das Hormon des Hypophysenvorderlappens—Darstellung, chemische Eigenschaften, biologische Wirkungen. *Klinische Wochenschrift*, **7**, 831–835.
10. Aschheim S., and Zondek, B. (1928) Die Schwangerschaftsdiagnose aus dem Harn durch Nachweis des Hypophysenvorderlappenhormons—I. Grundlagen und Technik der Methode. *Klinische Wochenschrift*, **7**, 1404–1411.
11. Aschheim, S., and Zondek, B. (1928) Die Schwangerschaftsdiagnose aus dem Harn durch Nachweis des Hypophysenvorderlappenhormons—II. Praktische und theoretische Ergebnisse aus den Harnuntersuchungen. *Klinische Wochenschrift*, **7**, 1453–1457.
12. Galli-Mainini, C. (1948) Pregnancy test using the male batrachia. *Journal of the American Medical Association*, **138**, 121–125.
13. Ruppert, H. (1950) Erfahrungen mit der Schwangerschaftsprobe nach Galli-Mainini. *Das Deutsche Gesundheitswesen*, **5**, 389–392.
14. Wide, L., and Gemzell, C.A. (1960) An immunological pregnancy test. *Acta Endocrinologica (Copenhagen)*, **25**, 261–267.
15. Brody, S., and Carlstroem, G. (1965) Human chorionic gonadotropin in abnormal pregnancy. Serum and urinary findings using various immunoassay techniques. *Acta Obstetricia et Gynecologica Scandinavica*, **44**, 32–44.
16. Stenman, U.H., Hotakainen, K., and Alfthan, H. (2008) Gonadotropins in doping: Pharmacological basis and detection of illicit use. *British Journal of Pharmacology*, **154**, 569–583.
17. Handelsman, D.J. (2006) Clinical review: The rationale for banning human chorionic gonadotropin and estrogen blockers in sport. *Journal of Clinical Endocrinology and Metabolism*, **91**, 1646–1653.

18. Valmu, L., Alfthan, H., Hotakainen, K., *et al.* (2006) Site-specific glycan analysis of human chorionic gonadotropin beta-subunit from malignancies and pregnancy by liquid chromatography—electrospray mass spectrometry. *Glycobiology*, **16**, 1207–1218.
19. Kicman, A.T., Parkin, M.C., and Iles, R.K. (2007) An introduction to mass spectrometry based proteomics-detection and characterization of gonadotropins and related molecules. *Molecular and Cellular Endocrinology*, **260–262**, 212–227.
20. Ramirez-Llanelis, R., Llop, E., Ventura, R., *et al.* (2008) Can glycans unveil the origin of glycoprotein hormones?—human chorionic gonadotrophin as an example. *Journal of Mass Spectrometry*, **43**, 936–948.
21. Gam, L.H., Tham, S.Y., and Latiff, A. (2003) Immunoaffinity extraction and tandem mass spectrometric analysis of human chorionic gonadotropin in doping analysis. *Journal of Chromatography B*, **792**, 187–196.
22. Liu, C.L., and Bowers, L.D. (1997) Mass spectrometric characterization of the β-subunit of human chorionic gonadotropin. *Journal of Mass Spectrometry*, **32**, 33–42.
23. Liu, C.L., and Bowers, L.D. (1996) Immunoaffinity trapping of urinary human chorionic gonadotropin and its high-performance liquid chromatographic-mass spectrometric confirmation. *Journal of Chromatography B*, **687**, 213–220.
24. Miyake, T., Kung, C.K., and Goldwasser, E. (1977) Purification of human erythropoietin. *Journal of Biological Chemistry*, **252**, 5558–5564.
25. Dordal, M.S., Wang, F.F., and Goldwasser, E. (1985) The role of carbohydrate in erythropoietin action. *Endocrinology*, **116**, 2293–2299.
26. Jelkmann, W. (2004) Molecular biology of erythropoietin. *Internal Medicine*, **43**, 649–659.
27. Jelkmann, W. (2003) Erythropoietin. *Journal of Endocrinological Investigation*, **26**, 832–837.
28. Jelkmann, W. (2007) Erythropoietin after a century of research: Younger than ever. *European Journal of Haematology*, **78**, 183–205.
29. Macdougall, I.C. (2006) Recent advances in erythropoietic agents in renal anemia. *Seminars in Nephrology*, **26**, 313–318.
30. Macdougall, I.C., and Eckardt, K.U. (2006) Novel strategies for stimulating erythropoiesis and potential new treatments for anaemia. *Lancet*, **368**, 947–953.
31. Franz, S.E. (2009) Erythropoiesis-stimulating agents: Development, detection, and dangers. *Drug Testing and Analysis*, **1**, 245–249.
32. Kochendoerfer, G.G., Chen, S-Y., Mao, F., *et al.* (2003) Design and chemical synthesis of a homogeneous polymer-modified erythropoiesis protein. *Science*, **299**, 884–887.
33. Videman, T., Lereim, I., Hemmingsson, P., *et al.* (2000) Changes in hemoglobin values in elite cross-country skiers from 1987–1999. *Scandinavian Journal of Medicine & Science in Sports*, **10**, 98–102.

34. Recny M.A., Scoble H.A., Kim Y. (1987) Structural characterization of natural human urinary and recombinant DNA-derived erythropoietin. Identification of des-arginine 166 erythropoietin. *Journal of Biological Chemistry* **262**: 17156–17163.
35. Sytkowski, A.J. (2004) *Erythropoietin: Blood, Brain and Beyond.* Wiley-VCH, Weinheim.
36. Hokke, C.H., Bergwerff, A.A., Van Dedem, G.W.K., *et al.* (1995) Structural analysis of the sialylated N- and O-linked carbohydrate chains of recombinant human erythropoietin expressed in Chinese hamster ovary cells. Sialylation patterns and branch location of dimeric N-acetyllactosamine units. *European Journal of Biochemistry*, **228**, 981–1008.
37. Kawasaki, N., Ohta, M., Hyuga, S., *et al.* (2000) Application of liquid chromatography/mass spectrometry and liquid chromatography with tandem mass spectrometry to the analysis of the site-specific carbohydrate heterogeneity in erythropoietin. *Analytical Biochemistry*, **285**, 82–91.
38. Kawasaki, N., Haishima, Y., Ohta, M., *et al.* (2001). Structural analysis of sulfated N-linked oligosaccharides in erythropoietin. *Glycobiology*, **11**, 1043–1049.
39. Sasaki, H., Bothner, B., Dell, A., and Fukuda, M. (1987) Carbohydrate structure of erythropoietin expressed in Chinese hamster ovary cells by a human erythropoietin cDNA. *Journal of Biological Chemistry*, **262**, 12059–12076.
40. Egrie, J.C., and Browne, J.K. (2001) Development and characterization of novel erythropoiesis stimulating protein (NESP). *British Journal of Cancer*, **84**, 3–10.
41. Zhou, G.H., Luo, G.A., Zhou, Y., *et al.* (1998) Application of capillary electrophoresis, liquid chromatography, electrospray-mass spectrometry and matrix-assisted laser desorption/ionization—time of flight—mass spectrometry to the characterization of recombinant human erythropoietin. *Electrophoresis*, **19**, 2348–2355.
42. Stübiger, G., Marchetti, M., Nagano, M., *et al.* (2005) Characterisation of intact recombinant human erythropoietins applied in doping by means of planar gel electrophoretic techniques and matrix-assisted laser desorption/ionisation linear time-of-flight mass spectrometry. *Rapid Communications in Mass Spectrometry*, **19**, 728–742.
43. Wide, L., and Bengtsson, C. (1990) Molecular charge heterogeneity of human serum erythropoiein. *British Journal of Haematology*, **76**, 121–127.
44. Wide, L., Bengtsson, C., Berglund, B., and Ekblom, B. (1995) Detection in blood and urine of recombinant erythropoietin administered to healthy men. *Medicine and Science in Sports and Exercise*, **27**, 1569–1576.
45. Caldini, A., Moneti, G., Fanelli, A., *et al.* (2003) Epoetin alpha, epoetin beta and darbepoetin alfa: Two-dimensional gel electrophoresis isoforms characterization and mass spectrometry analysis. *Proteomics*, **3**, 937–941.

46. Ohta, M., Kawasaki, N., Hyuga, S., *et al.* (2001) Selective glycopeptide mapping of erythropoietin by on-line high-performance liquid chromatography—electrospray ionization mass spectrometry. *Journal of Chromatography A*, **910**, 1–11.
47. Stanley, S.M., and Poljak, A. (2003) Matrix-assisted laser-desorption time-of flight ionisation and high-performance liquid chromatography-electrospray ionisation mass spectral analyses of two glycosylated recombinant epoetins. *Journal of Chromatography B, Analytical Technologies in the Biomedical and Life Sciences*, **785**, 205–218.
48. Groleau, P.E., Desharnais, P., Cote, L., and Ayotte, C. (2008) Low LC-MS/MS detection of glycopeptides released from pmol levels of recombinant erythropoietin using nanoflow HPLC-chip electrospray ionization. *Journal of Mass Spectrometry*, **43**, 924–935.
49. Nimtz, M., Martin, W., Wray, V., *et al.* (1993) Structures of sialylated oligosaccharides of human erythropoietin expressed in recombinant BHK—21 cells. *European Journal of Biochemistry*, **213**, 39–56.
50. Nimtz, M., Wray, V., Rüdiger, A., and Conradt, H.S. (1995) Identification and structural characterization of a mannose-6-phosphate containing oligomannosidic N-glycan from human erythropoietin secreted by recombinant BHK-21 cells. *FEBS Letters*, **365**, 203–208.
51. Guan F., Uboh C.E., Soma L.R., Birksz E., and Chen J. (2009) Identification of darbepoetin alfa in human plasma by liquid chromatography coupled to mass spectrometry for doping control. *International Journal of Sports Medicine*, **30**, 80–86.
52. Bell, P.H. (1954) Purification and structure of β-corticotropin. *Journal of the American Chemical Society*, **76**, 5565–5567.
53. Lee, T.H., Lerner, A.B., and Buettner-Janusch, V. (1961) On the structure of human corticotropin (adrenocorticotropic hormone). *Journal of Biological Chemistry*, **236**, 2970–2974.
54. Shepherd, R.G., Howard, K.S., Bell, P.H., *et al.* (1956) Studies with corticotropin. I. Isolation, purification and properties of β-corticotropin. *Journal of the American Chemical Society*, **78**, 5051–5058.
55. Howard, K.S., Shepherd, R.G., Eigner, E.A., *et al.* (1955) Structure of β-corticotropin: Final sequence studies. *Journal of the American Chemical Society*, **77**, 3419–3420.
56. Berson, S.A., and Yalow, R.S. (1968) Radioimmunoassay of ACTH in Plasma. *Journal of Clinical Investigation*, **47**, 2725–2751.
57. Bell, P.H., Howard, K.S., Shepherd, R.G., *et al.* (1956) Studies with corticotropin. II. Pepsin degradation of β-corticotropin. *Journal of the American Chemical Society*, **78**, 5059–5066.
58. Shepherd, R.G., Willson, S.D., Howard, K.S., *et al.* (1956) Studies with corticotropin. III. Determination of the structure of β-corticotropin and its active degradation products. *Journal of the American Chemical Society*, **78**, 5067–5076.

59. Kappeler, H., and Schwyzer, R. (1961) Die Synthese eines Tetracosapeptides mit der Aminosäuresequenz eines hochaktiven Abbauprodukts des β-Corticotropins (ACTH) aus Schweinehypophysen. *Helvetica Chimica Acta*, **44**, 1136–1141.

60. Toft, A.D., and Irvine, W.J. (1974) Hormonal effects of synthetic ACTH analogues. *Proceedings of the Royal Society of Medicine*, **67**, 749–750.

61. Schuler, W., Schar, B., and Desaulles, P. (1963) On the pharmacology of an ACTH-active, fully synthetic polypeptide, β^{1-24} corticotropin, Ciba 30920-Ba, Synacthen. *Schweizerische Medizinische Wochenschrift*, **93**, 1027–1030.

62. Eipper, B.A., and Mains, R.E. (1980) Structure and biosynthesis of pro-adrenocorticotropin/endorphin and related peptides. *Endocrine Reviews*, **1**, 1–27.

63. Simpson, E.R., and Waterman, M.R. (1988) Regulation of the synthesis of steroidogenic enzymes in adrenal cortical cells by ACTH. *Annual Review of Physiology*, **50**, 427–440.

64. Costa, J.L., Bui, S., Reed, P., *et al.* (2004) Mutational analysis of evolutionarily conserved ACTH residues. *General and Comparative Endocrinology*, **136**, 12–16.

65. Schöneshöfer, M., and Goverde, H.J.M. (1984) Corticotropin in human plasma. General considerations. *Survey of Immunologic Research*, **3**, 55–63.

66. Agwu, J.C., Spoudeas, H., Hindmarsh, P.C., *et al.* (1999) Tests of adrenal insufficiency. *Archives of Disease in Childhood*, **80**, 330–333.

67. Otto, H., Minneker, C., and Spaethe, R. (1966) [Rapid synacthen test for the evaluation of adrenocortical function]. *Deutsche medizinische Wochenschrift*, **91**, 934–939.

68. Crowley, S., Hindmarsh, P.C., Holownia, P., *et al.* (1991) The use of low doses of ACTH in the investigation of adrenal function in man. *Journal of Endocrinology*, **130**, 475–479.

69. Harnack, K. (1978) [Synacthen-depot–an equivalent to systemic glucocorticosteroid therapy]. *Dermatologische Monatsschrift*, **164**, 382–389.

70. Nelson, J.K., Neill, D.W., Montgomery, D.A., *et al.* (1968) Synacthen Depot–Adrenal response in normal subjects and corticotrophin-treated patients. *British Medical Journal*, **1**, 557–558.

71. Bridges, N.A., Hindmarsh, P.C., Pringle, P.J., *et al.* (1998) Cortisol, androstenedione (A4), dehydroepiandrosterone sulphate (DHEAS) and 17 hydroxyprogesterone (17OHP) responses to low doses of (1–24)ACTH. *Journal of Clinical Endocrinology and Metabolism*, **83**, 3750–3753.

72. Vogeser, M., Zachoval, R., and Jacob, K. (2001) Serum cortisol/cortisone ratio after Synacthen stimulation. *Clinical Biochemistry*, **34**, 421–425.

73. World Anti-Doping Agency (2009) *The 2009 Prohibited List*. Available at http://www.wada-ama.org/rtecontent/document/2009_Prohibited_List_ENG_Final_20_Sept_08.pdf. Accessed 02-01-2009.

74. Nicholl, R. (July 20, 1999) Cycling-Tour de France: Dierckxsens pays drugs penalty, in *The Independent*, London.
75. Arribas, C. (2006) El CSD pide ayuda internacional para analizar el "caso Eufemiano", in *El Pais*, Lyon.
76. Loo, J.A., Edmonds, C.G., and Smith, R.D. (1993) Tandem mass spectrometry of very large molecules. 2. Dissociation of multiply charged proline-containing proteins from electrospray ionization. *Analytical Chemistry*, **65**, 425–438.
77. Thevis, M., Bredehöft, M., Geyer H., *et al.* (2006) Determination of Synacthen in human plasma using immunoaffinity purification and liquid chromatography/tandem mass spectrometry. *Rapid Communications in Mass Spectrometry*, **20**, 3551–3556.
78. Schäfer, E.A. (1916) *The Endocrine Organs: An Introduction to the Study of Internal Secretion*. Longmans, Green, London.
79. von Mering, J., and Minkowski, O. (1889) Diabetes mellitus nach Pankreasextirpation. *Centralblatt für klinische Medicin*, **10**, 393–394.
80. Rosenfeld, L. (2002) Insulin: Discovery and controversy. *Clinical Chemistry*, **48**, 2270–2288.
81. DeGroot, L.J., and Jameson, J.L. (2001) Biosynthesis of insulin, in *Endocrinology* (eds L. J. DeGroot and J. L. Jameson) Elsevier, New York, pp. 671–727.
82. Newgard, C.B. (2001) Substrate control of insulin release, in *The Endocrine Pancreas and Regulation of Metabolism* (L. S. Jefferson and A. D. Cherrington) Oxford University Press, New York, pp. 125–151.
83. Tipton, K.D., Ferrando, A.A., Phillips, S.M., *et al.* (1999) Postexercise net protein synthesis in human muscle from orally administered amino acids. *American Journal of Physiology*, **276**, E628–634.
84. Tipton, K.D., Elliott, T.A., and Cree, M.G., *et al.* (2007) Stimulation of net muscle protein synthesis by whey protein ingestion before and after exercise. *American Journal of Physiology and Endocrinology and Metabolism*, **292**, E71–76.
85. Tipton, K.D., and Wolfe, R.R. (2001) Exercise, protein metabolism, and muscle growth. *International Journal of Sport Nutrition and Exercise Metabolism*, **11**, 109–132.
86. Manninen, A.H. (2006) Hyperinsulinaemia, hyperaminoacidaemia and post-exercise muscle anabolism: The search for the optimal recovery drink. *British Journal of Sports Medicine*, **40**, 900–905.
87. Ladriere, L., Louchami, K., Vinambres, C., *et al.* (1998) Insulinotropic action of the monoethyl ester of succinic acid. *General Pharmacology*, **31**, 377–383.
88. Garcia-Martinez, J.A., Zhang, T.M., Villanueva-Penacarrillo, M.L., *et al.* (1997) In vivo stimulation of insulin release by the monoethyl, monopropyl, monoisopropyl, monoallyl and diallyl esters of succinic acid. *Research Communications in Molecular Pathology and Pharmacology*, **95**, 209–216.

89. Valverde, I., Vicent, D., Villanueva-Penacarrillo, M.L., *et al.* (1997) Stimulation of insulin release in vivo by the methyl esters of succinic acid and glutamic acid. *Advances in Experimental Medicine and Biology*, **426**, 231–234.
90. Vicent, D., Villanueva-Penacarrillo, M.L., *et al.* (1994) In vivo stimulation of insulin release by succinic acid methyl esters. *Archives Internationales de Pharmacodynamie et de Thérapie*, **327**, 246–250.
91. Sonksen, P.H. (2001) Hormones and sport (insulin, growth hormone and sport). *Journal of Endocrinology*, **170**, 13–25.
92. Wolfe, R.R. (2005) Regulation of skeletal muscle protein metabolism in catabolic states. *Current Opinion in Clinical Nutrition and Metabolic Care*, **8**, 61–65.
93. International Diabetes Federation (2005) *Diabetes atlas*. Available at http://www.eatlas.idf.org/Prevalence/All_diabetes/. Accessed 04-30-2007.
94. International Diabetes Federation (2006). *Diabetes epidemic out of control.* Available at http://www.idf.org/home/index.cfm?unode=7F22F450-B1ED-43BB-A57C-B975D16A812D. Accessed 30-04-2007.
95. Fabris D., and Fenselau, C. (1999) Characterization of allosteric insulin hexamers by electrospray ioniziation mass spectrometry. *Analytical Chemistry*, **71**, 384–387.
96. Barnett, A.H., and Owens, D.R. (1997) Insulin analogues. *Lancet*, **349**, 47–51.
97. Rosak, C. (2001) Insulinanaloga: Struktur, Eigenschaften und therapeutische Indikationen (Teil 1: Kurzwirkende Insulinanaloga). *Der Internist*, **42**, 1523–1535.
98. Lindström, T., Hedman, C.A., and Arnqvist, H.J. (2002) Use of a novel double-antibody technique to describe the pharmacokinetics of rapid-acting insulin analogs. *Diabetes Care*, **25**, 1049–1054.
99. Plum, A., Agers, H., and Andersen, L. (2000) Pharmacokinetics of the rapid-acting insulin analog, insulin aspart, in rats, dogs, and pigs, and pharmacodynamics of insulin aspart in pigs. *Drug Metabolism and Disposition*, **28**, 155–160.
100. Becker, R.H., Frick, A.D., Burger, F., *et al.* (2005) Insulin glulisine, a new rapid-acting insulin analogue, displays a rapid time-action profile in obese non-diabetic subjects. *Experimental and Clinical Endocrinology & Diabetes*, **113**, 435–443.
101. Danne, T., Becker, R.H., Heise, T., *et al.* (2005) Pharmacokinetics, prandial glucose control, and safety of insulin glulisine in children and adolescents with type 1 diabetes. *Diabetes Care*, **28**, 2100–2105.
102. Becker, R.H., Frick, A.D., Burger, F., *et al.* (2005) A comparison of the steady-state pharmacokinetics and pharmacodynamics of a novel rapid-acting insulin analog, insulin glulisine, and regular human insulin in healthy volunteers using the euglycemic clamp technique. *Experimental and Clinical Endocrinology & Diabetes*, **113**, 292–297.

103. Campbell, R.K., White, J.R., Levien, T., and Baker, D. (2001) Insulin glargine. *Clinical Therapeutics*, **23**, 1938–1957.
104. Levien, T.L., Baker, D.E., White, J.R., and Campbell, R.K. (2002) Insulin glargine: A new basal insulin. *The Annals of Pharmacotherapy*, **36**, 1019–1027.
105. Reinhart, L., and Panning, C.A. (2002) Insulin glargine: A new long-acting insulin product. *American Society of Health-System Pharmacists*, **59**, 643–649.
106. Polonsky, K.S., and O'Meara, N.M. (2001) Secretion and metabolism of insulin, proinsulin and C-peptide, in *Endocrinology* (L. J. deGroot and J. L. Jameson) Saunders, Philadelphia, pp. 697–711.
107. Birkinshaw, V.J., Gurd, M.R., Randall, S.S., *et al.* (1958) Investigations in a case of murder by insulin poisoning. *British Medical Journal*, 463–468.
108. Marks, V. (1999) Murder by insulin. *Medico-Legal Journal*, **67**, 147–163.
109. Marks, V. (2009) Murder by insulin: Suspected, purported and proven—a review. *Drug Testing and Analysis*, **1**, 162–176.
110. Bauman, W.A., and Yalow, R.S. (1981) Insulin As a Lethal Weapon. *Journal of Forensic Sciences*, **26**, 594–598.
111. Haibach, H., Dix, J.D., and Shah, J.H. (1987) Homicide by insulin administration. *Journal of Forensic Sciences*, **32**, 208–216.
112. Biolo, G., and Wolfe, R.R. (1993) Insulin action on protein metabolism. *Baillière's Clinical Endocrinology and Metabolism*, **7**, 989–1005.
113. Biolo, G., Declan Fleming, R.Y., and Wolfe, R.R. (1995) Physiologic hyperinsulinemia stimulates protein synthesis and enhances transport of selected amino acids in human skeletal muscle. *Journal of Clinical Investigation*, **95**, 811–819.
114. Reverter, J.L., Tural, C., Rosell, A., *et al.* (1994) Self-induced insulin hypoglycemia in a bodybuilder. *Archives of Internal Medicine*, **154**, 225–226.
115. Evans, P.J., and Lynch, R.M. (2003) Insulin as a drug of abuse in body building. *British Journal of Sports Medicine*, **37**, 356–357.
116. Rich, J.D., Dickinson, B.P., Merriman, N.A., and Thule, P.M. (1998) Insulin use by bodybuilders. *Journal of the American Medical Association*, **279**, 1613.
117. Dawson, R., and Harrison, M. (1998) Use of insulin as an anabolic agent. *British Journal of Sports Medicine*, **31**, 259.
118. Young, J., and Anwar, A. (2007) Strong diabetes. *British Journal of Sports Medicine*, **41**, 335–336; discussion 336.
119. Hacke, D., Ludwig, U., Pfeil, G., and Wulzinger, M. (2006) *Inside the blood doping investigation*. Available at http://www.spiegel.de/international/spiegel/0,1518,425939,00.html. Accessed 04-17-07.
120. Chevenne, D., Trivin, F., and Porquet, D. (1999) Insulin assays and reference values. *Diabetes and Metabolism*, **25**, 459–476.

121. Yalow, R.S., and Berson, S.A. (1959) Radiobiology: Assay of plasma insulin in human subjects by immunological methods. *Nature*, **184**, 1648–1649.
122. Sapin, R., Galudec, V.L., Gasser, F., *et al.* (2001) Elecsys insulin assay: Free insulin determination and the absence of cross-reactivity with insulin lispro. *Clinical Chemistry*, **47**, 602–605.
123. Butter, N.L., Hattersley, A.T., and Clark, P.M. (2001) Development of a bloodspot assay for insulin. *Clinica Chimica Acta*, **310**, 141–150.
124. Thevis, M., Loo, J.A., Loo, R.R., and Schänzer, W. (2007) Recommended criteria for the mass spectrometric identification of target peptides and proteins (<8 kDa) in sports drug testing. *Rapid Communications in Mass Spectrometry*, **21**, 297–304.
125. March, R.E. (1997) An introduction to quadrupole ion trap mass spectrometry. *Journal of Mass Spectrometry*, **32**, 351–369.
126. Gordon, D.B., and Woods, M.D. (1993) Enlarging the stability domain for ion trap tandem mass-spectrometry (Ms Ms). *Rapid Communications in Mass Spectrometry*, **7**, 215–218.
127. Thevis, M., Thomas, A., Delahaut, P., *et al.* (2006) Doping control analysis of intact rapid-acting insulin analogues in human urine by liquid chromatography-tandem mass spectrometry. *Analytical Chemistry*, **78**, 1897–1903.
128. Thevis, M., Thomas, A., Delahaut, P., *et al.* (2005) Qualitative determination of synthetic analogues of insulin in human plasma by immunoaffinity purification and liquid chromatography-tandem mass spectrometry for doping control purposes. *Analytical Chemistry*, **77**, 3579–3585.
129. Thomas, A., Thevis, M., Delahaut, P., *et al.* (2007) Mass spectrometric identification of degradation products of insulin and its long-acting analogues in human urine for doping control purposes. *Analytical Chemistry*, **79**, 2518–2524.
130. Thevis, M., Ogorzalek-Loo, R.R., Loo, J.A., *et al.* (2003) Mass spectrometric identification of synthetic insulins, in *22nd Cologne Workshop on Doping Analysis* (eds A. Gotzmann, H. Geyer, U. Mareck, and W. Schänzer), Sport&Buch Strauß, Cologne, pp. 227–237.
131. Thevis, M., Horning, S., and Schänzer, W. (2007) Unpublished results.
132. Thevis, M., Thomas, A., and Schänzer, W. (2008) Mass spectrometric determination of insulins and their degradation products in sports drug testing. *Mass Spectrometry Reviews*, **27**, 35–50.
133. Thomas, A., Schänzer, W., Delahaut, P., and Thevis, M. (2009) Sensitive and fast identification of urinary human, synthetic and animal insulin by means of nano-UPLC coupled with high resolution/high accuracy mass spectrometry. *Drug Testing and Analysis*, **1**, 219–227.
134. Moffat, A.S. (1991) Three li'l pigs and the hunt for blood substitutes. *Science*, **253**, 32–34.
135. Cohn, S. (2000) Blood substitutes in surgery. *Surgery*, **127**, 599–602.

136. Greenburg, A., and Kim, H. (2004) Hemoglobin-based oxygen carriers. *Critical Care*, **8**, S61–S64.

137. Yu, Z., Friso, G., Miranda, J.J., *et al.* (1997) Structural characterization of human hemoglobin crosslinked by bis(3,5-dibromosalicyl) fumarate using mass spectrometry techniques. *Protein Science*, 2568–2577.

138. Bunn, F.H., Esham, W.T., and Bull, R.W. (1969) The renal handling of hemoglobin. I. Glomerular filtration. *Journal of Experimental Medicine*, **129**, 909–924.

139. Standl, T. (2005) [Autologous transfusion—from euphoria to reason: Clinical practice based on scientific knowledge. (Part IV). Artificial oxygen carriers: Cell-free hemoglobin solutions—current status 2004]. *Anaesthesiologie, Intensivmedizin, Notfallmedizin, Schmerztherapie*, **40**, 38–45.

140. Vandegriff, K., Malavalli, A., Wooldridge, J., *et al.* (2003) MP4, a new nonvasoactive PEG-Hb conjugate. *Transfusion*, **43**, 509–516.

141. Schumacher, Y.O., Schmid, A., Dinkelmann, S., *et al.* (2001) Artificial oxygen carriers-the new doping threat in endurance sport? *International Journal of Sports Medicine*, **22**, 566–571.

142. Schumacher, Y.O., and Ashenden, M. (2004) Doping with artificial oxygen carriers: An update. *Sports Medicine*, **34**, 141–150.

143. Goebel, C., Alma, C., Howe, C., *et al.* (2005) Methodologies for detection of hemoglobin-based oxygen carriers. *Journal of Chromatographic Science*, **43**, 39–46.

144. Gasthuys, M., Alves, S., Fenaille, F., Tabet, J.C. (2004) Simple identification of a cross-linked hemoglobin by tandem mass spectrometry in human serum. *Analytical Chemistry*, **76**, 6628–6634.

145. Gasthuys, M., Alves, S., and Tabet, J. (2005) N-terminal adducts of bovine hemoglobin with glutaraldehyde in a hemoglobin-based oxygen carrier. *Analytical Chemistry*, **77**, 3372–3378.

146. Thevis, M., Ogorzalek-Loo, R.R., Loo, J.A., and Schänzer, W. (2003) Doping control analysis of bovine hemoglobin-based oxygen therapeutics in human plasma by LC-electrospray ionization-MS/MS. *Analytical Chemistry*, **75**, 3287–3293.

147. Alma, C., Trout, G., Woodland, N., and Kazlauskas, R. (2002) Detection of haemoglobin based oxygen carriers, in *Recent Advances in Doping Analysis* (eds W. Schänzer, H. Geyer, A. Gotzmann, and U. Mareck) Sport&Buch Strauß, Cologne, pp. 169–178.

148. Thevis, M., and Schänzer, W. (2005) Examples of doping control analysis by liquid chromatography-tandem mass spectrometry: Ephedrines, beta-receptor blocking agents, diuretics, sympathomimetics, and cross-linked hemoglobins. *Journal of Chromatographic Science*, **43**, 22–31.

149. Li, C.H., and Papkoff, H. (1956) Preparation and properties of growth hormone from human and monkey pituitary glands. *Science*, **124**, 1293–1294.

150. Baumann, G. (1991) Growth hormone heterogeneity: Genes, isohormones, variants, and binding proteins. *Endocrine Reviews*, **12**, 424–449.
151. DeNoto, F.M., Moore D.D., and Goodman, H.M. (1981) Human growth hormone DNA sequence and mRNA structure: Possible alternative splicing. *Nucleic Acids Research*, **9**, 3719–3730.
152. Masuda, N., Watahiki, M., Tanaka, M., *et al.* (1988) Molecular cloning of cDNA encoding 20 kDa variant human growth hormone and the alternative splicing mechanism. *Biochimica et Biophysica Acta*, **949**, 125–131.
153. Lewis, U.J., Singh, R.N., Bonewald, L.F., *et al.* (1979) Human growth hormone: Additional members of the complex. *Endocrinology*, **104**; 1256–1265.
154. Kohler, M., Thomas, A., Puschel, K., *et al.* (2009) Identification of human pituitary growth hormone variants by mass spectrometry. *Journal of Proteome Research*, **8**, 1071–1076.
155. Stolar, M.W., Amburn, K., and Baumann, G. (1984) Plasma "big" and "big-big" growth hormone (GH) in man: An oligomeric series composed of structurally diverse GH monomers. *Journal of Clinical Endocrinology and Metabolism*, **59**, 212–218.
156. Brostedt, P., Luthman, M., Wide, L., *et al.* (1990) Characterization of dimeric forms of human pituitary growth hormone by bioassay, radioreceptor assay, and radioimmunoassay. *Acta Endocrinologica (Copenhagen)*, **122**, 241–248.
157. Brostedt, P., and Roos, P. (1989) Isolation of dimeric forms of human pituitary growth hormone. *Preparative Biochemistry*, **19**, 217–229.
158. Singh, R.N., Seavey, B.K., and Lewis, U.J. (1974) Heterogeneity of human growth hormone. *Endocrine Research Communications*, **1**, 449–464.
159. Baumann, G. (1999) Growth hormone heterogeneity in human pituitary and plasma. *Hormone Research*, **51**, 2–6.
160. Ranke, M.B., Orskov, H., Bristow, A.F., *et al.* (1999) Consensus on how to measure growth hormone in serum. *Hormone Research*, **51** Suppl 1, 27–29.
161. Radetti, G., Buzi, F., Tonini, G., *et al.* (2004) Growth hormone (GH) isoforms following acute 22-kDa GH injection: Is it useful to detect GH abuse? *International Journal of Sports Medicine*, **25**, 205–208.
162. Bidlingmaier, M., Wu, Z., and Strasburger, C.J. (2001) Doping with growth hormone. *Journal of Pediatric Endocrinology & Metabolism*, **14**, 1077–1084.
163. Wallace, J.D., Cuneo, R.C., Bidlingmaier, M., *et al.* (2001) Changes in non-22-kilodalton (kDa) isoforms of growth hormone (GH) after administration of 22-kDa recombinant human GH in trained adult males. *Journal of Clinical Endocrinology and Metabolism*, **86**, 1731–1737.
164. Wu, Z., Bidlingmaier, M., Dall, R., and Strasburger, C.J. (1999) Detection of doping with human growth hormone. *Lancet*, **353**, 895.

165. Bidlingmaier, M., Wu, Z., and Strasburger, C.J. (2000) Test method: GH. *Bailliere's Best Practice & Research Clinical Endocrinology and Metabolism*, **14**, 99–109.
166. Momomura, S., Hashimoto, Y., Shimazaki, Y., and Irie, M. (2000) Detection of exogenous growth hormone (GH) administration by monitoring ratio of 20kDa- and 22kDa-GH in serum and urine. *Endocrine Journal*, **47**, 97–101.
167. Baumann, G., Stolar, M.W., and Amburn, K. (1985) Molecular forms of circulating growth hormone during spontaneous secretory episodes and in the basal state. *Journal of Clinical Endocrinology and Metabolism*, **60**, 1216–1220.
168. Baumann, G., Stolar, M.W., and Buchanan, T.A. (1986) The metabolic clearance, distribution, and degradation of dimeric and monomeric growth hormone (GH): Implications for the pattern of circulating GH forms. *Endocrinology*, **119**, 1497–1501.
169. Becker, G.W., Bowsher, R.R., Mackellar, W.C., *et al.* (1987) Chemical, physical, and biological characterization of a dimeric form of biosynthetic human growth hormone. *Biotechnology and Applied Biochemistry*, **9**, 478–487.
170. Lewis, U.J., Peterson, S.M., Bonewald, L.F., *et al.* (1977) An interchain disulfide dimer of human growth hormone. *Journal of Biological Chemistry*, **252**, 3697–3702.
171. Nagatomi, Y., Ikeda, M., Uchida, H., *et al.* (2000) Reversible dimerization of 20 kilodalton human growth hormone (hGH). *Growth Hormone & IGF Research*, **10**, 207–214.
172. Zhan, X., Giorgianni, F., and Desiderio, D.M. (2005) Proteomics analysis of growth hormone isoforms in the human pituitary. *Proteomics*, **5**, 1228–1241.
173. Grigorian, A., Bustamante, J., Hernandez, P., *et al.* (2005) Extraordinarily stable disulfide-linked homodimer of human growth hormone. *Protein Science*, **14**, 902–913.
174. Singh, R.N.P., Seavey, B.K., and Lewis, U.J. (1983) Human growth hormone peptide 1–43: Isolation from pituitary glands. *Journal of Protein Chemistry*, **2**, 525–536.
175. Jenkins, P. (1999) Growth hormone and exercise. *Clinical Endocrinology (Oxf)*, **50**, 683–689.
176. Bidlingmaier, M., Wu, Z., and Strasburger, C. (2003) Problems with GH doping in sports. *Journal of Endocrinological Investigation*, **26**, 924–931.
177. Graham, M.R., Baker, J.S., Evans, P., *et al.* (2009) Potential benefits of recombinant human growth hormone (rhGH) to athletes. *Growth Hormone & IGF Research*, **19**, 300–307.
178. Graham, M.R., Baker, J.S., Evans, P., *et al.* (2007) Short-term recombinant human growth hormone administration improves respiratory function in

abstinent anabolic-androgenic steroid users. *Growth Hormone & IGF Research*, **17**, 328–335.

179. Graham, M.R., Baker, J.S., Evans, P., *et al.* (2008) Physical effects of short-term recombinant human growth hormone administration in abstinent steroid dependency. *Hormone Research*, **69**, 343–354.

180. Graham, M.R., Davies, B., Kicman, A., *et al.* (2007) Recombinant human growth hormone in abstinent androgenic-anabolic steroid use: Psychological, endocrine and trophic factor effects. *Current Neurovascular Research*, **4**, 9–18.

181. Graham, M.R., Evans, P., Davies, B., and Baker, J.S. (2008) AAS, growth hormone, and insulin abuse: Psychological and neuroendocrine effects. *Therapeutics and Clinical Risk Management*, **4**, 587–597.

182. Berggren, A., Ehrnborg, C., Rosen, T., *et al.* (2005) Short-term administration of supraphysiologic rhGH does not increase maximum endurance exercise capacity in healthy, active young men and women with normal GH-IGF-1 axes. *Journal of Clinical Endocrinology and Metabolism*, DOI: 10.1210/jc.2004–1209.

183. Ehrnborg, C., Ellegard, L., Bosaeus, I., *et al.* (2005) Supraphysiological growth hormone: Less fat, more extracellular fluid but uncertain effects on muscles in healthy, active young adults. *Clinical Endocrinology (Oxf)*, **62**, 449–457.

184. Healy, M.L., Gibney, J., Russell-Jones, D.L., *et al.* (2003) High dose growth hormone exerts an anabolic effect at rest and during exercise in endurance-trained athletes. *Journal of Clinical Endocrinology and Metabolism*, **88**, 5221–5226.

185. Wu, S., Jardine, I., Hancock, W., and Karger, B. (2004) A new and sensitive on-line liquid chromatography/mass spectrometric approach for top-down protein analysis: The comprehensive analysis of human growth hormone in an E. coli lysate using a hybrid linear ion trap/Fourier transform ion cyclotron resonance mass spectrometer. *Rapid Communications in Mass Spectrometry*, **18**, 2201–2207.

186. Giorgianni, F., Beranova-Giorgianni, S., and Desiderio, D.M. (2004) Identification and characterization of phosphorylated proteins in the human pituitary. *Proteomics*, **4**, 587–598.

187. Singh, R.N.P., Seavey, B.K., and Lewis, U.J. (1974) Heterogeneity of human growth hormone. *Endocrine Research Communications*, **1**, 449–464.

188. Brostedt, P., and Roos, P. (1989) Isolation of dimeric forms of human pituitary growth hormone. *Preparative Biochemistry*, **19**, 217–229.

189. Thevis, M., Ogorzalek-Loo, R.R., Loo, J.A., *et al.* (2005) Probing for characteristics of pituitary growth hormone by LC-MS/MS, in *Recent Advances in Doping Analysis* (eds W. Schänzer, G. Geyer, A. Gotzmann, and U. Mareck), Sport&Buch Strauss, Cologne, pp. 355–366.

190. Thevis, M., and Schänzer, W. (2005) Identification and characterization of peptides and proteins in doping control analysis. *Current Proteomics*, **2**, 191–208.
191. Lewis, U.J., Singh, R.N., Bonewald, L.F., and Seavey, B.K. (1981) Altered proteolytic cleavage of human growth hormone as a result of deamidation. *Journal of Biological Chemistry*, **256**, 11645–11650.
192. Wu, S.L., Jiang, H., Lu, Q., *et al.* (2009) Mass spectrometric determination of disulfide linkages in recombinant therapeutic proteins using online LC-MS with electron-transfer dissociation. *Analytical Chemistry*, **81**, 112–122.
193. Hepner, F., Cszasar, E., Roitinger, E., and Lubec, G. (2005) Mass spectrometrical analysis of recombinant human growth hormone (Genotropin®) reveals amino acid substitutions in 2% of the expressed protein. *Proteome Science*, **3**, 1.
194. Saugy, M., Cardis, C., Schweizer, C., *et al.* (1996) Detection of human growth hormone doping in urine: Out of competition tests are necessary. *Journal of Chromatography B*, **687**, 201–211.
195. Ankersen, M., Johansen, N.L., Madsen, K., *et al.* (1998) A new series of highly potent growth hormone-releasing peptides derived from ipamorelin. *Journal of Medicinal Chemistry*, **41**, 3699–3704.
196. Corpas, E., Harman, S.M., and Blackman, M.R. (1993) Human growth hormone and human aging. *Endocrine Reviews*, **14**, 20–39.
197. Bowers, C.Y., Reynolds, G.A., and Momany, F.A. (1984) New advances on the regulation of growth hormone (GH) secretion. *International Journal of Neurology*, **18**, 188–205.
198. Bowers, C.Y., Reynolds, G.A., Chang, D., *et al.* (1981) A study on the regulation of growth hormone release from the pituitaries of rats in vitro. *Endocrinology*, **108**, 1071–1080.
199. Momany, F.A., Bowers, C.Y., Reynolds, G.A., *et al.* (1981) Design, synthesis, and biological activity of peptides which release growth hormone in vitro. *Endocrinology*, **108**, 31–39.
200. Bowers, C.Y., Momany, F., Reynolds, G.A., *et al.* (1980) Structure-activity relationships of a synthetic pentapeptide that specifically releases growth hormone in vitro. *Endocrinology*, **106**, 663–667.
201. Momany, F.A., Bowers, C.Y., Reynolds, G.A., *et al.* (1984) Conformational energy studies and in vitro and in vivo activity data on growth hormone-releasing peptides. *Endocrinology*, **114**, 1531–1536.
202. Bowers, C.Y., Momany, F.A., Reynolds, G.A., and Hong, A. (1984) On the in vitro and in vivo activity of a new synthetic hexapeptide that acts on the pituitary to specifically release growth hormone. *Endocrinology*, **114**, 1537–1545.
203. Chapman, I.M., Pescovitz, O.H., Murphy, G., *et al.* (1997) Oral administration of growth hormone (GH) releasing peptide-mimetic MK-677 stimu-

lates the GH/insulin-like growth factor-I axis in selected GH-deficient adults. *Journal of Clinical Endocrinology and Metabolism*, **82**, 3455–3463.

204. Ghigo, E., Arvat, E., and Camanni, F. (1998) Orally active growth hormone secretagogues: State of the art and clinical perspectives. *Annals of Medicine*, **30**, 159–168.
205. Rinderknecht, E., and Humbel, R.E. (1976) Polypeptides with nonsuppressible insulin-like and cell-growth promoting activities in human serum: Isolation, chemical characterization, and some biological properties of forms I and II. *Proceedings of the National Academy of Sciences of the United States of America*, **73**, 2365–2369.
206. Rinderknecht, E., and Humbel, R.E. (1978) The amino acid sequence of human insulin-like growth factor I and its structural homology with proinsulin. *Journal of Biological Chemistry*, **253**, 2769–2776.
207. Rinderknecht, E., and Humbel, R.E. (1978) Primary structure of human insulin-like growth factor II. *FEBS Letters*, **89**, 283–286.
208. Raschdorf, F., Dahinden, R., Maerki, W., *et al.* (1988) Location of disulphide bonds in human insulin-like growth factors (IGFs) synthesized by recombinant DNA technology. *Biomedical and Environmental Mass Spectrometry*, **16**, 3–8.
209. Sato, A., Nishimura, S., Ohkubo, T., *et al.* (1993) Three-dimensional structure of human insulin-like growth factor-I (IGF-I) determined by 1H-NMR and distance geometry. *International Journal of Peptide and Protein Research*, **41**, 433–440.
210. Dubaquie, Y., Mortensen, D.L., Intintoli, A., *et al.* (2001) Binding protein-3-selective insulin-like growth factor I variants: Engineering, biodistributions, and clearance. *Endocrinology*, **142**, 165–173.
211. Abbaspour, A., Takata, S., Matsui, Y., *et al.* (2008) Continuous infusion of insulin-like growth factor-I into the epiphysis of the tibia. *International Orthopaedics*, **32**, 395–402.
212. Yakar, S., Rosen, C.J., Beamer, W.G., *et al.* (2002) Circulating levels of IGF-1 directly regulate bone growth and density. *Journal of Clinical Investigation*, **110**, 771–781.
213. Adams, G.R. (2000) Insulin-like growth factor in muscle growth and its potential abuse by athletes. *British Journal of Sports Medicine*, **343**, 412–413.
214. Adams, G.R., and McCue, S.A. (1998) Localized infusion of IGF-I results in skeletal muscle hypertrophy in rats. *Journal of Applied Physiology*, **84**, 1716–1722.
215. Weiss, J.M., Huang, W.Y., Rinaldi, S., *et al.* (2007) IGF-1 and IGFBP-3: Risk of prostate cancer among men in the Prostate, Lung, Colorectal and Ovarian Cancer Screening Trial. *International Journal of Cancer*, **121**, 2267–2273.

216. Gennigens, C., Menetrier-Caux, C., and Droz, J.P. (2006) Insulin-like growth factor (IGF) family and prostate cancer. *Critical Reviews in Oncology & Hematology*, **58**, 124–145.

217. Fletcher, O., Gibson, L., Johnson, N., *et al.* (2005) Polymorphisms and circulating levels in the insulin-like growth factor system and risk of breast cancer: A systematic review. *Cancer Epidemiology, Biomarkers and Prevention*, **14**, 2–19.

218. Sandhu, M.S., Dunger, D.B., and Giovannucci, E.L. (2002) Insulin, insulin-like growth factor-I (IGF-I), IGF binding proteins, their biologic interactions, and colorectal cancer. *Journal of the National Cancer Institute*, **94**, 972–980.

219. Chernausek, S.D., Backeljauw, P.F., Frane, J., *et al.* (2007) Long-term treatment with recombinant insulin-like growth factor (IGF)-I in children with severe IGF-I deficiency due to growth hormone insensitivity. *Journal of Clinical Endocrinology & Metabolism*, **92**, 902–910.

220. Bhatt, J.M., and Gordon, P.H. (2007) Current clinical trials in amyotrophic lateral sclerosis. *Expert Opinion in Investigational Drugs*, **16**, 1197–1207.

221. Bagley, C.J., May, B.L., Szabo, L., *et al.* (1989) A key functional role for the insulin-like growth factor 1 N-terminal pentapeptide. *Biochemical Journal*, **259**, 665–671.

222. Prosser, C.G., Davis, S.R., Hodgkinson, S.C., and Mohler, M.A. (1995) Pharmacokinetics and bioactivity of intact versus truncated IGF-I during a 24-h infusion into lactating goats. *Journal of Endocrinology*, **144**, 99–107.

223. King, R., Wells, J.R., Krieg, P., *et al.* (1992) Production and characterization of recombinant insulin-like growth factor-I (IGF-I) and potent analogues of IGF-I, with Gly or Arg substituted for Glu3, following their expression in Escherichia coli as fusion proteins. *Journal of Molecular Endocrinology*, **8**, 29–41.

224. Voorhamme, D., and Yandell, C.A. (2006) LONG R3IGF-I as a more potent alternative to insulin in serum-free culture of HEK293 cells. *Molecular Biotechnology*, **34**, 201–204.

225. Ballard, F.J., Knowles, S.E., Walton, P.E., *et al.* (1991) Plasma clearance and tissue distribution of labelled insulin-like growth factor-I (IGF-I), IGF-II and des(1-3)IGF-I in rats. *Journal of Endocrinology*, **128**, 197–204.

226. Tomas, F.M., Walton, P.E., Dunshea, F.R., and Ballard, F.J. (1997) IGF-I variants which bind poorly to IGF-binding proteins show more potent and prolonged hypoglycaemic action than native IGF-I in pigs and marmoset monkeys. *Journal of Endocrinology*, **155**, 377–386.

227. Francis, G.L., Ross, M., Ballard, F.J., *et al.* (1992) Novel recombinant fusion protein analogues of insulin-like growth factor (IGF)-I indicate the relative importance of IGF-binding protein and receptor binding for enhanced biological potency. *Journal of Molecular Endocrinology*, **8**, 213–223.

228. Ballard, F.J., Francis, G.L., Ross, M., *et al.* (1987) Natural and synthetic forms of insulin-like growth factor-1 (IGF-1) and the potent derivative, destripeptide IGF-1: Biological activities and receptor binding. *Biochemical & Biophysical Research Communications*, **149**, 398–404.

229. Bastian, S.E., Walton, P.E., Wallace, J.C., and Ballard, F.J. (1993) Plasma clearance and tissue distribution of labelled insulin-like growth factor-I (IGF-I) and an analogue LR3IGF-I in pregnant rats. *Journal of Endocrinology*, **138**, 327–336.

230. Loo, J.A., Edmonds, C.G., and Smith, R.D. (1993) Tandem mass spectrometry of very large molecules. 2. Dissociation of multiply charged proline-containing proteins from electrospray ionization. *Analytical Chemistry*, **65**, 425–438.

231. Casper, R.F. (1991) Clinical uses of gonadotropin-releasing hormone analogues. *Canadian Medical Association Journal*, **144**, 153–158.

232. DiLuigi, A.J., and Nulsen, J.C. (2007) Effects of gonadotropin-releasing hormone agonists and antagonists on luteal function. *Current Opinion in Obstetrics & Gynecology*, **19**, 258–265.

233. Myers, T.R., and Patonay, G. (2006) A new strategy utilizing electrospray ionization-quadrupole ion trap mass spectrometry for the qualitative determination of GnRH peptides. *Journal of Mass Spectrometry*, **41**, 950–959.

234. Matsuo, H., Baba, Y., Nair, R.M., *et al.* (2002) Structure of the porcine LH- and FSH-releasing hormone. I. The proposed amino acid sequence. 1991. *Journal of Urology*, **167**, 1011–1014.

235. Mulligan, T., Iranmanesh, A., Kerzner, R., *et al.* (1999) Two-week pulsatile gonadotropin releasing hormone infusion unmasks dual (hypothalamic and Leydig cell) defects in the healthy aging male gonadotropic axis. *European Journal of Endocrinology*, **141**, 257–266.

236. Tilbrook, A.J., and Clarke, I.J. (1995) Negative feedback regulation of the secretion and actions of GnRH in male ruminants. *Journal of Reproduction and Fertility Supplement*, **49**, 297–306.

237. van Breda, E., Keizer, H.A., Kuipers, H., and Wolffenbuttel, B.H. (2003) Androgenic anabolic steroid use and severe hypothalamic-pituitary dysfunction: A case study. *International Journal Sports Medicine*, **24**, 195–196.

238. Veldhuis, J.D., and Iranmanesh, A. (2004) Pulsatile intravenous infusion of recombinant human luteinizing hormone under acute gonadotropin-releasing hormone receptor blockade reconstitutes testosterone secretion in young men. *Journal of Clinical Endocrinology and Metabolism*, **89**, 4474–4479.

239. Peterson, J.E., and Nett, T.M. (1976) Clearance of gonadotropin-releasing hormone in beef heifers after intramuscular or intravenous administration. *Journal of Animal Science*, **43**, 1264–1269.

240. Yesalis, C.E., and Bahrke, M.S. (2000) Doping among adolescent athletes. *Bailliere's Best Practice & Research. Clinical Endocrinology and Metabolism*, **14**, 25–35.
241. Pascual, J., Belalcazar, V., de Bolos, C., *et al.* (2004) Recombinant erythropoietin and analogues: A challenge for doping control. *Therapeutic Drug Monitoring*, **26**, 175–179.
242. Apte, S.S. (2008) Blood substitutes—The polyheme trials. *McGill Journal of Medicine*, **11**, 59–65.
243. Napolitano, L.M. (2009) Hemoglobin-based oxygen carriers: First, second or third generation? Human or bovine? Where are we now? *Critical Care Clinics*, **25**, 279–301.
244. Thevis, M., and Schänzer, W. (2007) Mass spectrometry in sports drug testing: Structure characterization and analytical assays. *Mass Spectrometry Reviews*, **26**, 79–107.

6 Modern Mass Spectrometry-Based Analytical Assays

Mass spectrometry has become the method of choice in modern sports drug testing for numerous target analytes of low and high molecular weights and known as well as unknown structures. The continuous improvements of analytical instrumentations, chromatography, ion sources and interfaces, mass accuracy and resolution, as well as the sensitivity and speed of measurements concurrently allow the development of new or enhanced, comprehensive, fast, and sensitive doping control analytical assays. In the following, a selection of currently employed procedures that indicate the misuse of banned substances and/or enable the detection and identification of prohibited compounds in sports drug testing from blood or urine samples is described. Although doping control laboratories are required to follow strict rules as established by WADA and stipulated in respective documents such as the "International Standard for Laboratories (ISL),"[1] "Minimum Required Performance Limits for Detection of Prohibited Substances,"[2] or "Identification Criteria for Qualitative Assays Incorporating Chromatography and Mass Spectrometry,"[3] a certain autonomy with regard to analytical procedures is provided as long as the validity of an established procedure was demonstrated according to internationally accepted guidelines. Consequently, various different approaches toward comprehension, cost-effectiveness, sensitivity, specificity, etc., commonly designed to fit the equipment of respective laboratories, are found; however, all have to prove their "fitness for purpose" in frequent external quality assurance testing programs. Due to the enormous variety of methods available, only few typical examples are presented here.

Mass Spectrometry in Sports Drug Testing: Characterization of Prohibited Substances and Doping Control Analytical Assays, By Mario Thevis
Copyright © 2010 John Wiley & Sons, Inc.

6.1 GC-MS AND ISOTOPE RATIO MASS SPECTROMETRY

Although GC-MS has been losing its monopoly that it held before LC-MS(/MS) was routinely available to doping control laboratories, a number of detection assays still rely on this technique due to the considerable resolving power of GC columns, the robustness of EI sources, and the comparably low susceptibility to matrix effects, as well as the flexibility to interface various different analyzers to GC systems.[4,5]

6.1.1 Stimulants/Narcotics

One of the first assays including GC-MS was established to target stimulants[6] (see also Chapter 1) and was further developed to comprehensively and sensitively screen for nitrogen-containing drugs in general (commonly referred to as *Screening I* for volatile nitrogen-containing compounds).[7] The principle of extracting the basic alkaloids from urine at elevated pH into an organic solvent[8] (usually diethyl ether or *tert.*-butyl methyl ether, TBME)[9–12] or by means of a solid-phase extraction cartridge proved suitable for a variety of relevant and volatile compounds (later including also narcotics),[13–26] which have subsequently been measured either without or with derivatization using a dedicated GC-MS or GC-MS/nitrogen phosphorus detector (NPD) system.[27–31] The underlying considerations are still valid and various routine doping control laboratories measure a great share of stimulants as well as selected narcotics using GC-MS-based approaches.

A typical procedure aiming the analysis of drugs and metabolites excreted unconjugated into urine is shown in Scheme 6.1.[7] To a volume of 5 mL of urine, 10–25 µg of the internal standard(s) are added (e.g., dodecyl diisopropylamine [DIPA 12],[32] diphenylamine, or 7-ethyltheophylline), and the pH is adjusted to 13–14 by adding 0.5 mL of aqueous potassium hydroxide (5 M). Following the addition of 2 mL of TBME and approximately 3 g of sodium sulfate, the sample is shaken for 20 minutes and subsequently centrifuged at ca. 660 × *g* for 10 minutes. The organic layer is transferred to an autosampler vial and 3–5 µL are injected into GC-MS(/NPD) systems.[31] The analytical setup commonly consists of a GC oven interfaced to a quadrupole MS or NPD analyzer, which can also be used simultaneously in case of GC systems containing two analytical columns that connect the injection port with either of the detectors. A recently described GC-MS/NPD combination included two HP5MS columns (inner diameter: 0.25 mm, film thickness: 0.25 µm), one of which was 24 m (directed to the mass spectrometer) and the other 19 m (connected to the NPD) long.[33] The difference in length was

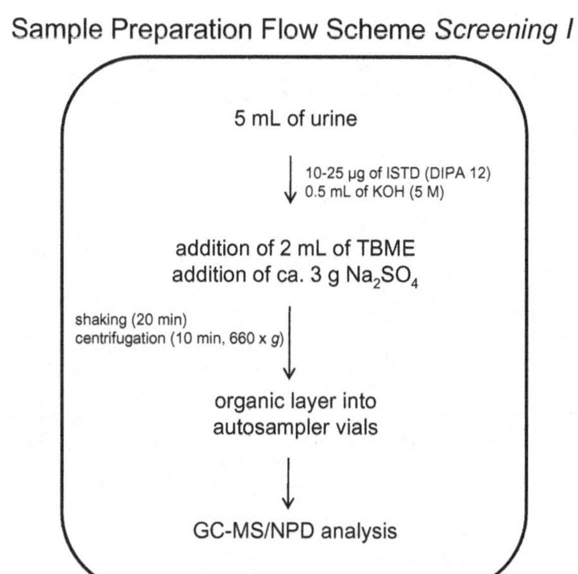

Scheme 6.1: Sample preparation flow scheme for the analysis of volatile drugs (mainly stimulants and narcotics) excreted unconjugated into urine. The screening procedure is commonly referred to as *Screening I*.[7]

due to different pressure conditions in MS and NPD in order to accomplish comparable retention times of the analytes. The GC carrier gas is usually helium operated at constant pressure (approximately 12 psi), and a temperature gradient of the GC commonly ranges from 85–330°C @ 28–35°C/min. The mass spectrometer is operated in EI mode with full scan analysis (*m/z* 40–400, 4 scans/s).

Due to the fact that most therapeutics are nitrogen-containing compounds, the combination of GC-MS and GC-NPD has proved considerable comprehensiveness also in doping controls. While many mass spectrometric procedures are optimized for selected target analytes (*vide infra*), the screening for known and unknown drugs particularly concerning the class of stimulants and narcotics is efficiently done using the combined MS/NPD detection system. Typical chromatograms obtained from a spiked quality control sample are depicted in Figure 6.1 with the upper window containing the NPD-signal and the lower window the total ion chromatogram recorded by the mass spectrometer. The mixture of stimulants (e.g., amphetamine, methamphetamine, ephedrine, nikethamide, phenmetrazine, pentylenetetrazol) yields abundant signals at the minimum required performance level (MRPL)

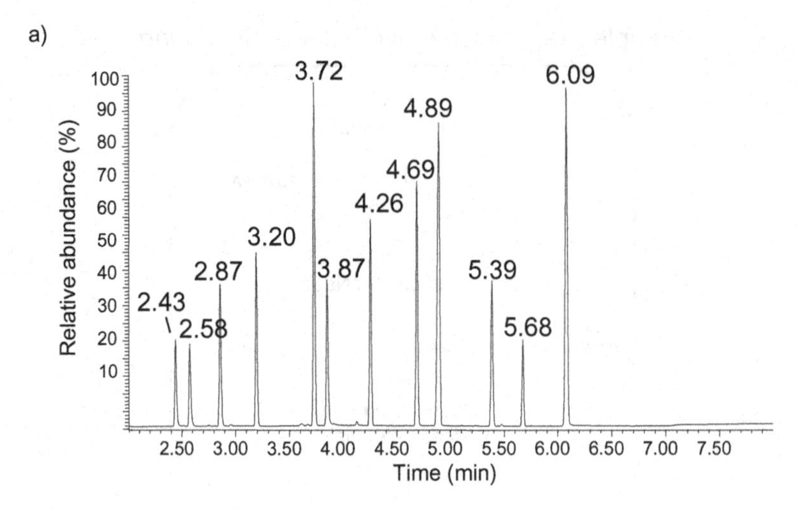

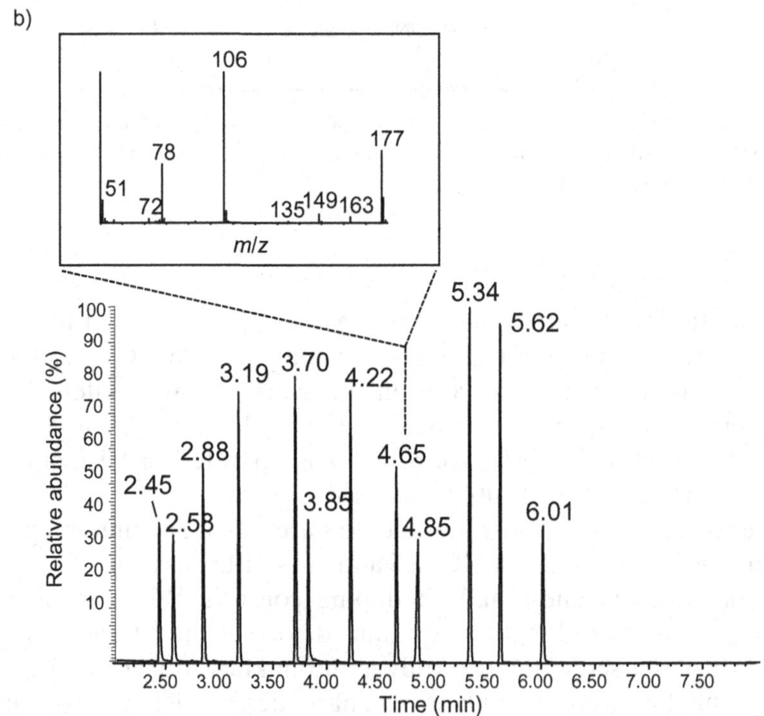

Figure 6.1: NPD (a) and MS (b) chromatograms of a quality control sample spiked with 11 analytes (heptaminol: 2.43 min; amphetamine: 2.58 min; methamphetamine: 2.87 min; N,N-dimethylamphetamine: 3.20 min; nicotine: 3.72 min; ephedrine: 3.87 min; phenmetrazine: 4.26 min; nikethamide 4.69 min; pentylenetetrazol: 4.89 min; fencamfamine: 5.39 min; and caffeine: 6.09 min) plus the internal standard DIPA 12 (5.68 min). The retention times of the added compounds align properly in both chromatograms, which allow the correlation of NPD signals with respective mass spectra.

of 500 ng/mL in both analysis modes, and full scan mass spectra enable the rapid identification of respective compounds. These and numerous additional target analytes, which include also a variety of metabolites, are detected as compiled e.g., in a review by Hemmersbach and de la Torre.[31]

In order to expand the group of target analytes toward heavy volatile substances and (conjugated) metabolites of target compounds, the simple and rapid screening procedure described above was modified to include hydrolysis of phase-II-metabolites (e.g., glucuronides and sulfates) before extraction from urine specimens, and chemical derivatization for improved chromatographic properties.[7] The hydrolysis of glucuronic acid and sulfate conjugates proved critical and complex; hence, two different approaches based on either enzymatic cleavage (using β-glucuronidase or a mixture of β-glucuronidase and aryl sulfatase) or acid hydrolysis are used. The latter option employs comparably harsh conditions, which, on the one hand, supposedly ensure efficient deconjugation but, on the other hand, might initiate degradation processes to target compounds. In contrast, the enzymatic cleavage is commonly referred to as "soft" but presumably does not guarantee comprehensive hydrolysis of phase-II-metabolites. Depending on the target analyte and type of conjugation, the issue has been controversial, e.g., for morphine and codeine in particular;[34,35] however, several strategies are employed complementary, which have been used as *Screening II* for several decades (Scheme 6.2).[7,31] Here, urine aliquots of 2.5–5 mL are fortified with an internal standard (ISTD) and 100 mg of cysteine (as antioxidant) and acidified with 0.5–1 mL of 6 M aqueous HCl. After incubation at 100°C for 30 minutes, acidic analytes (which were not of particular interest) are removed by extraction with 5 mL of ether, and the aqueous layer is subsequently adjusted to pH 9.6 using a solution of potassium hydroxide (6 M) and a mixture of potassium carbonate/sodium hydrogen carbonate. By means of 1 mL of *tert.*-butanol and 5 mL of TBME as well as approximately 3 g of sodium sulfate, the liberated phase-I-metabolites are extracted into the organic layer, which is subsequently evaporated to dryness (Scheme 6.2a).[7] Alternatively, the urine sample aliquot is not acid hydrolyzed but buffered to pH 5.2 and fortified with 50 μL of β-glucuronidase/aryl sulfatase from *Helix pomatia* and heated to 55°C for 2–3 hours. The aqueous layer is then subjected to either LLE (using the above described conditions) or SPE,[7,30,31] and the final extracts are evaporated to dryness. Derivatization of analytes is subsequently required, which either includes trimethylsilylation (TMS) only or selective derivatization including trimethylsilylation of hydroxyl functions and trifluoroacetylation (TFA) of amino residues

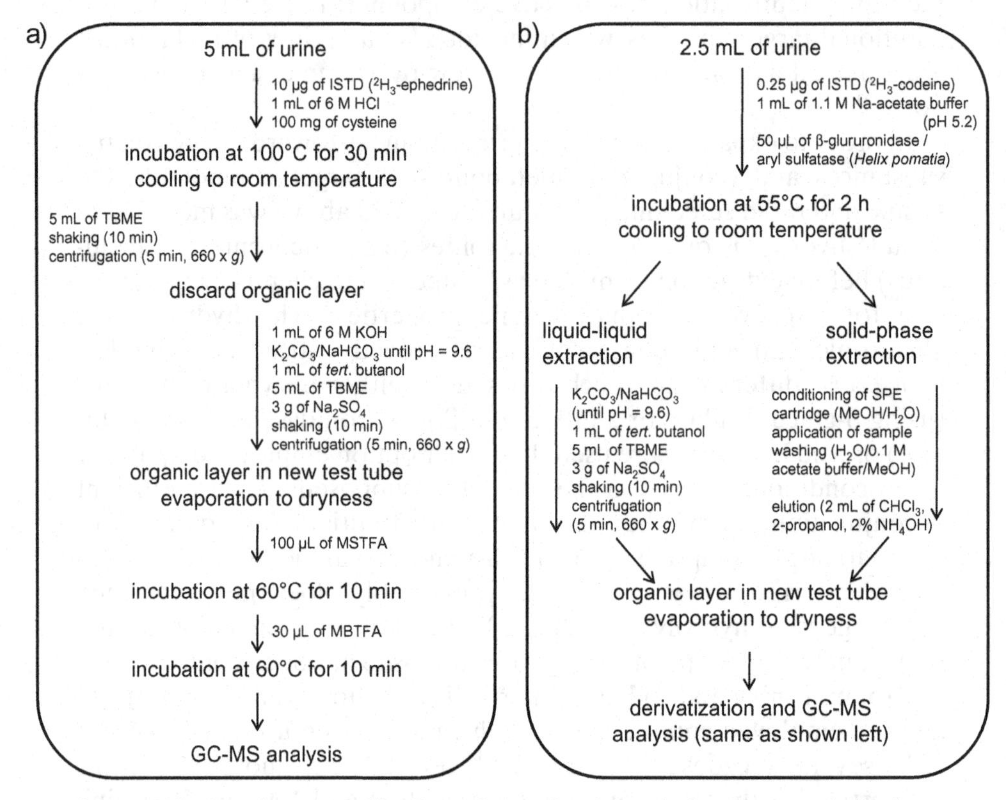

Scheme 6.2: Sample preparation flow schemes for the analysis of heavy volatile drugs (mainly stimulants, narcotics, and β-blockers) excreted conjugated into urine. The screening procedures have commonly been referred to as *Screening II* with a) acidic hydrolysis and b) enzymatic cleavage of conjugates.[7,31]

(Scheme 6.2b).[27–29] While TMS derivatives are readily formed by the addition of *N*-methyl-*N*-trimethylsilyl-trifluoroacetamide (MSTFA) and incubation for 10 minutes at 60°C, the substitution of TMS-groups by TFA residues at amino functions is accomplished by subsequent addition of *N*-methyl-bistrifluoroacetamide (MBTFA) and heating to 60°C for another 10 minutes.

The following GC-MS analysis is commonly done using a 12–16 minute HP5MS capillary GC column (inner diameter: 0.2 mm, film thickness: 0.33 µm) with helium as carrier gas operated at constant pressure (approximately 12 psi) with 0.8–1.0 mL/min. The temperature gradient of the GC usually starts from 100°C to 300°C with 15°C/min. The mass spectrometer is operated in EI mode with full scan or selected ion monitoring (SIM) detection.

Due to the option to omit derivatization of analytes when using LC-MS/MS procedures, several analytes of the above mentioned screening procedures were implemented in other, alternative assays; however, based on these sample preparation principles, also very recent GC-MS-based screening procedures were established, combining the goals of *Screening I, II*, and *IV* (see section 6.1.2).[36,37] A comprehensive approach covers more than 150 compounds at the required MRPL, and consists of two separate urine aliquot preparations that are combined prior to derivatization and GC-MS analysis. In detail, 1 and 3 mL of urine are prepared for analysis with the 1 mL-aliquot being treated in accordance to the procedure of Screening I, i.e., 1 mL of aqueous KOH (5 M) and 1 g of NaCl are added followed by extraction of the substances of interest into 1 mL of TBME. Rolling test tubes for 20 minutes is followed by centrifugation at 1200 × *g* and separation of the organic layer. The 3 mL aliquot is adjusted to pH 7 by addition of 1 mL of phosphate buffer, and conjugated compounds are hydrolyzed by means of 50 µL of β-glucuronidase from *E. coli* (K12) at 42°C overnight. The sample is further buffered to pH 9.5 and extracted with 5 mL of diethyl ether (20 minutes of rolling, 5 minutes of centrifugation at 1200 × *g*). Both organic layers are combined, evaporated to dryness, and derivatized using *in situ* generated trimethyliodosilane (TMIS, prepared from MSTFA, ammonium iodide, and ethanethiol) for 1 hour at 80°C (Scheme 6.3).[37] The employed GC-MS system consists of an Agilent 6890/5975 instrument equipped with a J&W Ultra-1 column (length: 17 m, inner diameter: 0.2 mm, film thickness: 0.11 µm) and helium as carrier gas provided at constant flow (0.6 mL/min). The temperature gradient starts at 70°C increasing to 100°C with 90°C/min, followed by an increase of 30°C/min until 180°C are reached, then 3°C/min to 232°C, and finally 40°C/min to 310°C. A total of 0.5 µL of the sample solution is injected in splitless mode, and the MS is operated using EI and SIM/scan options. The combined extraction of analytes with and without hydrolysis followed by derivatization allows for a comprehensive urine analysis, which is particularly informative due to the recording of full scan data. These can serve for retrospective analyses in cases when re-evaluations of urine samples regarding formerly unknown substances are required.

Case Vignette: 4-Methylhexan-2-amine The great advantage of analytical procedures such as Screening I, II, and methods having evolved from these is the general and comprehensive nature. This allowed for

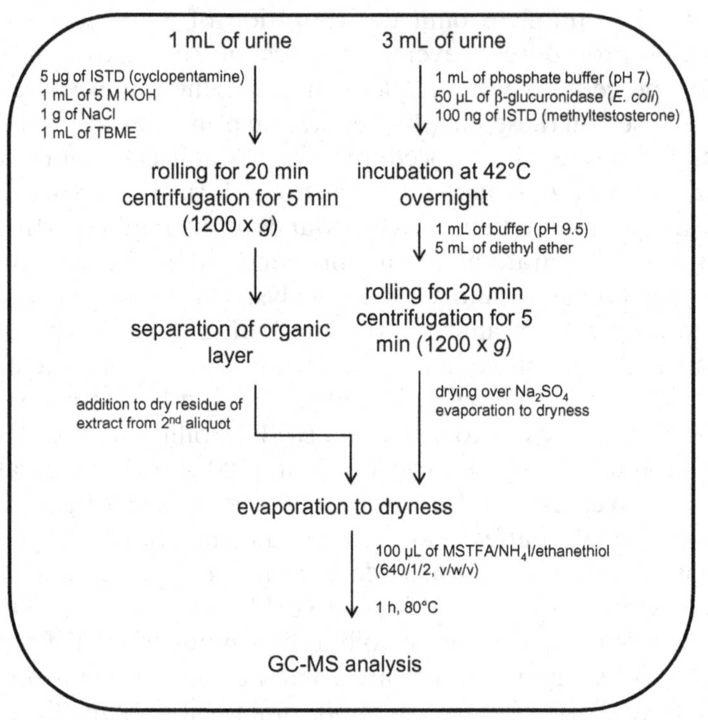

Scheme 6.3: Alternative sample preparation flow scheme for the combined analysis of various drugs (stimulants, narcotics, steroids, β_2-agonists, etc.) excreted conjugated and unconjugated into urine.[37]

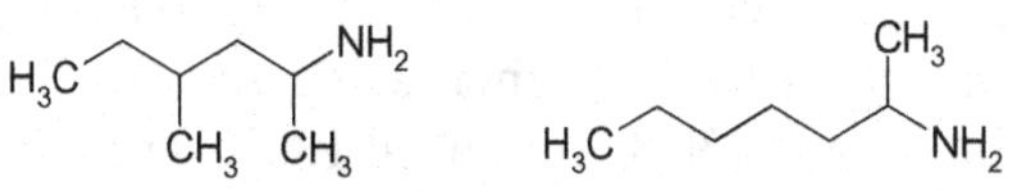

Figure 6.2: Structures of two banned stimulants 4-methylhexan-2-amine (**1**) and tuaminoheptane (**2**).

the detection of a drug termed 4-methylhexan-2-amine at several occasions in 2009. 4-Methylhexan-2-amine (geranamine, Fig. 6.2, **1**) is a natural product produced to a minor extent in *Pelargonium graveolens* (also referred to as *geranium or Pelargonium*), a plant that is indigenous particularly to South Africa. *Pelargonium* is largely cultivated due to the great interest in its foliage that is used for the preparation of various different scents, which are derived from an oily distillate that contains approximately 0.7% of 4-methylhexan-2-amine. The oil has been approved as food additive, and "*geranium* extracts" are frequently declared as ingredients of nutritional supplements and so-called party pills. In addition, 4-methylhexan-2-amine was synthetically obtained and prepared and patented in 1944 as pharmaceutical product, which was marketed as nasal decongestant (Forthane sulphate) and therapeutic agent for the treatment of hypertrophied gums. Although it was supposed to be less stimulating than drugs such as amphetamine or ephedrine, typical sympathomimetic effects such as tremor, excitement, or insomnia were reported. Its advantages over amphetamine and ephedrine were a greater volatility and reduced toxicity, and its efficacy was greater than that of heptylamine derivatives such as tuaminoheptane (Fig. 6.2, **2**).[38] In January 2007, WADA added substances such as tuaminoheptane to the Prohibited List, which was since monitored in all in-competition samples.

In 2009, several doping control urine specimens were found suspicious for a drug closely related to tuaminoheptane according to *Screening I* GC-MS and furthermore LC-MS/MS data. Retention times did not match the reference of tuaminoheptane, but due to the generic nature of the GC-MS/NP-based analytical method, the "unknown" drug was observed, and further studies led to the identification of 4-methylhexan-2-amine using authentic reference material and dedicated detection assays based on GC-MS[33] and LC-MS/MS.[39] In brief, the analyte of interest was extracted from urine according to *Screening I* and subsequently derivatized to Schiff-bases by adding a methanolic solution of benzaldehyde to the final ether extract. After 30 minutes without further treatment, samples were injected into the GC-MS/NPD system yielding abundant signals for 4-methylhexan-2-amine at 4.27 and 4.32 minutes, indicating the presence of diastereomers as also reported earlier.[39] The mass spectrometric fragmentation of 4-methylhexan-2-amine as benzaldehyde derivative was in accordance with earlier reported dissociation pathways.[33] Imines such as the derivatized target analytes are likely ionized by EI at the nitrogen atom due to the electron-donating nature of the amino function.[40] The resulting radical cations can subsequently undergo isomerization triggered by intramolecular hydrogen abstraction and formation of so-called distonic ions.[41] Hence, besides commonly observed and characteristic α-cleavage products,[42] complex cascades of rearrangements were described to precede the dissociation of amines, which allowed for the explanation of frequently detected additional fragment ions in EI mass spectra of aliphatic amines. The molecular ion of the condensation product of 4-methylhexan-2-amine at m/z 203 is hardly visible, which is a well

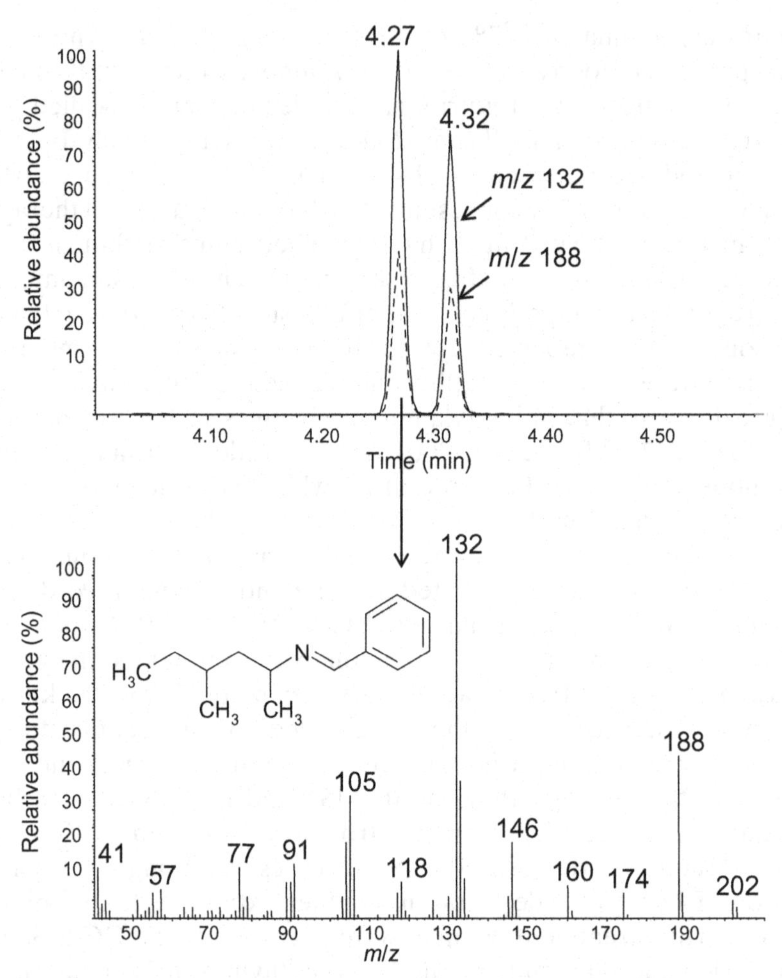

Figure 6.3: Extracted ion chromatogram (top) of a urine sample tested positive for the two stereoisomers of 4-methylhexan-2-amine as Schiff-base derivatives using benzaldehyde (benzylidene-(1,3-dimethylpentyl)-amine) with corresponding EI mass spectrum (bottom, mol wt = 203).

known issue of EI-MS-based assays for primary and secondary amines. However, abundant and diagnostic ions originating from α-cleavages are found at *m/z* 188 and 132 (Fig. 6.3), suggested to represent the cations of benzylidene-(3-methyl-pentyl)-amine and benzylidene-ethylamine, respectively. These are complemented by losses of alkyl radicals such as ethyl-, propyl- and butyl-residues from the molecular ion to yield the fragments at *m/z* 174, 160, and 146, which support the MS-based identification of 4-methylhexan-2-amine.

6.1.2 Anabolic Androgenic Steroids

The class of anabolic androgenic steroids has probably been the most challenging subset of analytes in doping controls due to the existence of dozens of fully synthetic but also natural steroids, both of which have been subject of misuse for many decades.[43] The major goals for the detection of xenobiotic steroids are utmost retrospect (e.g., by measuring long-term metabolites) and sensitivity; however, there are also misused synthetic counterparts of endogenous steroids (referred to as "natural steroids" in the following) such as testosterone, dehydroepiandrosterone (DHEA), dihydrotestosterone (DHT) or androstenedione, and the crucial aspect is to differentiate between the structurally identical compounds of endogenous and exogenous origin. Except for a few steroids, including stanozolol, tetrahydrogestrinone, and trenbolone, most AAS and, particularly, their metabolites, are still preferably analyzed by GC-MS-based assays.[44–50]

6.1.2.1 Synthetic Anabolic Androgenic Steroids The limited oral bioavailability of testosterone[51] as well as the need to differentiate desired anabolic from undesired androgenic effects[52] and corresponding health issues[53–55] have led to the investigation of thousands of potential AAS in the past,[56] but only a minor percentage of these has finally advanced to pharmaceutical and medicinal applications. In addition to numerous chemical modifications applied to emphasize the aimed anabolic effects and improve pharmacological profiles, alternative routes for testosterone administration, e.g., intramuscular injection or transdermal application, have been investigated and marketed since they demonstrated favorable medicinal properties.[57] However, a serious problem has arisen from an enormous variety of old and new AAS that are now available as performance-enhancing drugs, mostly illegally and without clinical approval,[58–62] and the majority of these must be considered relevant for doping controls as they possess considerable potential for misuse in sports.

The development of analytical assays enabling the comprehensive detection of AAS in sports drug testing samples has required the consideration of the commonly rather extensive metabolism of these xenobiotics.[62,63] Principle metabolic reactions of synthetic anabolic steroids are closely related to those reported for testosterone (see section 6.1.2.2), with major pathways controlled by various enzymes such as members of the cytochrome P450 system, 17β-hydroxysteroid dehydrogenase, 3α-/3β-hydroxysteroid dehydrogenase and 5α-reductases, which are primarily responsible for the conversion of testosterone into active or inactive metabolic products.[64] These systems as well as additional metabolizing enzymes also affect most synthetic AAS and yield mostly typical metabolic products that are commonly used as target analytes in doping controls as presented in Table 6.1;[62,63] however, also common metabolites of different steroids were found due to shared metabolic pathways. A generalized overview of the most common metabolic pathways involving the A/B-ring of 3-oxo-steroids and the D-ring of 17-hydroxysteroids is depicted in Scheme 6.4. The reduction of the 3-oxo-residue and the adjacent double bond between C-4 and C-5 of the A-ring commonly yields three out of four possible combinations (Scheme 6.4a). In contrast, 3-oxo-1,4-diene steroids are preferably transferred to 3α,5β-oriented metabolites (Scheme 6.4b). The D-ring of 17-hydroxylated steroids is frequently subjected to oxidation and reduction reactions yielding 17-oxo-steroids if no 17-alkylation is present. In addition, hydroxylation at C-16 is observed that results also in 16-oxo-metabolites (Scheme 6.4c). Following the phase-I-metabolism, most products undergo phase-II metabolic reactions and are renally eliminated as glucuronic acid or sulfate conjugates. Besides these major pathways, various steroids undergo additional and more individual metabolic degradation processes, leading to specific and, in some cases, long-term metabolites that enable the retrospective detection of steroid abuse even weeks after cessation of the drug.[44,45,65–67]

Screening for xenobiotic AAS in doping control urine samples has been a complex task, which was first approached by radioimmunoassays (RIAs), complemented and later substituted entirely by MS-based methods (see also Chapter 1). The metabolism of AAS was already accounted for in seminal procedures for the detection of AAS in urine specimens by means of GC-MS such as those described by Ward,[68] Donike,[46] or Catlin,[69] and their corresponding colleagues, which commonly consisted of an initial SPE followed by enzymatic hydrolysis of the dried and reconstituted eluate and subsequent LLE of liberated AAS and/or their phase-I-metabolites. In order to increase the gas chromatographic properties of the analytes and, thus, the sensitivity of the assays, derivatization of steroids was conducted, predominantly to corresponding trimethylsilylated compounds.

The option to efficiently extract steroid conjugates by SPE using, e.g., Amberlite XAD-2 polystyrene resin, was reported in the late 1960s,[70] which proved essential for the enzymatic cleavage of steroid glucuronides by means of β-glucuronidase. Various natural inhibitors of β-glucuronidase were identified in human urine including low and high molecular weight compounds, which are removed by the preceding SPE, thus ensuring complete liberation of phase-I-metabolites from

TABLE 6.1: Selected Synthetic Anabolic Steroids and Respective Target Analytes Used in Commonly Employed Doping Control Procedures[a]

Anabolic Steroid	Target Analyte(s) in Urine
1-Androstenedione	1-androstenedione
	3α-hydroxy-5α-androstan-17-one
	3α-hydoxy-5α-androst-1-en-17-one
	17β-hydroxy-5α-androst-1-en-3-one
	5α-androst-1-ene-3α,17β-diol
1-Androstenediol	*See* 1-androstendione
Bolasterone	7α,17α-dimethyl-5β-androstane-3α,17β-diol
Boldenone	boldenone
	5β-androst-1-en-17β-ol-3-one
	5β-androst-1-en-3α-ol-17-one
Boldione	*See* boldenone
Calusterone	7β,17α-dimethyl-5β-androstane-3α,17β-diol
	17β-hydroxy-7β,17α-dimethyl-androst-4-en-3-one
	7β,17α-dimethyl-5α-androstane-3α,17β-diol
Clostebol	4-chloro-androst-4-en-3α-ol-17-one
	4-chloro-5α-androstan-3α-ol-17-one
	4-chloro-5β-androstan-3α-ol-17-one
	4-chloro-androst-4-ene-3α,17β-diol
Danazol	2α-hydroxymethylethisterone
Dehydrochlormethyltestosterone	6β-hydroxy-4-chloro-1,2-dehydro-17α-methyltestosterone
	4ξ-chloro-3α,6β,17β-trishydroxy-17α-methyl-5β-androst-1-en-16-one
Dehydro-17-methylclostebol	4-chloro-17α-methyl-androst-4-ene-3ξ,17β-diol
Desoxymethyltestosterone	Desoxymethyltestosterone (17α-methyl-5α-androst-2-en-17β-ol)
Drostanolone	2α-methyl-5α-androstan-3α-ol-17-one
Ethylestrenol	17α-ethyl-5β-estrane-3α,17β-diol
Fluoxymesterone	9α-fluoro-18-nor-17,17-dimethyl-androst-4,13-dien-11β-ol-3-one
	9α-fluoro-17α-methyl-androst-4-ene-3α,6β,11β,17β-tetra-ol
Formebolone	2-hydroxymethyl-17α-methyl-androsta-1,4-diene-11α,17β-diol-3-one
Furazabol	16β-hydroxyfurazabol
4-Hydroxytestosterone	4-hydroxyandrost-4-ene,3,17-dione
Mestanolone	17α-methyl-5α-androstane-3α,17β-diol
	17β-methyl-5α-androstane-3α,17α-diol
	18-nor-17,17-dimethyl-5α-androst-13-en-3α-ol
Mesterolone	1α-methyl-5α-androstan-3α-ol-17-one
Metandienone	17-epimetandienone
	6β-hydroxymetandienone
	17α-methyl-5β-androstane-3α,17β-diol
	17β-methyl-5β-androst-1-ene-3α,17α-diol
	18-nor-17,17-dimethyl-5β-androsta-1,13-dien-3α-ol
	18-nor-17β-hydroxymethyl,17α-methyl-androst-1,4,13-trien-3-one

TABLE 6.1: (*Continued*)

Anabolic Steroid	Target Analyte(s) in Urine
Metenolone	1-methylene-5α-androstan-3α-ol-17-one
Methandriol	17α-methyl-5β-androstane-3α,17β-diol
Methasterone	Methasterone (2α,17α-dimethyldihydrotestosterone)
	2α,17α-dimethyl-5α-androstane-3α,17β-diol
6α-Methyl-androstendione	3α-hydroxy-6α-methyl-5β-androstan-17-one
Methyldienolone	Methyldienolone
17-Methylnortestosterone	17α-methyl-5β-estrane-3α,17β-diol
	17α-methyl-5α-estrane-3α,17β-diol
18-Methyl-19-nortestosterone	13β-ethyl-3α-hydroxy-5α-gonan-17-one
	13β-ethyl-3α-hydroxy-5β-gonan-17-one
Methyltestosterone	17α-methyl-5α-androstane-3α,17β-diol
	17α-methyl-5β-androstane-3α,17β-diol
	17β-methyl-androst-4,6-dien-17α-ol-3-one
Methyl-1-testosterone	Methyl-1-testosterone
	17α-methyl-5α-androst-1-en-3α,17β-diol
	17β-methyl-5α-androst-1-en-3α,17α-diol
	17α-methyl-5α-androstane-3α,17β-diol
Methyltrienolone	Methyltrienolone
Mibolerone	*7α,17α-dimethyl-5β-estrane-3α,17β-diol*
Nandrolone	5α-estran-3α-ol-17-one
	5β-estran-3α-ol-17-one
19-Norandrostenediol	*See* nandrolone
19-Norandrostenedione	*See* nandrolone
Norbolethone	13β,17α-diethyl-3α,17β-dihydroxy-5β-gonane
Norclostebol	4-chloro-4-estren-3α,17β-diol
Norethandrolone	17α-ethyl-5β-estrane-3α,17β-diol
Oxabolone	4-hydroxynorandrostendione
Oxandrolone	17-epioxandrolone
Oxymesterone	Oxymesterone
Oxymetholone	17α-methyl-5α-androstane-3α,17β-diol
Prostanozol	Prostanozol
	17-oxo-5α-androstano[3,2-c]pyrazole
Quinbolone	*See* boldenone
Stanozolol	3′-hydroxystanozolol
	3′-hydroxy-17-epistanozolol
	16β-hydroxystanozolol
	4β-hydroxystanozolol
Stenbolone	3α-hydroxy-2-methyl-5α-androst-1-ene-17-one
1-Testosterone	*See* 1-androstenedione
Tetrahydrogestrinone	Tetrahydrogestrione
Tibolone	3α-hydroxytibolone
	3β-hydroxytibolone
Trenbolone	Trenbolone
	17-epitrenbolone

[a] Refs. 210–211.

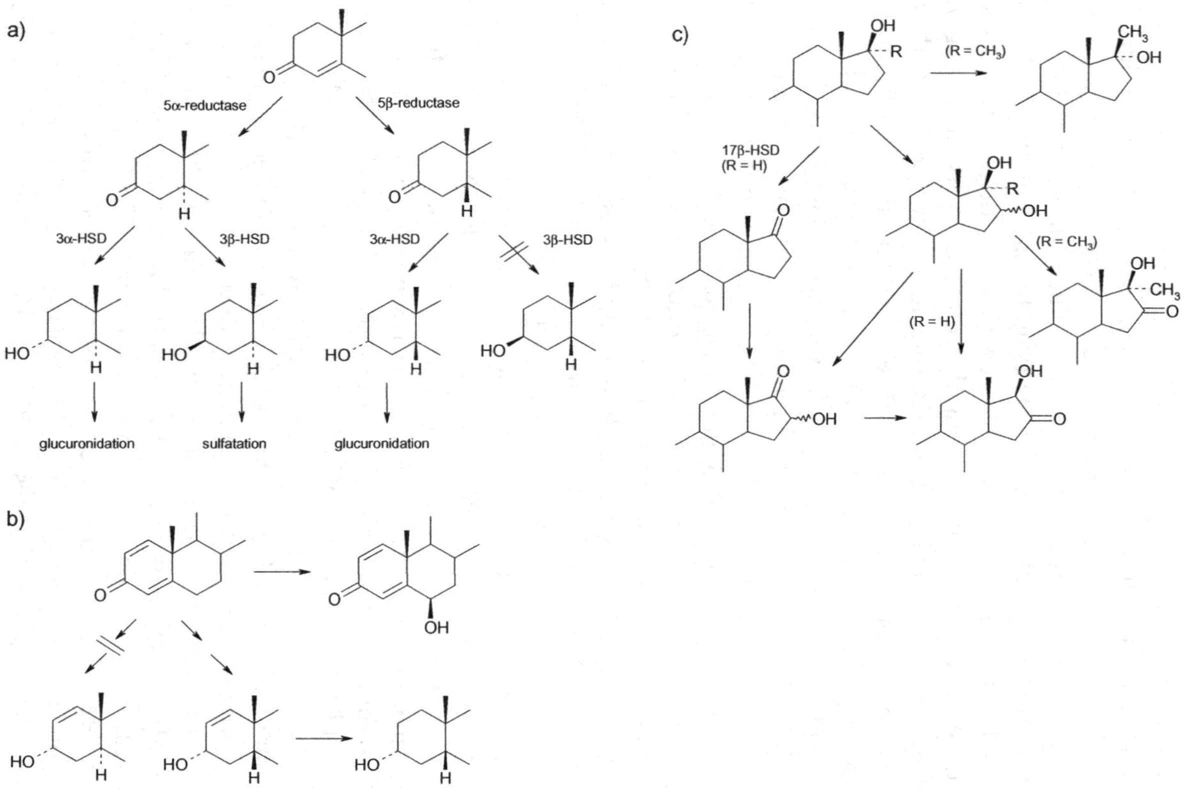

Scheme 6.4: Common metabolism pathways of anabolic androgenic steroids (HSD = hydroxysteroid dehydrogenase).

the glucuronic acid moiety.[71] Consequently, isolation of steroid conjugates from urine prior to enzymatic hydrolysis was mandatory,[72] until reliable tools to control the completeness of deconjugation in urine were established. In 1997, Geyer and associates introduced the use of isotopically labeled steroid glucuronides as internal standards in routine doping control procedures for AAS,[73] which enabled the determination of the conversion of intact to glucuronides to corresponding aglycons by means of 2,2,3,4,4-$^{2}H_5$-androsterone glucuronide and, thus, the so-called "direct hydrolysis" of sports drug testing samples. Based on these facts, detection methods for AAS and endogenous steroids were developed, which are commonly referred to as *Screening IV* (Scheme 6.5).[7,73] A typical protocol includes the use of 2 mL of urine, which are fortified with a mixture of deuterium-labeled steroids ([2,2,4,4-$^{2}H_4$]-etiocholanolone,[16,16,17-$^{2}H_3$]-testosterone,[16,16,17-$^{2}H_3$]-epitestosterone, [2,2,4,4-$^{2}H_4$]-11β-hydroxyandrosterone, and [2,2,3,4,4-$^{2}H_5$]-androsterone glucuronide) and methyltestosterone. The sample is buffered to pH 7.0 with 0.75 mL of a 0.8 M phosphate buffer (Na_2HPO_4: NaH_2PO_4, 1:2, w:w), 25 µL of β-glucuronidase from *E. coli* are added, and the specimen is incubated at 50°C for 1 hour. After cooling to room temperature, the pH is adjusted to 9.6 by adding 0. 5 mL of an aqueous solution containing potassium carbonate and potassium bicarbonate (20%, 1:1, w:w), and target analytes are partitioned into 5 mL of *tert.*-butyl methyl ether by shaking mechanically for 5 minutes. After centrifugation at 600 × *g* for 10 minutes, the organic layer is transferred to a fresh glass tube, evaporated to dryness at 50°C under reduced pressure, and further stored over phosphorus pentoxide *in vacuo* for an additional 60 minutes. The dry residue is subsequently derivatized using MSTFA-ammonium iodide-ethanethiol (500:1:3, v/w/v) for 10 minutes at 60°C and subjected to GC-MS(/MS) analysis. Routine GC-MS analyses are commonly performed on HP Ultra1 GC columns (length 17 m, inner diameter 0.2 mm, film thickness 0.11 µm) with an injection volume of 3 µL (splitless injection). A frequently used GC oven temperature program starts at 180°C increasing with 3°C/min to 229°C and 40°C/min to 310°C using helium as carrier gas (0.96 mL/min at 180°C, constant pressure). The injector temperature is set to 300°C, the interface temperature to 320°C and the ion source temperature to 230°C. Ionization is accomplished using EI (70 eV), and selected ion monitoring is employed at dwell times of 20 ms. Up to four ions are recorded per analyte to ensure its identity and enhance the specificity of the initial testing (i.e., screening). Typical extracted ion chromatograms of six compounds added to a quality control urine specimen are depicted in Figure 6.4.

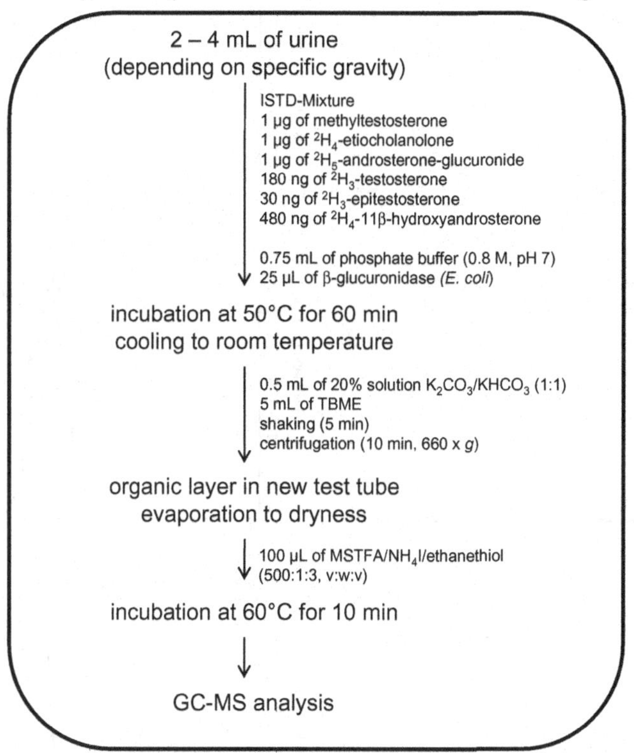

Sample Preparation Flow Scheme Screening IV
2 – 4 mL of urine
(depending on specific gravity)
ISTD-Mixture
1 µg of methyltestosterone
1 µg of 2H4-etiocholanolone
1 µg of 2H5-androsterone-glucuronide
180 ng of 2H3-testosterone
30 ng of 2H3-epitestosterone
480 ng of 2H4-11β-hydroxyandrosterone
0.75 mL of phosphate buffer (0.8 M, pH 7)
25 µL of β-glucuronidase (E. coli)
incubation at 50°C for 60 min
cooling to room temperature
0.5 mL of 20% solution K2CO3/KHCO3 (1:1)
5 mL of TBME
shaking (5 min)
centrifugation (10 min, 660 x g)
organic layer in new test tube
evaporation to dryness
100 µL of MSTFA/NH4I/ethanethiol
(500:1:3, v:w:v)
incubation at 60°C for 10 min
GC-MS analysis

Scheme 6.5: Sample preparation flow scheme for the analysis of AAS and endogenous steroids. The screening procedure is commonly referred to as *Screening IV*.[7,73]

The option of quadrupole MS was complemented by so-called high sensitivity instruments, which were introduced in 1993[74] with high resolution magnetic sector mass spectrometers and became mandatory in sports drug testing in 1996.[75] It has considerably enhanced detection and identification capabilities in doping controls, especially with regard to long-term excreted metabolites.[44,66] Primarily due to the high financial burden, alternatives were sought and found in tandem mass spectrometers commonly composed by ion trap or triple quadrupole devices. The gain in sensitivity as accomplished by MS/MS (/MS) experiments has also allowed the detection of the specified analytes 3′-hydroxystanozolol, 17α-methyl-5β-androstane-3α,17β-diol, 17β-methyl-5β-androst-1-ene-3α,17α-diol, 19-norandrosterone, and clenbuterol at required detection limits.[49,76,77] Those compounds have been particular targets in sports drug testing due to their frequent

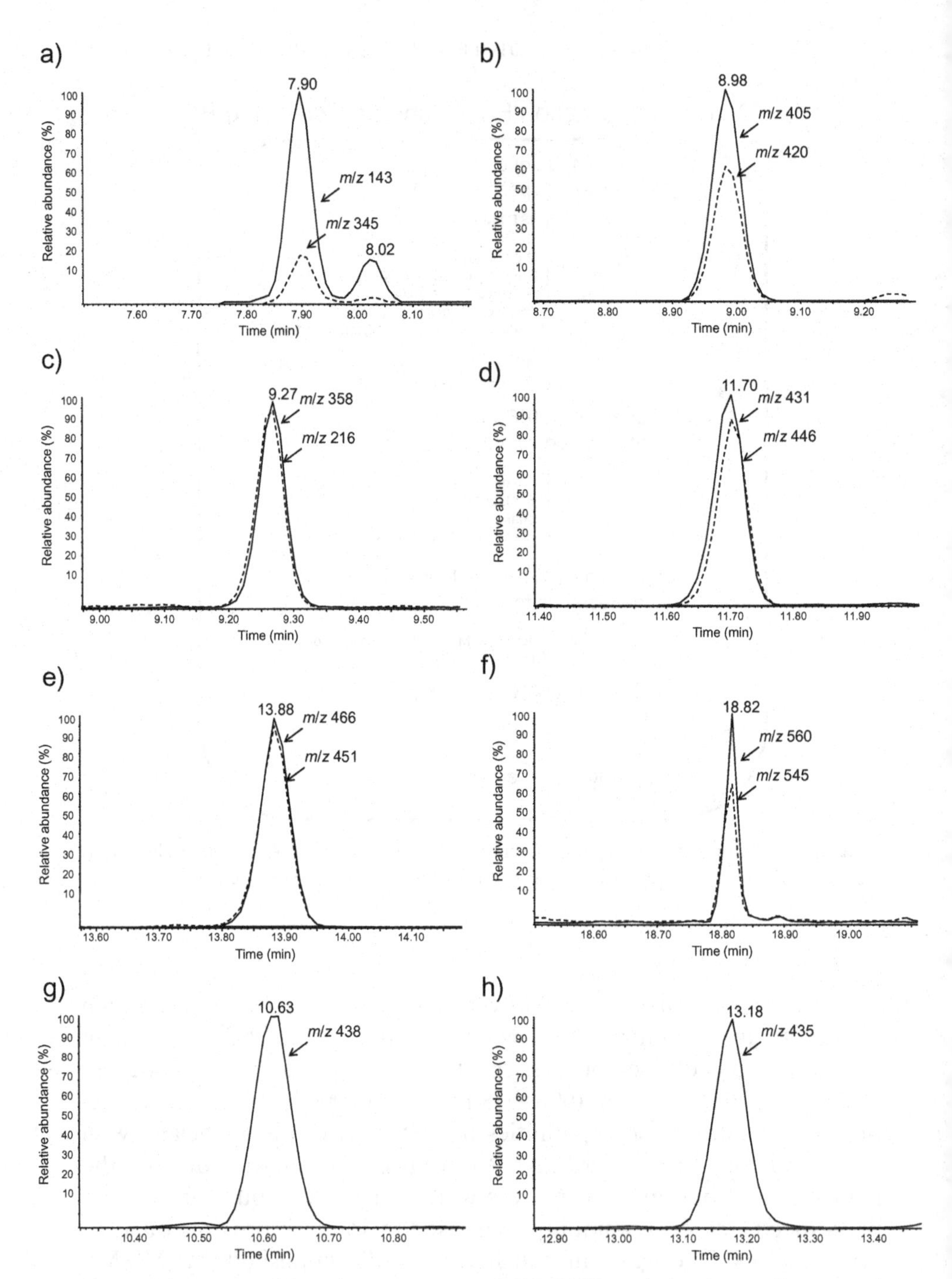

Figure 6.4: Extracted ion chromatograms of a GC-MS analysis of a quality control sample containing, among others, 6 analytes ([a] desoxymethyltestosterone: 7.90 min; [b] 19-norandrosterone: 8.98 min; [c] 18-nor-17,17-dimethyl-5β-androsta-1,13-dien-3α-ol: 9.27 min; [d] 1-methylene-5α-androstan-3α-ol-17-one: 11.70 min; [e] 4-chloro-androst-4-en-3α-ol-17-one: 13.88 min; and [f] 3′OH-stanozolol: 18.82 min) plus the internal standards $^{2}H_{4}$-etiocholanolone ([g], 10.63 min) and $^{2}H_{3}$-testosterone ([h], 13.18 min).

misuse; hence, long-term traceability has been of great interest and research focused on improved retrospective is an ongoing pursuit, and excellent alternatives for stanozolol metabolites and clenbuterol were found in LC-MS/MS procedures (see section 6.2).

6.1.2.1.1 Case Vignette: Long-term Metabolite Identification for Metandienone. A recent example for "old drugs and new targets" was found in 18-nor-17β-hydroxymethyl,17α-methyl-5β-androsta-1,13-dien-3-one, a newly characterized long-term metabolite of metandienone. The EI-mass spectrum of the trimethylsilylated analyte and respective extracted ion chromatograms are illustrated in Figure 6.5.

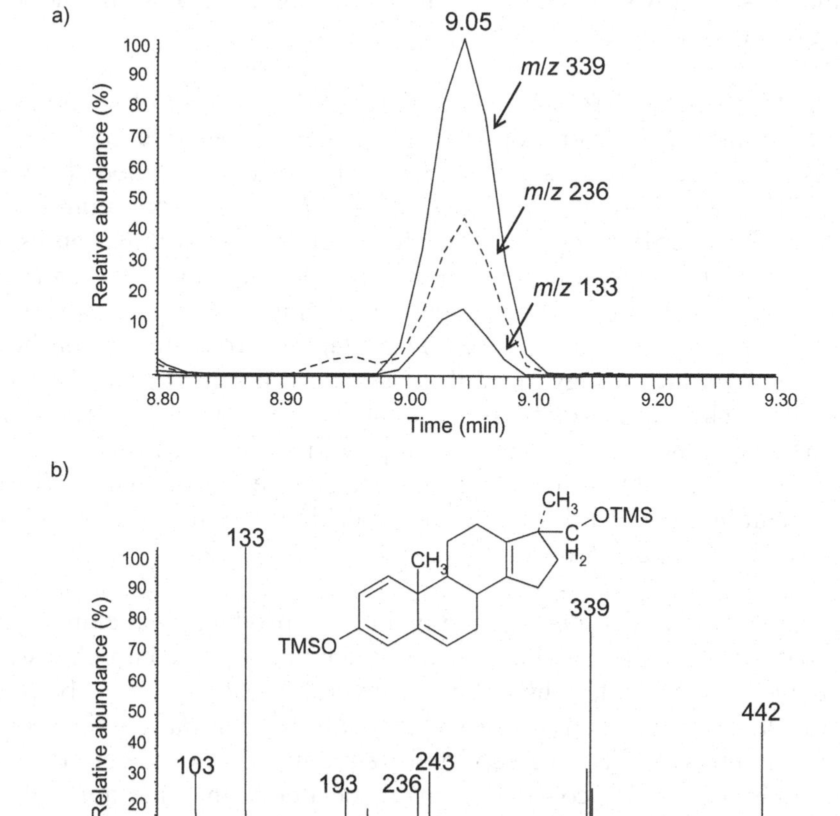

Figure 6.5: (a) Extracted ion chromatograms of a GC-MS/MS analysis (precursor ion: *m/z* 442) regarding the long-term metabolite 18-nor-17β-hydroxymethyl,17α-methyl-5β-androsta-1,13-dien-3-one resulting from metandienone application, and (b) the corresponding EI mass spectrum obtained from an authentic doping control specimen.

The data were obtained from a urine specimen tested positive after administration of metandienone. The utility of this comparably new long-term target in sports drug testing[45] has become evident with an increase of adverse analytical findings associated with the misuse of metandienone of more than 300% in the Cologne doping control laboratory in 2006.[65] Out of approximately 11,000 urine samples, 68 (0.6%) were tested positive for 18-nor-17β-hydroxymethyl,17α-methyl-5β-androsta-1,13-dien-3-one, thus proving the misuse of metandienone. In comparison to the years 2003–2005, where the total amount of doping control urine specimens ranged also between 10,000 and 12,000 per year, only 12–15 adverse analytical findings were reported. Moreover, in 2006 a total of 126 doping violations with metandienone were determined, 54% of which were identified utilizing the above mentioned long-term metabolite.

6.1.2.2 "Natural" Steroids In contrast to the misuse of xenobiotic AAS, the illicit administration of synthetic steroids with identical structures compared to those naturally occurring in the human body (in the following referred to as "natural" steroids, e.g., testosterone, epitestosterone, DHT, DHEA, etc.) is difficult to detect and requires sophisticated analytical approaches. Those include the so-called steroid profiling,[78] which has been developed for doping control purposes over a period of approximately 25 years by adapting strategies established in clinical endocrinology, as well as isotope ratio mass spectrometry (IRMS).[79] The analysis of naturally occurring urinary steroids is accomplished using the same procedure as reported above for synthetic AAS; here, the use of numerous deuterium-labeled internal standards allows the quantification of relevant steroids, which is essential for assessing the obtained steroid profiles.

6.1.2.2.1 Steroid Profiling. The human body produces and eliminates a great variety of steroidal hormones, which undergo a comprehensive biosynthetic and metabolic pathway. Several of these steroids belong to the class of androgens possessing anabolic and/or androgenic properties, and misuse has been reported frequently. Due to the enormous variability of urinary concentrations associated with urine flow rates, the use of hormone ratios for the detection of "natural" steroid administration to athletes was introduced in 1979, initially with regard to the misuse of testosterone (T).[80] The original assay required the determination of the ratio of urinary T and luteinizing hormone (LH), with the latter one necessitating an immunological method. Due to the preferred use of GC-MS-based test methods, an alternative approach was suggested in 1982 utilizing the ratio of urinary T and epitestosterone (EpiT),[78] which was found to be sensitive to administrations of T as the T/EpiT value significantly increases from basal levels. Deduced from population-based reference values, a threshold level T/EpiT > 6 was installed and later corrected to T/EpiT > 4 to indicate the misuse of T or related prohormones. Further to the determination of T and EpiT, metabolites of T, namely androsterone (A), etiocholanolone (E), DHT, 5α-androstane-3α,17β-diol (Adiol), and 5β-androstane-3α,17β-diol (Bdiol), as well as the precursor DHEA were quantified to compose the basis of the urinary steroid profiling program (Table 6.2).[81] This has been complemented by a variety of additional urinary steroids, and the entirety shall support, facilitate, and substantiate the interpretation of atypical findings using either absolute concentrations (commonly adjusted for a specific gravity of 1.020) or ratios (Table 6.3) that indicate the misuse of particular natural steroids (Table 6.4).

Due to their considerable intra-individual stability, the ratios T/EpiT, A/E, A/T, and Adiol/Bdiol in particular are relevant for steroid profiling, as these are demonstrably not influenced by intense and/or sustained exercise, menstrual cycle, and circadian or annual rhythms.[82] In contrast, the administration of steroids including AAS and "natural" compounds impairs the sensitively balanced system of steroid biosynthesis, and indications for steroid abuse are obtained and followed-up using, e.g., IRMS-based approaches (see section 6.1.2.2.2). In addition to the initial idea of applying population-based reference values as established for T, EpiT, A, E, DHT, Adiol, and Bdiol (Table 6.2), the much more sensitive nature of subject-based (or individual) reference ranges due to a rather low variability within the biosynthesis of steroids was suggested as early as 1994.[83,84] Improved statistical considerations[85,86] and the comparison of sequential measurements of biomarkers of individuals enabled by sophisticated and Internet-based databases collecting individual doping control analytical results (Anti-Doping Administration & Management System—*ADAMS*) have been of great importance for modern sports drug testing activities.

6.1.2.2.1.1 EFFECTS OF TESTOSTERONE AND EPITESTOSTERONE ADMINISTRATION. The injection or oral intake of testosterone significantly influences the steroid profile, especially the urinary concentration of T and, thus, the T/EpiT and A/T ratios (Table 6.4).[78,87] Moreover, the increased urinary elimination of metabolites derived from T (including the glucuronide and sulfate of T and the glucuronic acid conjugates of A, E, Adiol, and Bdiol) combined with a reduced excretion of E and 5-androstene-3β,17α-diol was described, which can further substantiate

TABLE 6.2: Analytes Relevant for Steroid Profiling and Rrespective Reference Ranges/Thresholds Where Determined or Applicable[a]

Steroid	Abbreviation	Reference Ranges (ng/mL)*				Threshold (ng/mL)*
		Male (low)	Male (high)	Female (low)	Female (high)	
Testosterone	T	2	137	1	57	200
Epitestosterone	EpiT	5	112	1	42	200
Androsterone	A	867	6703	404	6439	10000
Etiocholanolone	E	674	5294	473	6107	10000
5α-androstane-3α,17β-diol	Adiol	14	167	5	91	—
5β-androstane-3α,17β-diol	Bdiol	20	550	9	366	—
Dihydrotestosterone	DHT	0.2	21	0.2	18	—
5β-pregnane-3α,20α-diol	PD	73	951	83	3089	—
Dehydroepiandrosterone	DHEA	—	—	—	—	100
3α,5-cyclo-5α-androstan-6β-ol-17-one	3α,5-cyclo	—	—	—	—	—
5β-androstane-3α,17α-diol	17-epi-Bdiol	—	—	—	—	—
5α-androstane-3α,17α-diol	17-epi-Adiol	—	—	—	—	—
5β-androstane-3,17-dione	Bdione	—	—	—	—	—
5β-androst-1-en-17β-ol-3-one	B-M1	—	—	—	—	—
5α-estran-3α-ol-17-one	NorA	—	—	—	—	2
5β-estran-3α-ol-17-one	NorE	—	—	—	—	—
5-androstene-3β,17α-diol	—	—	—	—	—	—
5-androstene-3β,17β-diol	—	—	—	—	—	—
Epiandrosterone	EpiA	—	—	—	—	—
5α-androstane-3,17-dione	Adione	—	—	—	—	—
5α-androstane-3β,17β-diol	Trans-Adiol	—	—	—	—	—
Boldenone	B	—	—	—	—	—
11β-hydroxyandrosterone	11-OH-A	—	—	—	—	—
11β-hydroxyetiocholanolone	11-OH-E	—	—	—	—	—
5β-pregnane-3β,17α,20β-triol	PT	—	—	—	—	—

[a] Ref. 82.

*Urinary concentrations adjusted for a specific gravity of 1.020.

TABLE 6.3: Ratios of Urinary Steroids Relevant for Profile Interpretations[a]

Ratio	Reference Ranges				Threshold
	Male (low)	Male (high)	Female (low)	Female (high)	
T/EpiT	0.1	5.2	0.1	6.3	4
A/E	0.6	2.9	0.4	2.2	
A/T	22.5	1164	47.9	2184	
A/EpiT	27.7	406	65.3	939	
E/T	16.7	819	37.0	2397	
E/EpiT	20.5	346	60.5	1205	
Adiol/Bdiol	0.1	1.6	0.1	1.2	
DHT/EpiT	0.01	0.73	0.02	2.3	
DHT/E*	0.1	8.2	0.1	8.5	

[a] Ref. 82.
*Value of DHT/E multiplied by 1000.

suspicious T/EpiT ratios > 4.[88] Cheating athletes, however, reportedly co-administered T and EpiT to artificially decrease this threshold-furnished ratio and consequently prevent further in-depth investigations;[89,90] hence, a maximum allowed urinary concentration of 200 ng/mL of EpiT was established, and additional parameters potentially indicating the application of EpiT such as increased urinary concentrations of 5β-androstane-3α,17α-diol (17-epi-Bdiol) and 5α-androstane-3α,17α-diol (17-epi-Adiol) are now monitored (Table 6.4). In contrast to orally or intramuscularly administered T, the use of transdermal applications yield steroid profiles being noticeable by the increased ratios Adiol/EpiT, A/EpiT, and T/EpiT, resulting from both reduced EpiT and elevated Adiol and T excretion;[91] however, these alterations were of minor intensity and difficult to observe using population-based reference ranges.

6.1.2.2.1.2 EFFECTS OF DIHYDROTESTOSTERONE AND 5α-ANDROSTANEDIONE ADMINISTRATION. Due to the significant increase of the T/EpiT ratio after administration of T, the misuse of DHT was predicted as an "alternative" for cheating athletes. DHT does not influence the T/EpiT ratio, but several other indicators were found in elevated urinary concentrations of DHT (exceeding 21 and 18 ng/mL for men and women, respectively, Table 6.2), Adiol (exceeding 204 and 89 ng/mL for men and women, respectively), and A (exceeding 9103 and 7562 ng/mL for men and women, respectively),[82] and ratios of

TABLE 6.4: Factors Influenced by Administration of "Natural" Steroids[a]

Steroid	Factors Influenced (concentration and/or ratio)	
	Increased	Decreased
Testosterone (oral, injection)	T T/EpiT	A/T
Testosterone gel (transdermal)	T Adiol T/EpiT Adiol/EpiT A/EpiT	EpiT
Dihydrotestosterone or 5α-androstanedione	DHT Adiol A Trans-Adiol EpiA DHT/E DHT/EpiT Adiol/Bdiol A/E	
4-Androstene-3,17-dione	A E T DHT EpiT 17-epi-Bdiol 17-epi-Adiol T/EpiT	5-androstene-3β,17α-diol
Epitestosterone	EpiT 17-epi-Bdiol 17-epi-Adiol	T/EpiT
Dehydroepiandrosterone	DHEA A E 5-androstene-3β,17β-diol 3α,5-cyclo	

[a] Ref. 82.

DHT/EpiT, DHT/E, Adiol/Bdiol, A/E (Table 6.4).[92–95] The administration of 5α-androstanedione (as an intermediate product of the T metabolism) caused comparable but less pronounced alterations to the steroid profile as observed after DHT application.

6.1.2.2.1.3 EFFECTS OF DEHYDROEPIANDROSTERONE, 4-ANDROSTENE-3,17-DIONE, AND 4-ANDROSTENEDIOL ADMINISTRATION. The use of so-called prohormones of T such as DHEA, 4-androstene-3,17-dione and 4-androstene-3,17-diol induces various changes to the urinary steroid profile. Particularly the concentrations of A, E, T, and DHT, as well as the T/EpiT ratio are affected in case of 4-androstene-3,17-dione intake,[96–99] all of which demonstrated unusually high values in administration study urine samples. Moreover, the detection of 6-hydroxylated metabolites such as 6α-hydroxyandrostenedione, 6β-hydroxyandrosterone, 6β-hydroxyetiocholanolone, and 6β-hydroxyepiandrosterone were reported, and increased urinary concentrations of EpiT, 17-epi-Bdiol and 17-Epi-Adiol, as well as decreased 5-androstene-3β,17α-diol levels were found.[100] Hence, the combined presence of 6α-hydroxyandrostenedione and elevated ratios of 17-epi-Bdiol/EpiT and 17-epi-Adiol/EpiT were considered indicative for 4-androstene-3,17-dione applications.

In case of DHEA, only minimal effects on the T/EpiT ratio were detected in administration study urine samples.[101,102] Hence, the urinary concentration of its glucuronide was considered the most relevant parameter, and a threshold value of 300 ng/mL (corresponding to the unconjugated DHEA) was installed and recently lowered to 100 ng/mL. Nevertheless, the search for more discriminating parameters has been pursued and the utility of 3α,5-cyclo (a major metabolite of DHEA, Table 6.2) as indicator for a DHEA misuse was recently reported.[103] Urinary concentrations exceeding 140 ng/mL were considered suspicious, and the combined application of DHEA and 3α,5-cyclo threshold levels might support the efforts to economize routine doping controls[104] as all urine samples being suspicious for the administration of "natural" steroids are recommended to be analyzed by IRMS.

6.1.2.2.1.4 OTHER FACTORS INFLUENCING THE STEROID PROFILE. Besides the fact that the administration of steroids alters the "normal" urinary steroid profile, other aspects affecting the concentration and/or analysis of steroids need consideration in doping control analyses as compiled in several recent reviews.[48,50,82] One of the most important issues are the genotype-dependent variations of the T/EpiT ratio, which have been reported for different ethnic groups, mainly associated with a polymorphism of the UGT2B17 gene.[105] The UGT2B17 enzyme plays a significant role in the glucuronidation of T and EpiT, and comparing the UGT2B17 genotypes with corresponding urinary levels of T, persons homozygous for the UGT2B17 depletion genotype eliminated significantly reduced amounts of T. The genetic phenomenon is prevalent especially in the Asian population,[106] which caused concerns about false-negative test results as Asian athletes would hardly exceed T/EpiT ratios of 4 when administering T or related prohormones.

Another factor that impacts urinary steroid profiles is the acute consumption of alcohol.[107–109] Urinary T/EpiT values increase and A/T levels decrease under the influence of ethanol intake, which necessitates analyses for ethanol (and/or its glucuronic acid conjugate) in urine samples being suspect due to the above mentioned alterations in the steroid profile.[110]

In addition, the influence of oral contraceptives as well as 5α-reductase inhibitors (common active principles in drugs used for the treatment of benign prostate hyperplasia as well as alopecia) was described. The first mentioned class of drugs leads to an increase of the T/EpiT ratio due to the suppression of EpiT elimination; however, other parameters such as A/E and Adiol/Bdiol remain stable. The use of 5α-reductase inhibitors strongly interferes with steroid profile interpretations as particularly A and Adiol are significantly reduced and, thus, the formerly mentioned intra-individually stable ratios of A/E and Adiol/Bdiol are considerably modified.[111]

In terms of the threshold-substance 5α-estran-3α-ol-17-one (NorA), several aspects need consideration, e.g., pregnancy and bacterial activity. Urinary levels > 2 ng/mL represent doping violations, but several studies outlined the elevated production and elimination of NorA throughout gestation[112,113] as well as the possibility to form NorA from A by demethylation.[114] Hence, careful interpretations of steroid profiles and tests for bacterial activity (for instance, by means of deuterium-labeled A) are carried out.

Several additional factors that are either technical or drug-related reportedly influence urinary steroid profiles and must be eliminated to ensure adequate interpretations of test results. Consequently, confirmatory analyses substantiating the indications obtained from steroid profiling are highly desirable and found in IRMS as outlined in the following text.

6.1.2.2.2 GC/C/IRMS. Isotope ratio mass spectrometry (IRMS) allows the determination of isotopic signatures of compounds by measuring different (stable) isotopes of an element. The principle of this strategy relies on the variation of isotopic contents of analytes, resulting from fractionation processes that cause discrimination of one isotope of an element against another. In sports drug testing, carbon isotope ratio (CIR) mass spectrometry is the most frequently applied technique to demonstrate whether a urinary steroid originates from human biosynthesis or from exogenous sources.[79] Here, the $\delta^{13}C$ value of target compounds such as T, EpiT, A, E, Adiol, Bdiol, NorA, etc. are determined, which describes the abundance of ^{13}C of the substance in part-per-thousand (per-mille, ‰) in relation to a reference standard termed Vienna Pee Dee Belemnite (VPDB) using the formula

$$\delta^{13}C(‰) = (\text{ratio}_{\text{Sample}} / \text{ratio}_{\text{Reference}} - 1) \times 1000 \qquad (6.1)$$

Natural variations in the abundance of ^{13}C reflect the process of (bio) synthesis and/or transformation of a molecule that is commonly accompanied by isotopic fractionation, which causes a small but significant enrichment or depletion of ^{13}C. These differences can be measured and visualized by quantitative combustion of compounds into CO_2 followed by high-precision IRMS employing a magnetic sector and three Faraday cups that are positioned to collect the ions *m/z* 44, 45, and 46 corresponding to $^{12}C^{16}O^{16}O^{+\bullet}$, $^{13}C^{16}O^{16}O^{+\bullet}$, and $^{12}C^{16}O^{18}O^{+\bullet}$, respectively. The ratio of *m/z* 45 to *m/z* 44 is determined for target compounds and corrected for contributions to the abundance of *m/z* 45 caused by $^{12}C^{16}O^{17}O^{+\bullet}$. While pharmaceutically produced steroids are commonly derived from soy stigmasterol or sitosterol comprising a reduced ^{13}C content, endogenously produced steroids mirror the individual's diet, which is usually composed of various different sources of carbon and, thus, contains a comparably elevated ^{13}C proportion. Reference ranges for urinary steroid metabolites were determined between –15‰ to –25‰, and numerous pharmaceutical steroidal products yielded $\delta^{13}C$ values of –30.1‰ ± 2.6‰ and –25.9‰ to –32.8‰.[115,116] To account for the fact that diets vary considerably and, thus, also $\delta^{13}C$ values of urinary steroids, endogenous reference compounds (ERCs), i.e., endogenously produced steroidal compounds that are not involved in the androgen metabolic pathway, were included in doping control IRMS methods. Frequently used ERCs are e.g., pregnanediol (PD) and 5α-androst-16-en-3α-ol, and the difference of an ERC $\delta^{13}C$ values from a urinary steroid (metabolite) (i.e., the $\Delta\delta^{13}C$ value) has been considered the most efficient and reliable means to uncover the administration of "natural" steroids for doping purposes. Exceeding a $\Delta\delta^{13}C$ threshold level specified for particular pairs of analytes constitutes a doping rule violation,[117] these were evaluated in recent studies[118,119] and considerably enhance the level of confidence as well as the discrimination power of IRMS compared to earlier approaches that used a generic threshold of 3‰.

The analysis of $\Delta\delta^{13}C$ values of steroids found in doping control urine samples has been accomplished by various different sample preparation strategies. First approaches were published in 1994 and since, two principal routes (without and/or with analyte derivatization) were pursued providing the required symmetry and purity of GC peaks. As an example, a method utilized in the Cologne doping control laboratory is described in more detail in the following text.

Conditions to hydrolyze and extract urinary steroids from doping control specimens were mostly adapted from methods established for conventional GC-MS screening procedures targeting AAS. Deconjugated and extracted steroids are, however, not trimethylsilylated but subjected to HPLC purification. For that purpose, the dried residues are commonly reconstituted in 50 µL of methanol and injected onto a LiChrospher 100 RP18 (250 × 4 mm, 5 µm particle size) analytical column using a flow rate of 1 mL/min. Solvents used are water (A) and acetonitrile (B) employing a gradient starting at 30% B, increasing linearly to 100% B in 25 minutes, and maintaining at 100% B for another 5 minutes. Fractions containing the desired steroids are collected in time windows (Table 6.5) that were previously defined by injecting respective reference compounds, and resulting aliquots are evaporated to dryness. Fractions I (containing 11-OH-A and 11-OH-E), and IV (containing A and E) are measured using GC/C/IRMS without further treatment, while the fractions II (T), III (EpiT, ADiol, Bdiol, and DHEA), V (PD), and VI (5α-androst-16-en-3α-ol) are acetylated using acetic anhydride and pyridine, followed by another HPLC purification step using either the same setup as before or, due to significantly decreased polarities of target analytes, an adjusted gradient

TABLE 6.5: HPLC Fractions and Respective Steroids Isolated for GC/C/IRMS Analysis

Fraction No.	Fractionation 1		Retention Time (min)	Fractionation 2
	Retention Time (min)	Analytes		Analytes (acetylated)
I	9.0–10.5	11-OH-A 11-OH-E		
II	11.3–12.5	T	19.0–21.0	T
III	12.5–14.9	EpiT Adiol Bdiol DHEA	10.0–11.2	EpiT
			13.5–15.1	DHEA
			23.2–24.2	Bdiol
			24.2–25.2	Adiol
IV	14.9–17.0	E A		
V	17.0–18.2	PD		
VI	26.8–28.0	5α-androst-16-en-3α-ol		

starting at 70% B and increasing linearly to 100% B in 33 minutes. Collected fractions are evaporated to dryness and ready for GC/C/IRMS analysis.

The GC/C/IRMS instrument is composed of an Agilent 6890 GC connected to a Delta C IRMS using a GC Combustion Interface. The GC is equipped with an OPTIMA δ3 column (length: 20 m, inner diameter: 0.25 mm, film thickness: 0.25 μm) and a retention gap (length: 1 m, inner diameter: 0.53 mm). Following cool-on-column injection at 50°C, the oven temperature is maintained for 0.5 min and subsequently ramped to 250°C at 30°C/min, to 270°C at 3°C/min and finally to 295°C at 15°C/min. Helium (purity grade 5.0) is used as carrier gas at constant flow (2.2 mL/min), and the combustion furnace is operated at 940°C. Generated water is removed in a cryotrap and carbon dioxide of target compounds analyzed for its $\delta^{13}C$ values. Due to the introduction of additional carbon into several analytes by derivatization, the measured values must be corrected for the acetate moieties according to the formula

$$n_{cd}\delta^{13}C_{cd} = n_c\delta^{13}C_c + n_d\delta^{13}C_{dcorr} \tag{6.2}$$

with n = number of moles of carbon, c = analyte of interest (e.g., T), d = derivatization residue, and cd = derivatized compound. $\delta^{13}C_d$ must be empirically determined.

With the obtained data, the desired $\Delta\delta^{13}C$ values are calculated and doping rule violations either verified or falsified. The formerly valid generic threshold of 3‰ represented a means to identify suspicious differences in carbon isotope signatures of biosynthetic precursors and corresponding metabolites; however, the "customized" thresholds for defined pairs of precursors/metabolites substantially enhance the level of confidence as well as the discrimination power of IRMS.

6.1.3 Diuretics

Although procedures to detect diuretic agents by means of GC-MS are almost entirely obsolete, various strategies were tested in the past to efficiently screen for this heterogeneous class of compounds.[120] Because most diuretics require derivatization to be compatible with GC, sophisticated procedures facilitating and accelerating the sample preparation were established.[121] These have commonly been based on alkylation of diuretic agents following or accompanying SPE or LLE. The considerable differences in physicochemical properties of diuretic agents (e.g., acidic, neutral, or alkaline character, presence of carboxyl, amino, and/or sulfonamide residues, etc.) necessitated the adjustment of the urine's pH to different values and multiple extractions of target analytes into organic solvents such as ethyl acetate or diethyl ether followed by combined evaporation to ensure a comprehensive recovery.[122] However, the joint isolation of diuretics at pH 9.5 was accomplished in a single-step LLE using ethyl acetate, which utilized the supportive effect of additional sodium chloride (salting-out effect) that substantially increased the partitioning of acidic diuretic agents from the alkaline aqueous phase into the organic layer.[123] Alternatively, SPE has been employed using either C-18, C-8, polystyrene, or mixed-mode resins,[124,125] which yielded a reasonable compromise for all relevant diuretics regarding respective extraction efficiencies. Both extracts from LLE and SPE were methylated prior to GC-MS analysis using either the so-called flash methylation approach,[126–128] which commonly employs quaternary ammonium hydroxides (such as tetramethylammonium hydroxide or trimethylaniline hydroxide) or conventional alkylation in acetone or acetonitrile using iodomethane and potassium carbonate (as catalyst).[124,129] The first mentioned method utilizes the elevated temperatures of the GC injection port to rapidly alkylate extracted analytes with the highly reactive methylation reagents, while the latter requires 2–3 hours of incubation at 50–60°C but commonly generates higher yields of target derivatives.

Complementary to these procedures of extraction and subsequent methylation, extractive alkylation was described as an efficient means to isolate and derivatize target compounds for GC-MS analysis.[130,131] Here, quaternary ammonium salts such as tetrahexylammonium hydroxide act as phase-transfer reagents supporting the extraction of ionic species such as organic acids from the aqueous phase into an aprotic organic solvent (e.g., toluene) containing the alkylation reagent such as iodomethane employing ion-pairing properties. The alkylation reaction occurs at room temperature, and derivatized analytes were subsequently purified from phase-transfer reagents to minimize interferences resulting from pyrolysis of the reagents or secondary derivatization reactions occurring in the injection port of the GC.[130–132]

The reported assays allow the detection of numerous diuretic agents at LODs as low as 0.05 ng/mL, and various other analytes are commonly included in the screening procedure that is frequently referred to as *Screening V* (Scheme 6.6).[7] A typical sample preparation and analysis procedure consists of the preconditioning of an SPE cartridge prepared with approximately 30 mg of XAD-2 polystyrene resin using 2 mL of methanol and 2 mL of water, followed by loading a total of 2 mL of urine enriched with 100 ng/mL of the ISTD (mefruside). The urine is passed

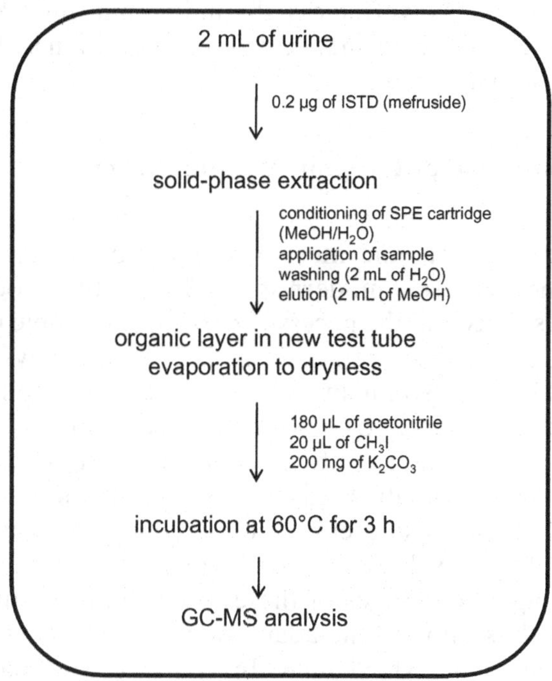

Scheme 6.6: Sample preparation flow scheme for the analysis of diuretics and masking agents. The screening procedure is commonly referred to as *Screening V*.[7]

through the SPE column by gravity flow, the resin washed with 2 mL of water, and the retained compounds are eluted into a test tube using 2 mL of methanol. The eluate is evaporated to dryness and reconstituted in 200 μL of acetonitrile containing 10% of iodomethane. About 200 mg of anhydrous potassium carbonate are added to the sample solution prior to incubation at 50°C for 3 hours in a heating block. After cooling to room temperature, the supernatant is transferred to autosampler vials and 2 μL are injected into the GC-MS system. The analytical instrumentation consists of an Agilent 6890/5973 GC-MS apparatus, equipped with an HP5-MS GC column (length: 17 m, inner diameter: 0.25 mm, film thickness: 0.25 μm). A temperature gradient starting at 100°C increasing to 320°C within 12 minutes is employed, and helium is used with a constant flow of 1.5 mL/min as carrier gas. The mass spectrometer is operated in SIM mode with 20 ms dwell time per ion.

Due to the superior sensitivity of LC-MS(/MS) systems over most GC-MS approaches, excellent compatibility of diuretic agents with LC and ESI or APCI, as well as the option to omit any derivatization using hazardous chemicals, *Screening V* now includes LC-MS(/MS) rather than GC-MS analysis. Nevertheless, the unique information provided by EI-MS spectra of selected analytes after methylation is still valuable for structural characterization and/or confirmatory measurements.

6.1.4 β_2-Agonists and β-Receptor Blocking Agents

In agreement with the analysis of diuretic agents, the detection of β-receptor blocking agents and most β_2-agonists by means of GC-MS methods does not represent modern sports drug testing procedures due to a limited sensitivity and the necessity of complex sample preparation procedures. In the absence of adequate LC-MS(/MS) systems in the late 1980s and early 1990s, many studies were conducted to convert β_2-agonists and β-blockers into GC-compatible compounds that provide sufficient mass spectrometric information under EI conditions. Most β-blockers were conveniently measured using a modified *Screening II* approach (*vide supra*) using enzymatic deconjugation and selective derivatization of hydroxyl and amino functions.[7,28,29,31,133,134] The comparably high therapeutic dosages of these drugs did not require utmost sensitivity and, thus, allowed the adaptation of established procedures providing sufficient analytical data. In contrast, the analysis of β_2-agonists proved more challenging due to low pharmacologically relevant dosages and greatly varying physicochemical properties of the analytes with serious implications particularly regarding multi-residue extraction procedures and derivatization strategies.[135] Several specialized methods targeting a subset of β_2-agonists were established,[136–141] and a representative screening method for doping control purposes consisting of enzymatic hydrolysis, SPE, and trimethylsilylation of target compounds allowed the simultaneous detection of eight β_2-agonists or respective metabolic products at LODs of 0.5–5 ng/mL.[142] A total of 2 mL of urine is enriched with 100 ng of the ISTD penbutolol and 50 µL of β-glucuronidase/arylsulfatase from *Helix pomatia*. After adjusting the specimen to pH 5.2 (using 1.1 M acetate buffer), an incubation step of 2 hours at 55°C is conducted, followed by the addition of 100 µL of a 5.3 M ammonium chloride solution (pH adjusted to 9.5 using ammonium hydroxide) that buffers the pH of the urine specimen to 9.5. The sample is then applied to a Bond Elut Certify SPE cartridge preconditioned with 2 mL of methanol and 2 mL of water. After the urine passed through, the resin is washed consecutively with 2 mL of water, acetate buffer (pH = 4), and methanol, prior to the elution using 2 × 2 mL of chloroform/isopropanol (4:1, *v/v*) containing 2% ammonium hydroxide. After evaporation to dryness, the residue is derivatized

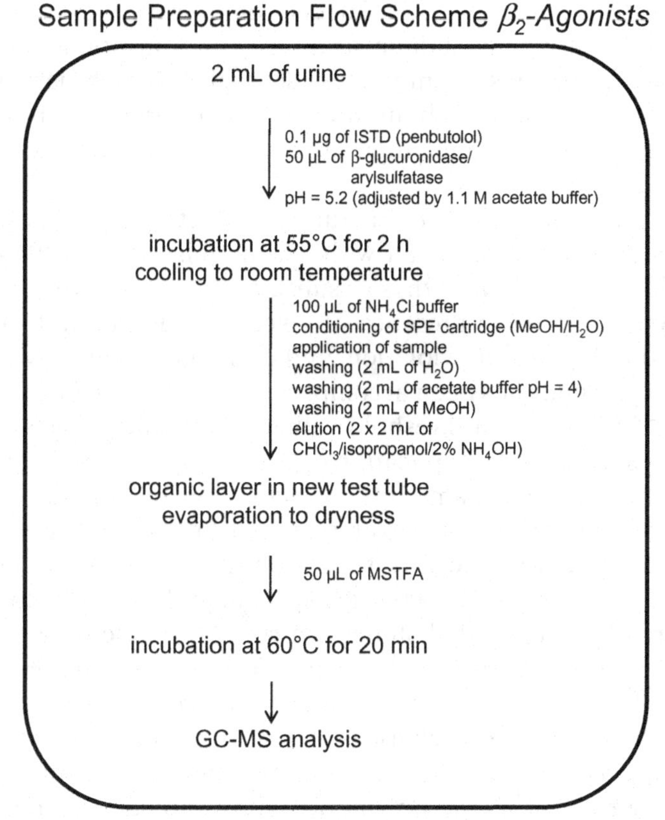

Scheme 6.7: Sample preparation flow scheme for the analysis of β_2-agonists.[142]

by means of 50 µL of MSTFA (20 minutes at 60°C) and subjected to GC-MS analysis (Scheme 6.7).[142] The analytical apparatus consists of a HP5890/5970 GC-MS system operated in SIM mode monitoring 3 diagnostic ions per analyte. A temperature gradient was used starting at 100°C for 2 min, increasing to 190°C at 30°C/min, then to 300°C at 20°C/min, and helium was used with a flow of 0.7 mL/min as carrier gas.

6.1.5 Carbohydrate-Based Agents

Carbohydrate-based agents are relevant for different classes of substances prohibited in elite sports such as glycerol and hydroxyethyl starch (as plasma-volume expanding and hyperhydratation agents) as well as mannitol (as masking agent and osmotic diuretic). Their polar nature complicates their extraction from urine, but the need of large doses to achieve beneficial effects from intravenous or oral (glycerol and mannitol only) administration results in abundant urinary concentrations of these compounds. Several different derivatization strategies were tested to obtain all relevant information required to identify the drugs, determine their polymeric structure, and/or perform quantification.

Glycerol and mannitol are natural products that are commonly found in urine specimens even without administration of therapeutic formulations containing these substances as active principles. Consequently, the mere presence of glycerol and mannitol cannot be considered a doping rule violation, and although quantitative analysis of these compounds has been accomplished by means of GC-MS-based assays,[143,144] the lack of a threshold value or any other means to differentiate the surreptitious application of these drugs for doping purposes from naturally occurring amounts does not allow the unequivocal identification of a misuse yet; however, an upper limit of 200 μg of glycerol per mL of urine was recently suggested, based on the quantitative analysis of approximately 1000 urine specimens. In contrast to glycerol and mannitol, hydroxyethyl starch is a semi-synthetic polysaccharide, which is detected in urine only after infusion of the plasma-volume expander.[145–147] Hence, the qualitative analysis of this drug and/or its hydroxyethylated monosaccharides serves the purpose of screening for an illicit application. For a confirmation of a finding regarding the high molecular weight drug, the polymeric nature should be demonstrated by generating partially methylated alditol acetates[148,149] followed by GC-MS measurement.

A generic screening procedure was established to monitor HES, glycerol, and mannitol requiring 20 μL of urine (Scheme 6.8).[150] The aliquot is treated with 0.5 mL of 3 M HCl for 30 minutes at 100°C, and the solution is evaporated to dryness. The dry residue is subsequently trimethylsilylated using a mixture of 25 μL of pyridine and 75 μL of *in situ* generated trimethyliodosilane, and the sample injected into a GC-MS system consisting of the same setup as formerly used for the analysis of diuretic agents (see section 6.1.3). The only difference is the fact that the analysis is performed in full scan mode.[150] Per-trimethylsilylated hydroxyethylated glucose is readily observed in GC-MS analyses by means of diagnostic fragment ions (e.g., *m/z* 235, 248 , and 261), and suspicious test results are subsequently confirmed in a second analysis with partially methylated alditol acetates derived from HES.[151] This confirmation procedure consists of the evaporation of 20 μL of urine, which is followed by permethylation by adding 200 μL of a suspension of sodium hydroxide in anhydrous dimethylsulfoxide

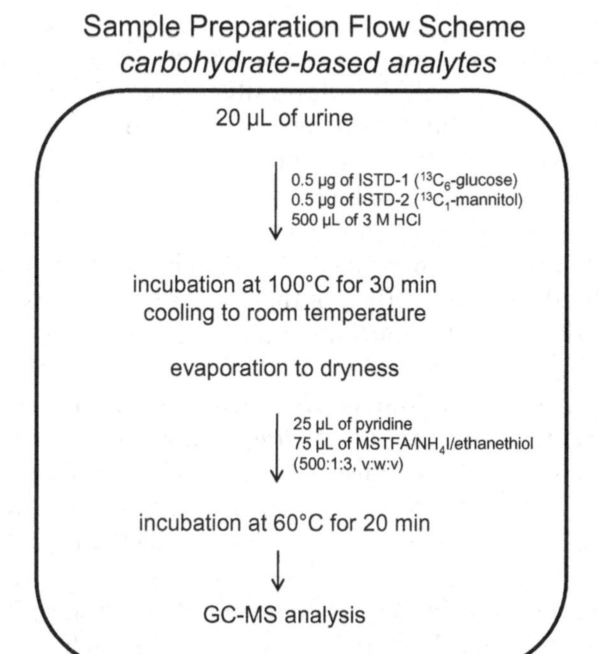

Sample Preparation Flow Scheme
carbohydrate-based analytes
20 µL of urine
0.5 µg of ISTD-1 ($^{13}C_6$-glucose)
0.5 µg of ISTD-2 ($^{13}C_1$-mannitol)
500 µL of 3 M HCl
incubation at 100°C for 30 min
cooling to room temperature
evaporation to dryness
25 µL of pyridine
75 µL of MSTFA/NH_4I/ethanethiol
(500:1:3, v:w:v)
incubation at 60°C for 20 min
GC-MS analysis

Scheme 6.8: Sample preparation flow scheme for the analysis of the polysaccharide-based plasma-volume expander HES, glycerol, and mannitol.[150]

(DMSO) containing 10% of iodomethane. After 10 minutes at room temperature, all free hydroxyl functions are alkylated, and the excess of sodium hydroxide, DMSO, and iodomethane is removed by partitioning the permethylated polysaccharide between 2 mL of water and 2 mL of chloroform. The organic layer is dried and the dry residue subjected to acidic hydrolysis using 0.5 mL of 3 M HCl (1 hour, 100°C). After cooling to room temperature, the solution is evaporated, obtained partially methylated monosaccharides are reduced to corresponding alditols by adding 200 µL of a 0.5% solution of sodium borohydride in methanol and 0.03 M NaOH (1:3, *v/v*) and incubation overnight. The addition of 20 µL of glacial acetic acid is followed by evaporation to dryness, and partially methylated alditols are acetylated. Therefore, a total of 350 µL of a solution composed of acetonitrile, acetic anhydride, and pyridine (3:3:1, *v/v/v*) is added and incubated at 80°C for 2 hours. The solution is subsequently cooled to ambient temperature, 2 mL of water are added, and peracetylated analytes are extracted into 2 mL of chloroform. After evaporation of the organic layer, the residue is reconstituted in 100 µL of isopropanol and analyzed by GC-MS.

In case of atypically high levels of mannitol or glycerol in routine screening analyses, an accurate quantitation is performed to substantiate the result. For glycerol, an isotopically labeled ISTD ($^2H_5$-glycerol) is employed, and the routine screening procedure is modified omitting the acidic hydrolysis.[143] With mannitol it is mandatory to ensure its separation from its "non-prohibited" stereoisomers allitol, altritol, dulcitol, iditol, and sorbitol, the first mentioned of which is employed as internal standard. The confirmatory analysis is based on the peracetylation of a dried residue of 20 μL of the suspicious urine using the same approach as described for partially methylated alditols (see above).[144]

6.1.5.1 Case Vignette: Systematic Misuse of Hydroxyethyl Starch in the Finnish Cross-Country Skiing Team The above reported approaches to screen and confirm the presence of HES was successfully applied during the 2001 Nordic World Championships (WC) held in Lahti, Finland. Developed between 1998 and 2000, the 2001 WC were the first big sporting event where the newly released procedures for HES were part of routine doping controls. Already the first batch of samples received from the very first competitions yielded a suspicious result for HES in the screening procedure based on acidic hydrolysis, trimethylsilylation, and GC-MS analysis. Repetitive measurements using the screening and confirmation approach proved the presence of approximately 10 mg HES per mL of urine in that particular doping control specimen. Reported to the Fédération Internationale de Ski (FIS), the B-sample analysis was requested the following day. After confirmation of the A-sample analysis result, the athlete convicted of the misuse of HES confessed the infusion of HEMOHES, a plasma-volume expander containing the detected compound; however, he claimed that none of his team colleagues or physicians was involved in the doping rule violation. Nevertheless, out-of-competition controls of male and female team members were conduced. These yielded another 5 adverse analytical findings with HES, although urine sampling was conducted up to 7 days post-administration.[152] Moreover, the team doctor's bag was later found at a gas station containing several infusion units of HEMOHES.

6.2 LC-MS/MS

The utility of LC-MS(/MS) for the detection of a great number of drugs relevant for doping controls has been demonstrated in various methods developed and employed during the last decade.[153] The constantly improving quality of LC columns regarding separation power as accomplished, e.g., by monolithic columns or sub-2 μm particle size, in addition to increasing scan speed, polarity switching capability of MS instruments, high resolution/high accuracy analyzers, etc., has allowed taking alternative routes to detect comprehensively many different categories of prohibited drugs in doping control specimens.[154–165] Assays are commonly not particularly dedicated to a single group of analytes but substances belonging to different groups are measured in one single run, and many different combinations and compilations of categories and analytes were reported in the past. In agreement with GC-MS developments, also here numerous approaches have been established, and the methods described in more detail in the following are intended to serve as one example out of many accepted options applied currently in sports drug testing laboratories.

6.2.1 Stimulants/narcotics/β_2-Agonists/β-Receptor Blocking Agents

Numerous methods have been established in the past that allow the detection of stimulants,[166,167] narcotics,[168,169] β_2-agonists,[170,171] or beta-receptor blocking agents[162,172] using LC-MS(/MS). In addition, several approaches were designed to comprehensively screen for most of their representatives simultaneously with adequate sensitivity and specificity to meet the requirements for sports drug testing programs as established by WADA.[2] Due to the considerably different physicochemical properties of several analytes, the isolation as well as ionization of all compounds has been a great challenge and required sophisticated sample preparation as well as measurement strategies. A recently published screening protocol allows for the detection of 197 different compounds relevant for doping controls including the above mentioned classes of banned substances.[164] In order to fulfill the demands of appropriate sports drug testing, a two-step SPE methodology followed by two consecutive analytical runs (in positive and negative ESI mode) was suggested using the following conditions: One mL of urine is enriched with two ISTDs (10 ng of each dibenzepin and $^2H_4$-cortisol) and subjected to enzymatic hydrolysis by adding 0.375 mL of 0.8 M sodium/potassium phosphate buffer (pH = 7) and 15 μL of β-glucuronidase from *E. coli*. The solution is incubated for 1 hour at 50°C, cooled to room temperature, and loaded onto a mixed-mode cation exchange SPE cartridge (Isolute IST HCX, 130 mg) preconditioned with 1 mL of methanol and 1 mL of 0.1 M ammonium acetate buffer (pH = 6). The SPE column is washed twice, first with 2 mL of water, and second with 2 mL 0.01 M aqueous HCl. Both fractions are collected, combined, and later used for further analyte extraction. The compounds retained on the HCX resin are eluted using 2 × 1 mL of a mixture prepared from methanol and ammonium hydroxide (98:2, *v/v*) yielding the elution volume **1**. The combined wash fractions are adjusted to pH 7 by adding 0.6 mL of a mixture of 1 M ammonium acetate and 0.05 M NaOH, and then applied to a HAX mixed-mode anion exchange SPE cartridge (130 mg), preconditioned with 1 mL of methanol and 1 mL of a mixture of 1 M ammonium acetate and 0.05 M NaOH (pH = 7). The HAX resin is subsequently washed with 1 mL of 0.1 M ammonium acetate (pH = 7) and 1 mL of 50% methanol, followed by the elution of target compounds using 2 × 1 mL of a mixture of methanol and acetic acid (98:2, *v/v*) yielding the elution volume **2** (Scheme 6.9).[164] Both elution volumes are combined, evaporated to dryness, and reconsti-

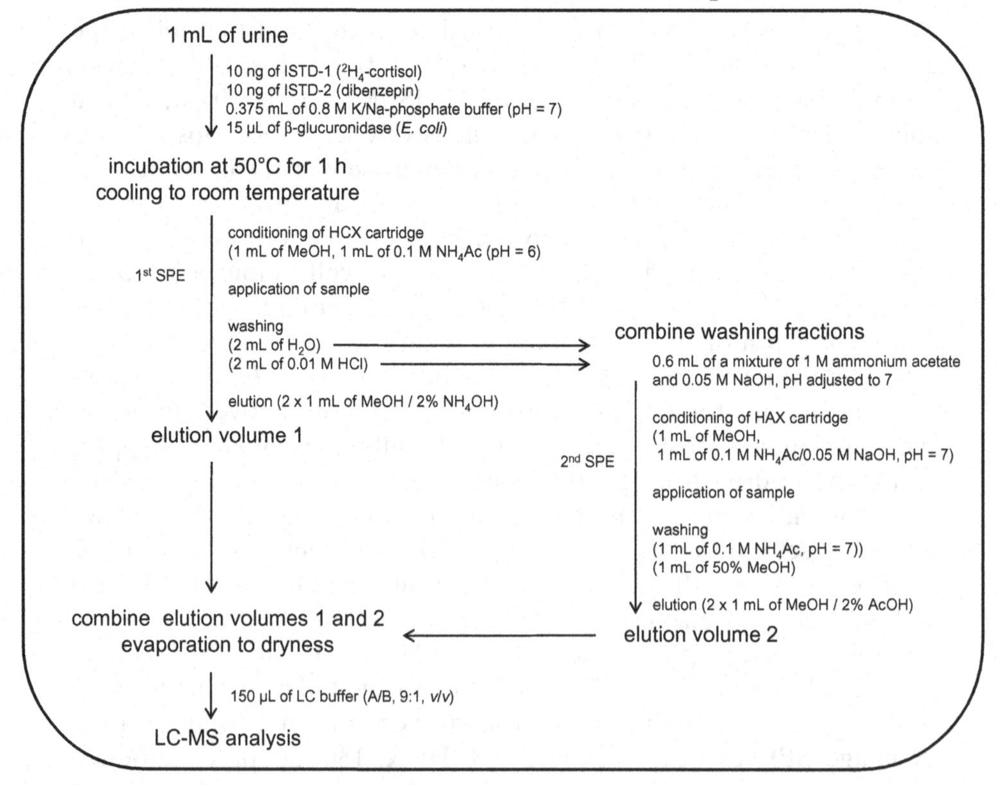

Scheme 6.9: Sample preparation flow scheme for the combined analysis of stimulants, narcotics, β_2-agonists, and β-blockers.[164]

tuted in 150 μL of the mobile phases A and B (9:1, *v/v*), which are composed of 2.5 mM ammonium formate/0.1% formic acid (A) and 2.5 mM ammonium formate/0.1% formic acid in 90% acetonitrile (B).

The analysis of the urine extract is conducted using an Agilent 1200 series rapid resolution LC system equipped with a Zorbax Eclipse Plus rapid resolution HT C-18 column (50 × 2.1 mm, particle size 1.8 μm). The injection volume is 3 μL, the flow rate is 0.4 mL/min, and gradient elution of analytes starts with a 1 minute isocratic step at 90% A, followed by a linear decrease to 60% A in 2 minutes, to 30% A in 1 minute, and finally to 10% A in 2 minutes. The composition is maintained at 10% A for another 0.5 min before re-equilibration. The employed mass spectrometer is a Bruker Daltonics micrOTOF operated either in positive ESI (capillary voltage = 4000 V) or negative ESI (capillary voltage = −3400 V) mode, covering the full scan mass ranges *m/z* 50–600 allowing for an average resolution of 11,000 for the ISTD dibenzepin at *m/z* 296.

Due to the obtained purity of the urine extracts combined with the high resolution and accurate mass capability of the analytical instrumentation, 72 stimulants and metabolites, 21 narcotics and metabolites, 10 β_2-agonists, and 31 β-blockers are, besides numerous additional compounds, simultaneously detected at minimum required performance levels.[164]

6.2.2 Anabolic Androgenic Steroids/Corticosteroids/Hormone Antagonists and Modulators

The lipophilic character of most steroidal agents has resulted in various successful attempts to efficiently combine the extraction and analysis of AAS and corticosteroids as well as other banned compounds such as hormone antagonists and modulators.[159,173–175] Especially the class of corticosteroids and a subset of AAS (including e.g., trenbolone, stanozolol, gestrinone, and tetrahydrogestrinone) has been challenging for GC-MS systems primarily due to their poor GC properties, and LC-MS(/MS) was found to readily provide the required sensitivity and specificity for these analytes in sports drug testing. Dedicated methods are available to screen and confirm (cortico)steroids,[157,176,177] aromatase inhibitors,[178] antiestrogens,[179] hormone modulators, etc.;[155] however, modified versions of the above mentioned *Screening IV* (see section 6.1.2.1) protocol are commonly used to allow for the combined extraction of these drugs from urine specimens, and sufficient LODs were found in several different studies using LC-MS/MS detection. Hence, also the splitting of conventional *Screening IV* sample extracts into

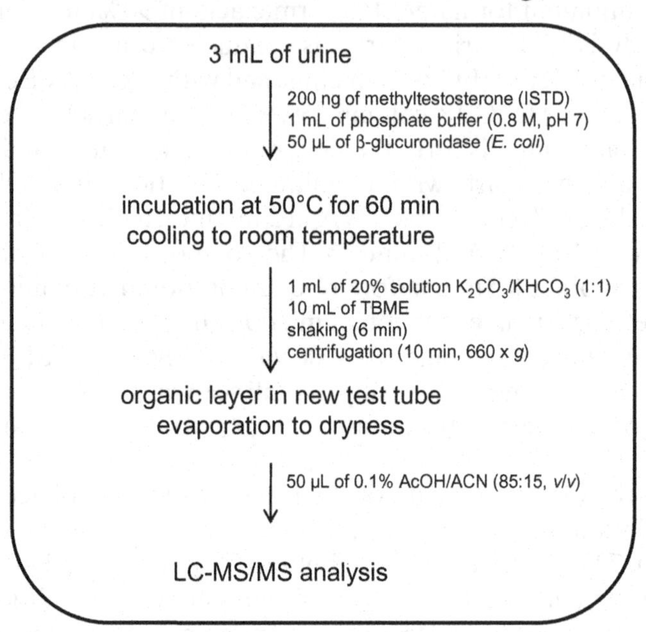

Scheme 6.10: Sample preparation flow scheme for the combined analysis of corticosteroids, selected AAS, and hormone antagonists and modulators.[173]

LC-MS/MS and GC-MS aliquots prior to the derivatization step has been successfully employed.[175] A representative approach[173] is reported in brief in the following (Scheme 6.10).[173]

To a 3 mL aliquot of urine, 1 mL of phosphate buffer (pH = 7), 50 μL of β-glucuronidase, and 200 ng of methyltestosterone (ISTD) are added, and the solution is incubated at 50°C for 1 hour. After cooling to room temperature, the mixture is adjusted to pH = 9 using 1 mL of carbonate buffer, and analytes are extracted into 10 mL of *tert.*-butyl methyl ether. The organic layer is separated, evaporated to dryness, and the residue reconstituted in 50 μL of 0.1% acetic acid and acetonitrile (85:15, *v/v*). The LC-MS/MS analysis is conducted using an Agilent 1100 Series HPLC equipped with a Zorbax C-18 column (2.1 × 50 mm, particle size 1.8 μm) using 0.1% acetic acid (solvent A) and acetonitrile (containing 0.1% acetic acid, solvent B). Following an injection of 5 μL, a gradient elution is accomplished starting at 15% B, increasing to 60% B within 5 minutes and subsequently to 100% B in another 2 minutes, using a flow rate of 0.3 mL/min. The effluent is directed to an API4000 triple quadrupole MS employing positive ESI and MRM. Two ion transitions per analyte are used to provide sufficient specificity and sensitivity for the detection of 15 corticosteroids, 3 hormone antagonists, and 4 AAS at required MRPLs.

6.2.3 Masking Agents (Including Diuretics)

One of the first classes of substances transferred from GC-MS to LC-MS(/MS) analysis was the heterogeneous group of diuretics.[180] Driven by the aim to circumvent the time-consuming derivatization that necessitated the use of hazardous chemicals, established sample extraction steps (either LLE or SPE) were used to recover the target compounds from urine specimens. The dried extracts were dissolved in LC buffers, and the underivatized substances measured by LC-MS(/MS), preferably by multiple reaction monitoring.[181,182] The simplicity of the approach and the comprehensiveness of the extraction allowed the inclusion of many additional analytes into the screening procedure such as probenecid, efaproxiral,[183] the cocaine metabolite benzoyl ecgonine, selective androgen receptor modulators,[184] and many more.

With the introduction of ultra-performance LC (UPLC), fast polarity switching MS instruments, and/or information-dependent data acquisition, the direct injection of urine specimens following dilution was established and yielded comprehensive detection methods, particularly for diuretics, masking agents, etc.[156,185] The convenience of skipping nearly all sample preparation steps is challenged only by the increased risk of ion suppression due to the significantly elevated amounts of matrix components that are not removed by any purification step; however, a recently reported assay demonstrated the fitness-for-purpose of a screening procedure for 130 target compounds using automated urine dilution followed directly by injection of the specimens.[156] The assay consists of a dilution step where 100 μL of the urine samples is added to 100 μL of an ammonium acetate buffer (20 mM) containing 2 μg/mL of the ISTD (4-phenoxy-3-[1H-pyrrol-1-yl]-5-sulfamoylbenzoic acid) in a 96-well microtiter plate, which is accomplished by a Xiril Robotic Workstation pipetting robot. Subsequently, the samples are subjected to LC-MS/MS analysis injecting 5 μL into a Waters Acquity UPLC system. LC is done on a BEH Shield RP-18 column (50 × 2.1 mm, particle size 1.7 μm) at 60°C using 10 mM ammonium acetate (solvent A) and methanol (solvent B). The flow rate is set to 0.4 mL/min and starts with 5% B for 0.5 minute. A linear gradient increases B to 40% at 4 minutes followed by another linear step to 95% B at 5 minutes. After maintaining at 95% for 1 minute, re-equilibration is conducted for 1.5 minutes, allowing to complete a single run within 7.5 minutes in total. The LC effluent is directed to a Waters Quattro Premier triple-quadrupole MS interfaced via ESI source. The MS is operated in fast polarity switching mode and MRM for targeted analysis allowing dwell times of 5 ms per ion transition. The MS method is split into 30 groups to enable sufficient data point recording (>10 data points per peak, 2 ion transitions per analyte), and more than 35 diuretics, 20 narcotics, 65 stimulants, and several additional analytes are measured at required MRPLs.

6.2.4 Peptide Hormones <10 kDa Including Insulins, Synacthen, Gonadorelin, IGF-1

One of the most important contributions of LC-MS(/MS) to doping controls has been the ability to detect peptide hormones with superior selectivity and specificity (and, thus, confidence) compared to many immunoassays. With the significant improvements in sensitivity of LC-MS(/MS) instruments as accomplished by the introduction of nano-flow LC systems interfaced by nanospray ion sources to state-of-the-art mass spectrometers, minute amounts of banned peptide hormones such as insulins, synacthen, gonadorelin, and IGF-1 as well as its derivatives were successfully detected in plasma and urine specimens.[186–194] A central aspect has been the isolation of target analytes from doping control samples using immunoaffinity purification prior to the MS-based analysis, in order to provide a specimen of adequate quality to allow nanoscale analytical methods. Initially achieved by means of SPE and/or immunoaffinity chromatography, more recently the utility of secondary antibody-coated magnetic beads has demonstrated excellent performance and ease of use.[190] Moreover, the option to extend an existing sample preparation to new (classes of) analytes by adding appropriate primary antibodies to the immunoaffinity purification step allows the flexible expansion of established methods without the need of additional sample preparations and corresponding specimen volume consumption.

A comprehensive assay allowing the detection of synthetic (rapid- as well as long-acting) and animal insulins, gonadorelin, IGF-1, and synacthen in human urine was recently described requiring a total of 5 mL of urine.[195] The sample is enriched with 500 μL of acetonitrile and 500 fmol of the ISTD (bovine insulin) and subsequently subjected to SPE using an Oasis HLB cartridge preconditioned with 2 mL of acetonitrile and 2 mL of water. Loading of the urine sample is followed by washing of the resin with 2 mL of water and elution of the retained material into an Eppendorf tube using 1.4 mL of 80% acetonitrile. After evaporation of the eluate, the residue is reconstituted in 500 µL of phosphate-buffered saline (PBS) and extraction of target analytes is conducted by adding 35 µL of a suspension of secondary antibody-coated magnetic beads and adequate amounts of monoclonal or polyclonal primary antibodies directed against insulins, synacthen, as well as gonadorelin. Following a 2 hour incubation period at 4°C, the nanoparticles carrying the antibody-antigen complexes are immobilized by a magnetic separator device, washed repeatedly with 300 µL of PBS, and analytes are finally eluted using 50 µL of acetic acid (2%). The immunoaffinity-purified sample is subsequently forwarded to nanoLC-MS/MS analysis (Scheme 6.11). The apparatus consists of an Acquity nanoflow UPLC (nanoUPLC) system equipped with a BEH-130 C-18

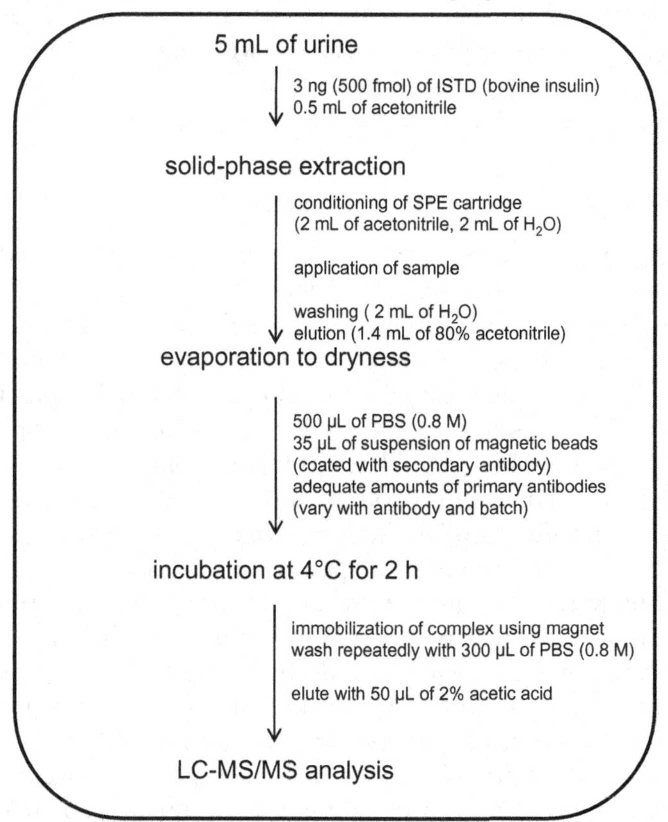

Scheme 6.11: Sample preparation flow scheme for the detection of peptide hormones including insulins, LH-RH, synacthen, and IGF-1.

analytical column (75 µm × 100 mm, 1.7 µm particle size) protected by a Symmetry C-18 trapping column (180 µm × 20 mm, 5 µm particle size), and a LTQ-Orbitrap hybrid mass spectrometer. Solvents used are ultra-pure water (A) and acetonitrile (B), both containing 0.1% of formic acid. After injecting 1 µL of the sample, analytes are concentrated on the trapping column at a flow rate of 5 µL and a solvent composition of 97% A / 3% B. After 3 minutes, the flow is reduced to 750 nL/min and diverted via the analytical column to the mass spectrometer employing gradient elution. Following an isocratic step of 1 minute at 97% A, the percentage of A is reduced linearly to 10% within 21 minutes, and re-equilibration is required at 97% A for 13 minutes prior to the next injection. The mass spectrometer is operated in two different modes including high resolution/high accuracy full scan analysis (resolving power 30,000) and low resolution tandem mass spectrometry for pre-selected precursor ions of target compounds.

Using the selected approach, a multi-analyte screening procedure for peptide hormones has been established that allows the identification of a variety of drugs in urine specimens with detection limits ranging from 1–5 pg/mL. A chromatogram of a spiked urine specimen containing 14 different analytes is depicted in Figure 6.6.

6.2.5 Hemoglobin-Based Oxygen Carriers

For hemoglobin-based oxygen carriers (HBOCs), a MS-based screening procedure has not been required due to the literally visible influence of HBOCs on the color of human serum or plasma.[196] Only in case of suspicion, plasma or serum samples are subjected to further investigation, which can be accomplished either by non-mass spectrometric methods[197–200] or assays using bottom-up sequencing approaches employing LC-MS/MS.[199,201–204] The latter commonly consist of the enzymatic degradation of plasma proteins and the detection of either bovine hemoglobin-specific peptides, the presence of which constitute a doping offence, or the determination of peptide ratios derived from human hemoglobin, which indicate the presence of lysine-located crosslinkers. A representative assay was reported by Goebel and associates in 2005, which consists of the enzymatic hydrolysis of 100 µL of plasma by the addition of 10 µL of 0.5 M NH_4HCO_3 and 20 µg of trypsin. After incubation at 37°C for 4.5 hours, the hydrolysis is terminated by the addition of 20 µL of 0.5 M hydrochloric acid (Scheme 6.12).[199] Ten µL of the final solution are injected onto a Phenomenex Jupiter Proteo 90A column (2.1 × 150 mm, particle size 4 µm), which is protected by a Phenomenex Security Guard Max-RP column (2 × 4 mm). Solvents for gradient elution are 0.2% formic acid (A) and acetonitrile (B), starting at 95%

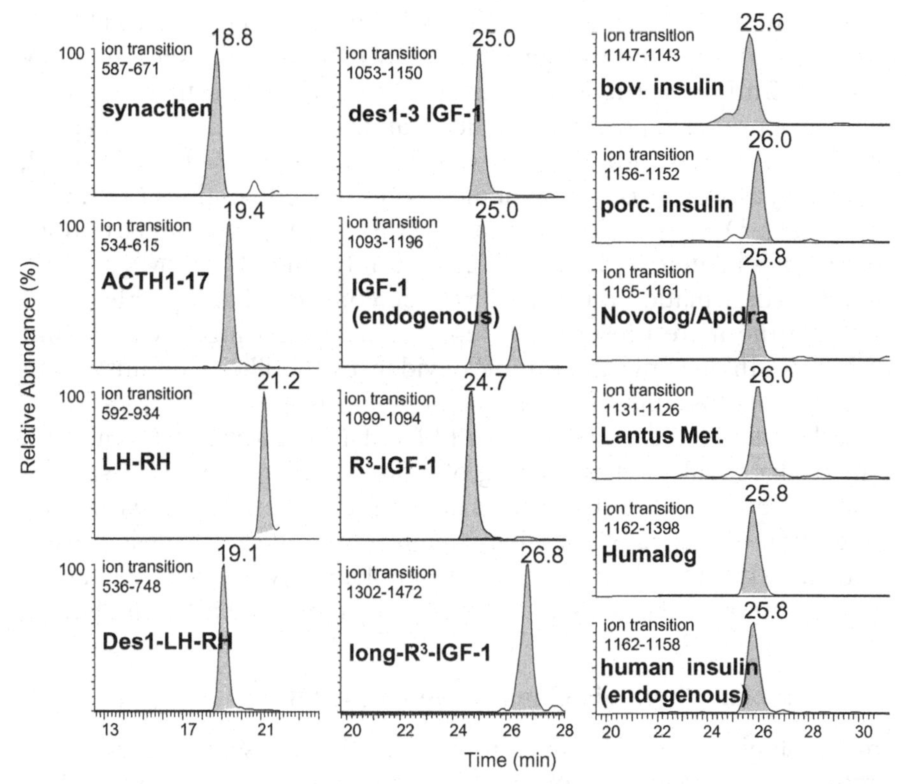

Figure 6.6: Extracted ion chromatograms of a urine sample fortified with synacthen (7 fmol/mL), ACTH 1-17 (ISTD, 10 fmol/mL), LH-RH (20 fmol/mL), des1-LH-RH (ISTD, 20 fmol/mL), des1-3-IGF-1 (15 fmol/mL), R^3-IGF-1 (15 fmol/mL), long-R^3-IGF-1 (15 fmol/mL), bovine insulin (ISTD), porcine insulin, Novolog, Apidra, a Lantus metabolite and Humalog (20 fmol/mL each), measured on a Waters nanoAcquity/Thermo LTQ instrument.

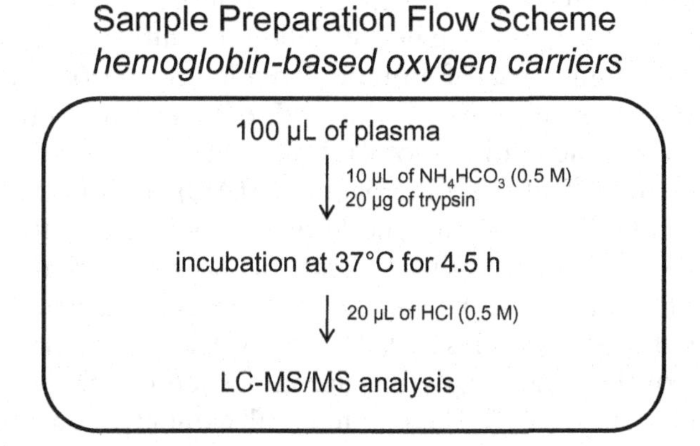

Scheme 6.12: Sample preparation flow scheme for the detection of hemoglobin-based oxygen carriers.[199]

A for 5 minutes. The percentage of A is linearly decreased to 60% A within 60 minutes and further reduced to 10% in 80 minutes at a flow rate of 0.2 mL/min. The MS (Micromass Quattro micro triple quadrupole) is operated in positive ESI mode, and precursor-product ion pairs are generated to identify human (aT12, bT1), bovine (aT8, aT9, aT12, and bT1), as well as common hemoglobin-derived peptides (aT3, aT5, and aT11). Due to the comparably fast and simple screening options for HBOCs in human plasma, the comparably long LC-MS/MS analysis time for confirmation purposes has been considered acceptable.

While the mere presence of peptides characterizing bovine hemoglobin in athletes' plasma samples evidences the illicit administration of HBOCs, cross-linked human hemoglobin is identified by ratios of peptide abundances, especially of aT12 and aT3 as well as bT1 and aT3. Due to the covalent cross-linking of alpha- and beta-chains at specific lysine residues, trypsin digestion is hampered at these locations and the generation of respective peptides significantly reduced. This is considered indicative for the presence of intramolecularly cross-linked human hemoglobin as present, e.g., in case of PolyHeme (see also Chapter 5).

6.2.6 Human Chorionic Gonadotrophin (hCG)

Immunological methods are commonly used in the screening and confirmation of hCG;[205–207] however, in 2003 an MS-based method was reported that allows the qualitative and quantitative detection of hCG in human urine using a bottom-up approach, which was designed to serve as confirmatory procedure in case of adverse analytical findings.[208] Although a differentiation between natural and recombinant hCG is not accomplished, the identification and quantitation at threshold levels is achieved (Scheme 6.13).[208] The method requires a total of 11 mL of urine, which is centrifuged to remove particulate matter. Before loading 10 mL of urine in 2 mL-aliquots onto an immunoaffinity purification column (2 mL of antibody-coupled Sepharose 4B gel), the antibody-coated resin is flushed with 6 mL of water and 5 mL of 0.01 M PBS (pH = 7.2). Each volume of 2 mL of urine is allowed to incubate with the immunoaffinity chromatography (IAC) gel for 20 minutes by closing the exit of the column, and after removal of the urine aliquot, a washing step is conducted using 2 mL of the above mentioned PBS buffer. When the entire urine sample (5 × 2 mL) is extracted, the resin is washed with 7 bed volumes of 0.1% Tween 20 in 0.1 M PBS (pH = 7.2), and elution of the target analyte is accomplished by means of 9 mL of citric acid (1 M, pH = 2.2). The eluate is subsequently concentrated on an SPE cartridge, and hCG is recovered by 0.1% formic acid in 70%

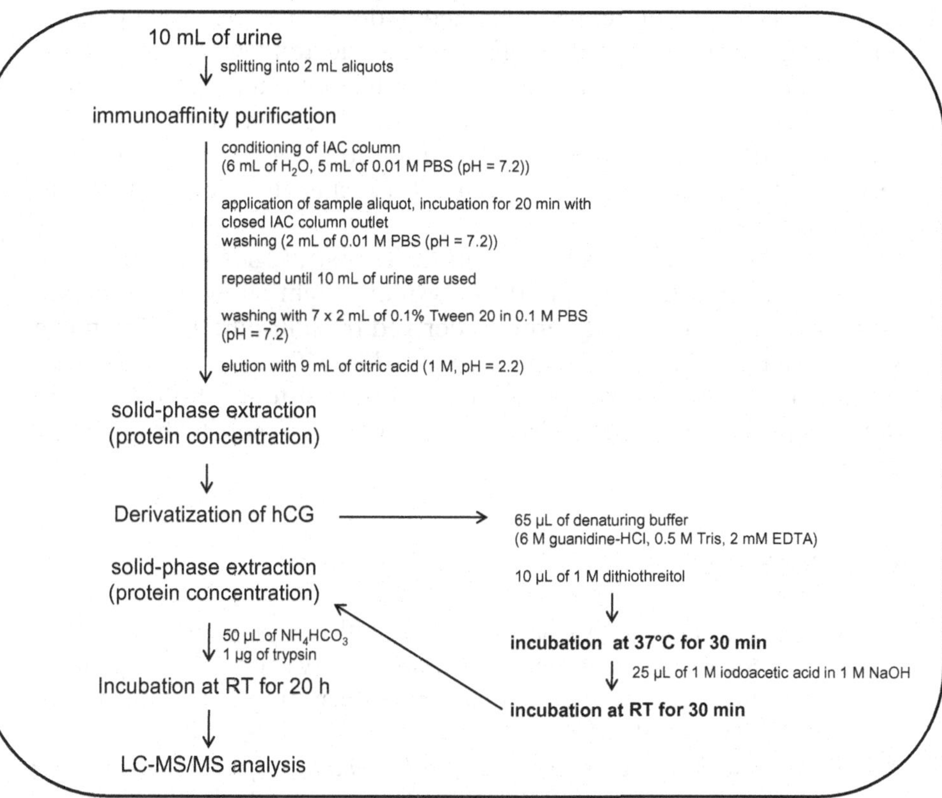

Scheme 6.13: Sample preparation flow scheme for the determination of hCG in human urine.[208]

acetonitrile. After evaporation to dryness, the protein is reconstituted in 65 µL of a denaturing buffer (6 M guanidine-HCl, 0.5 M Tris, 2 mM EDTA, pH = 8.6), and 10 µL of 1 M dithiothreitol (DTT) are added before incubation at 37°C for 30 minutes. Further, 25 µL of 1 M iodoacetic acid in 1 M NaOH are added followed by incubation of the solution at room temperature. After removal of excess reagents by another SPE (according to the conditions described above), the derivatized hCG is hydrolyzed in 50 µL of 50 mM NH_4HCO_3 containing 1 µg of trypsin. After 20 hours at room temperature, 2 µL of the sample is subjected to LC-MS/MS analysis, which is focused particularly on one distinct peptide (bT5) that allows the differentiation of hCG and LH (see also Chapter 5). The LC-MS/MS instrument is composed of an Agilent 1100 Series HPLC equipped with a Vydac RP C-18 column

(1.0 × 250 mm, 5 μm particle size). Gradient elution (linear from 10% B to 60% B in 60 minutes) is employed to determine the target peptide bT5 using the solvents 0.05% TFA (A) and acetonitrile (B, containing 0.05% TFA). The flow rate is set to 50 μL/min, and the effluent is directed to a Finnigan MAT LCQ Cassic MS, interfaced to the LC system by ESI in positive mode. Precursor ions for tryptic peptides of hCG are preselected and programmed for inclusion in the data-dependent analysis.

More modern immunoaffinity purification strategies (e.g., based on magnetic nanoparticles) and UPLC systems might allow the detection and quantitation of target peptides derived from urinary hCG in much shorter analytical run times, and MALDI-MS/MS might serve as alternative to LC-ESI-MS/MS also;[209] however, until today, no other validated approach has been presented with regard to sports drug testing applications.

REFERENCES

1. World Anti-Doping Agency (2009) *The World Anti-Doping Code—International Standard for Laboratories*. Available at http://www.wada-ama.org/rtecontent/document/International_Standard_for_Laboratories_v6_0_January_2009.pdf. Accessed 11-24-2008.
2. World Anti-Doping Agency (2009) *Minimum Required Performance Limits for Detection of Prohibited Substances*. Available at http://www.wada-ama.org/rtecontent/document/MINIMUM_REQUIRED_PERFORMANCE_LEVELS_TD_v1_0_January_2009.pdf. Accessed 11-24-2008.
3. World Anti-Doping Agency (2004) *Identification Criteria for Qualitative Assays Incorporating Chromatography and Mass Spectrometry*. Available at http://www.wada-ama.org/rtecontent/document/criteria_1_2.pdf. Accessed 07-27-2006.
4. Thevis, M., and Schänzer, W. (2005) Mass spectrometry in doping control analysis. *Current Organic Chemistry*, **9**, 825–848.
5. Thevis, M., and Schänzer, W. (2007) Mass spectrometry in sports drug testing: Structure characterization and analytical assays. *Mass Spectrometry Reviews*, **26**, 79–107.
6. Beckett, A.H., Tucker, G.T., and Moffat, A.C. (1967) Routine detection and identification in urine of stimulants and other drugs, some of which may be used to modify performance in sport. *Journal of Pharmacy and Pharmacology*, **19**, 273–294.
7. Donike, M., Geyer, H., Gotzmann, A., *et al.* (1987) Dope analysis, in *International Athletic Foundation—World Symposium on Doping in Sport*

(eds P. Bellotti, G. Benzi, and A. Ljungqvist) Arti Grafiche Danesi, Florence, pp. 53–80.

8. Richter, D. (1938) A colour reaction for benzedrine. *Lancet*, **234**, 1275.
9. Donike, M. (1970) Stickstoffdetektor und Temperaturprogrammierte Gas-Chromatographie, ein Fortschritt für die Routinemäßige Dopingkontrolle. *Sportarzt und Sportmedizin*, **21**, 27–30.
10. Donike, M. (1970) Temperature programmed gas chromatographic analysis of nitrogen-containing drugs: The reproducibility of retention times (I). *Chromatographia*, **3**, 422–424.
11. Donike, M., Jaenicke, L., Stratmann, D., and Hollmann, W. (1970) Gas chromatographic detection of nitrogen-containing drugs in aqueous solutions by means of the nitrogen detector. *Journal of Chromatography*, **52**, 237–250.
12. Donike, M., and Stratmann, D. (1974) Temperaturprogrammierte gas-chromatographische Analyse stickstoffhaltiger Pharmaka: Die Reproduzierbarkeiten der Retentionszeiten und der Mengen bei automatischer Injektion (II) "Die Screeningprozedur für flüchtige Dopingmittel bei den Olympischen Spielen der XX. Olympiade München 1972". *Chromatographia*, **7**, 182–189.
13. Jacobsen, E., and Gad, I. (1940) Die Ausscheidung des β-Phenylisopropylamins bei Menschen. *Archiv für Experimentelle Pathologie und Pharmakologie*, **196**, 280–289.
14. Brodie, B.B., and Udenfriend, S. (1945) The estimation of basic organic compounds and a technique for the appraisal of specificity. *Journal of Biological Chemistry*, **158**, 705–714.
15. Beyer, K.H., and Skinner, J.T. (1940) The detoxication and excretion of beta phenylisopropylamine (benzedrine). *Journal of Pharmacology and Experimental Therapeutics*, **68**, 419–432.
16. McNally, W.D., Bergman, W.L., and Polli, J.F. (1944) The quantitative determination of amphetamine. *Journal of Laboratory and Clinical Medicine*, **32**, 913–917.
17. Schoen, K. (1944) A rapid and simple method for the determination of ephedrine. *Journal of the American Pharmaceutical Association*, **33**, 116–118.
18. Keller, R.E., and Ellenbogen, W.C. (1952) The determination of d-amphetamine in body fluids. *Journal of Pharmacology and Experimental Therapeutics*, **106**, 77–82.
19. Axelrod, J. (1953) Studies on sympathomimetic amines. II. The biotransformation and physiological disposition of D-amphetamine, D-P-hydroxyamphetamine and D-methamphetamine. *Journal of Pharmacology and Experimental Therapeutics*, **110**, 315–326.
20. Munier, R., and Macheboeuf, M. (1949) Microchromatographie de partage des alcaloides et de diverses bases azotees biologiques. *Bulletin De La Societe De Chimie Biologique*, **31**, 1144–1162.

21. Munier, R., and Macheboeuf, M. (1951) Microchromatographie de partage sur papier des alcaloides et de diverses bases azotees biologiques. 3. Exemples de separations de divers alcaloides par la technique en phase solvante acide (familles de latropine, de la cocaine, de la nicotine, de la sparteine, de la strychnine et de la corynanthine). *Bulletin De La Societe De Chimie Biologique*, **33**, 846–856.
22. Mannering, G.J., Dixon, A.C., Carrol, N.V., and Cope, O.B. (1954) Paper chromatography applied to the detection of opium alkaloids in urine and tissues. *Journal of Laboratory and Clinical Medicine*, **44**, 292–300.
23. Jatzkewitz, H. (1953) Ein klinisches Verfahren zur Bestimmung von basischen Suchtmitteln um Harn. *Hoppe-Seylers Zeitschrift fur Physiologische Chemie*, **292**, 94–100.
24. Kaiser, H., and Jori, H. (1954) Beiträge zum toxikologischen Nachweis von Dromoran "Roche", Morphin, Dilaudid, Cardiazol Coramin und Atropin mit Hilfe der Papierchromatographie. *Arch Pharm Ber Dtsch Pharm Ges*, **287**, 224–242; contd.
25. Kaiser, H., and Jori, H. (1954) Beiträge zum toxikologischen Nachweis von Dromoran "Roche", Morphin, Dilaudid, Cardiazol Coramin und Atropin mit Hilfe der Papierchromatographie. *Arch Pharm Ber Dtsch Pharm Ges*, **287**, 253–258; concl.
26. Vidic, E. (1955) Eine Methode zur Identifizierung papierchromatographisch isolierter Arzneistoffe. *Archiv für Toxikologie. Fuehner Wielands Sammlung von Vergiftungsfällen*, **16**, 63–73.
27. Donike, M. (1973) Acylierung mit Bis(Acylamiden); N-Methyl-Bis (Trifluoracetamid) und Bis(Trifluoracetamid), zwei neue Reagenzien zur Trifluoracetylierung. *Journal of Chromatography*, **78**, 273–279.
28. Donike, M. (1975) N-Trifluoracetyl-O-trimethylsilyl-phenolalkylamine—Darstellung und massenspezifischer gaschromatographischer Nachweis. *Journal of Chromatography*, **103**, 91–112.
29. Donike, M., Derenbach, J. (1976) Die Selektive Derivatisierung Unter Kontrollierten Bedingungen: Ein Weg zum Spurennachweis von Aminen. *Zeitschrift für analytische Chemie*, **279**, 128–129.
30. Solans, A., Carnicero, M., de la Torre, R., and Segura, J. (1995) Comprehensive screening procedure for detection of stimulants, narcotics, adrenergic drugs, and their metabolites in human urine. *Journal of Analytical Toxicology*, **19**, 104–114.
31. Hemmersbach, P., and de la Torre, R. (1996) Stimulants, narcotics and β-blockers: 25 years of development in analytical techniques for doping control. *Journal of Chromatography B*, **687**, 221–238.
32. Nolteernsting, E., Opfermann, G., and Donike, M. (1996) 1-(N,N-Diisopropylamino)-n-alkanes: A new reference system for systematical identification of nitrogen containing substances by gas-chromatography and nitrogen specific or mass spectrometrical detection, in *Recent Advances*

in Doping Analysis (eds M. Donike, H. Geyer, A. Gotzmann, and U. Mareck-Engelke), Sport&Buch Strauss, Cologne, pp. 369–388.

33. Thevis, M., Sigmund, G., Koch, A., and Schänzer, W. (2007) Determination of tuaminoheptane in doping control urine samples. *European Journal of Mass Spectrometry*, **13**, 213–221.
34. Delbeke, F.T., and Debackere, M. (1993) Influence of hydrolysis procedures on the urinary concentrations of codeine and morphine in relation to doping analysis. *Journal of Pharmaceutical and Biomedical Analysis*, **11**, 339–343.
35. Jennison, T.A., Wozniak, E., and Nelson, G. (1991) The quantitative conversion of morphine 3-β-D-glucuronide to morphine using β-glucuronidase obtained from *Patella vulgata* as compared to acid hydrolysis. *Journal of Analytical Toxicology*, **17**, 208–210.
36. Van Thuyne, W., Van Eenoo, P., and Delbeke, F.T. (2007) Comprehensive screening method for the qualitative detection of narcotics and stimulants using single step derivatisation. *Journal of Chromatography. B, Analytical Technologies in the Biomedical and Life Sciences*, **857**, 259–265.
37. Van Thuyne, W., Van Eenoo, P., and Delbeke, F.T. (2008) Implementation of gas chromatography combined with simultaneously selected ion monitoring and full scan mass spectrometry in doping analysis. *Journal of Chromatography A*, **1210**, 193–202.
38. Marsh, D.F., Howard, A., and Herring, D.A. (1951) The comparative pharmacology of the isomeric nitrogen methyl substituted heptylamines. *Journal of Pharmacology and Experimental Therapeutics*, **103**, 325–329.
39. Perrenoud, L., Saugy, M., and Saudan, C. (2009) Detection in urine of 4-methyl-2-hexaneamine, a doping agent. *Journal of Chromatography B*, **877**, 3767–3770.
40. Gohlke, R.S., and McLafferty, F.W. (1962) Mass spectrometric analysis—Aliphatic amines. *Analytical Chemistry*, **34**, 1281–1287.
41. Hammerum, S. (1988) Distonic radical cations in gaseous and condensed phase. *Mass Spectrometry Reviews*, **7**, 123–202.
42. Hammerum, S., Norrman, K., Solling, T.I., *et al.* (2005) Competing simple cleavage reactions: The elimination of alkyl radicals from amine radical cations. *Journal of the American Chemical Society*, **127**, 6466–6475.
43. Todd, T. (1987) Anabolic steroids: The gremlins of sport. *Journal of Sports History*, **14**, 87–107.
44. Schänzer, W., Delahaut, P., Geyer, H., *et al.* (1996) Long-term detection and identification of metandienone and stanozolol abuse in athletes by gas chromatography- high- resolution mass spectrometry. *Journal of Chromatography B*, **687**, 93–108.
45. Schänzer, W., Geyer, H., Fusshöller, G., *et al.* (2006) Mass spectrometric identification and characterization of a new long-term metabolite of metandienone in human urine. *Rapid Communications in Mass Spectrometry*, **20**, 2252–2258.

46. Donike, M., Zimmermann, J., Bärwald, K.R., *et al.* (1984) Routinebestimmung von Anabolika in Harn. *Deutsche Zeitschrift für Sportmedizin*, **35**, 14–24.
47. Ayotte, C., Goudreault, D., and Charlebois, A. (1996) Testing for natural and synthetic anabolic agents in human urine. *Journal of Chromatography B, Biomedical Applications*, **687**, 3–25.
48. Kicman, A.T., Gower, D.B. (2003) Anabolic steroids in sport: Biochemical, clinical and analytical perspectives. *Annals of Clinical Biochemistry*, **40**, 321–356.
49. Marcos, J., Pascual, J.A., de la Torre, X., and Segura, J. (2002) Fast screening of anabolic steroids and other banned doping substances in human urine by gas chromatography/tandem mass spectrometry. *Journal of Mass Spectrometry*, **37**, 1059–1073.
50. Saugy, M., Cardis, C., Robinson, N., and Schweizer, C. (2000) Test methods: Anabolics. *Baillière's Best Practice and Research Clinical Endocrinology and Metabolism*, **14**, 111–133.
51. Kochakian, C.D. (1976) Metabolic effects of anabolic-androgenic steroids in experimental animals, in *Anabolic-Androgenic Steroids* (ed C.D. Kochakian) Springer, Berlin-Heidelberg-New York, pp. 5–44.
52. Potts, G.O., Arnold, A., and Beyler, A.L. (1976) Dissociation of the androgenic and other hormonal activities from the protein anabolic effect of steroids, in *Anabolic-Androgenic Steroids* (ed C.D. Kochakian) Springer, Berlin-Heidelberg-New York, pp. 361–406.
53. Casavant, M.J., Blake, K., Griffith, J., *et al.* (2007) Consequences of use of anabolic androgenic steroids. *Pediatric Clinics of North-America*, **54**, 677–690, x.
54. Kerr, J.M., and Congeni, J.A. (2007) Anabolic-androgenic steroids: Use and abuse in pediatric patients. *Pediatric Clinics of North-America*, **54**, 771–785, xii.
55. Kicman, A.T. (2008) Pharmacology of anabolic steroids. *British Journal of Pharmacology*, **154**, 502–521.
56. Maisel, A.Q. (1965) *The Hormone Quest*. Random House, New York.
57. Hong, B.S., and Ahn, T.Y. (2007) Recent trends in the treatment of testosterone deficiency syndrome. *International Journal of Urology*, **14**, 981–985.
58. Geyer, H., Parr, M.K., Koehler, K., *et al.* (2008) Nutritional supplements cross-contaminated and faked with doping substances. *Journal of Mass Spectrometry*, **43**, 892–902.
59. Catlin, D.H., Ahrens, B.D., and Kucherova, Y. (2002) Detection of norbolethone, an anabolic steroid never marketed, in athletes' urine. *Rapid Communications in Mass Spectrometry*, **16**, 1273–1275.
60. Catlin, D.H., Sekera, M.H., Ahrens, B.D., *et al.* (2004) Tetrahydrogestrinone: Discovery, synthesis, and detection in urine. *Rapid Communications in Mass Spectrometry*, **18**, 1245–1249.

61. Sekera, M.H., Ahrens, B.D., Chang, Y.C., *et al.* (2005) Another designer steroid: Discovery, synthesis, and detection of "madol" in urine. *Rapid Communications in Mass Spectrometry*, **19**, 781–784.
62. Van Eenoo, P., Delbeke, F.T. (2006) Metabolism and excretion of anabolic steroids in doping control—New steroids and new insights. *Journal of Steroid Biochemistry and Molecular Biology*, **101**, 161–178.
63. Schänzer, W. (1996) Metabolism of anabolic androgenic steroids. *Clinical Chemistry*, **42**, 1001–1020.
64. Rommerts, F.F.G. (2004) Testosterone: An overview of biosynthesis, transport, metabolism and non-genomic actions, in *Testosterone: Action, Deficiency, Substitution* (eds E. Nieschlag and H.M. Behre),Cambridge University Press, Cambridge, pp. 1–38.
65. Fusshöller, G., Mareck, U., Schmechel, A., and Schänzer, W. (2007) Long-term detection of metandienone abuse by means of the new metabolite 17β-hydroxymethyl-17α-methyl-18-norandrost-1,4,13-trien-3-one, in *Recent Advances in Doping Analysis* (eds W. Schänzer, H. Geyer, A., Gotzmann, and U. Mareck) Sport&Buch Strauss, Cologne, pp. 393–396.
66. Schänzer, W., Horning, S., Opfermann, G., and Donike, M. (1996) Gas chromatography/mass spectometry identification of long-term excreted metabolites of the anabolic steroid 4-chloro-1,2-dehydro-17alpha-methyltestosterone in humans. *Journal of Steroid Biochemistry and Molecular Biology*, **57**, 363–376.
67. Pozo, O.J., Van Eenoo, P., Deventer, K., *et al.* (2009) Detection and characterization of a new metabolite of 17{alpha}-methyltestosterone. *Drug Metabolism and Disposition*, **37**, 2153–2162.
68. Ward, R.J., Shackleton, C.H., and Lawson, A.M. (1975) Gas chromatographic—mass spectrometric methods for the detection and identification of anabolic steroid drugs. *British Journal of Sports Medicine*, **9**, 93–97.
69. Catlin, D.H., Kammerer, R.C., Hatton, C.K., *et al.* (1987) Analytical chemistry at the Games of the XXIIIrd Olympiad in Los Angeles, 1984. *Clinical Chemistry*, **33**, 319–327.
70. Bradlow, H.L. (1968) Extraction of steroid conjugates with a neutral resin. *Steroids*, **11**, 265–272.
71. Graef, V., Furuya, E., and Nishikaze, O. (1977) Hydrolysis of steroid glucuronides with beta-glucuronidase preparations from bovine liver, Helix pomatia, and E. coli. *Clinical Chemistry*, **23**, 532–535.
72. Vestergaard, P. (1978) The hydrolysis of conjugated neutral steroids in urine. *Acta Endocrinologica (Copenhagen)*, **Suppl. 217**, 96–120.
73. Geyer, H., Schänzer, W., Mareck-Engelke, U., *et al.* (1997) Screening procedure for anabolic steroids-The control of the hydrolysis with deuterated androsterone glucuronide and studies with direct hydrolysis, in *Recent Advances in Doping Analysis* (eds W. Schänzer, H. Geyer, A. Gotzmann, and U. Mareck-Engelke) Sport und Buch Strauss, Cologne, pp. 99–102.

74. Horning S., Donike M. (1994) High resolution GC/MS. In: Donike, M., Geyer, H., Gotzmann, A., Mareck-Engelke, U., and Rauth, S. *Recent advances in doping analysis*. Cologne. Sport&Buch Strauß, p 155–161.
75. Bowers, L.D. (1997) Analytical advances in detection of performance-enhancing compounds. *Clinical Chemistry*, **43**, 1299–1304.
76. Mateus-Avois, L., Mangin, P., and Saugy, M. (2005) Use of ion trap gas chromatography-multiple mass spectrometry for the detection and confirmation of 3′hydroxystanozolol at trace levels in urine for doping control. *Journal of Chromatography B, Analytical Technologies in the Biomedical and Life Sciences*, **816**, 193–201.
77. Munoz-Guerra, J., Carreras, D., Soriano, C., *et al.* (1997) Use of ion trap gas chromatography-tandem mass spectrometry for detection and confirmation of anabolic substances at trace levels in doping analysis. *Journal of Chromatography B*, **704**, 129–141.
78. Donike, M., Bärwald, K.R., Klostermann, K., *et al.* (1983) The detection of exogenous testosterone, in *Leistung und Gesundheit, Kongressbd, Dtsch. Sportärztekongress* (eds H. Heck, W. Hollmann, and H. Liesen), Deutscher Ärtze-Verlag, Köln, pp. 293–298.
79. Cawley, A.T., and Flenker, U. (2008) The application of carbon isotope ratio mass spectrometry to doping control. *Journal of Mass Spectrometry*, **43**, 854–864.
80. Brooks, R.V., Jeremiah, G., Webb, W.A., and Wheeler, M. (1979) Detection of anabolic steroid administration to athletes. *Journal of Steroid Biochemistry*, **11**, 913–917.
81. Rauth, S. (1994) *Referenzbereiche von urinären Steroidkonzentrationen und Steroidquotienten—Ein Beitrag zur Interpretation des Steroidprofils in der Routinedopinganalytik*. Institut für Biochemie, Deutsche Sporthochschule Köln. Dissertation.
82. Mareck, U., Geyer, H., Opfermann, G., *et al.* (2008) Factors influencing the steroid profile in doping control analysis. *Journal of Mass Spectrometry*, **43**, 877–891.
83. Donike, M., Rauth, S., Mareck-Engelke, U., *et al.* (1993) Evaluation of longitudinal studies: The determination of subject-based reference ranges of the testosterone/epitestosterone ratio, in *Recent Advances in Doping Analysis* (eds M. Donike, H. Geyer, A. Gotzmann, U. Mareck-Engelke, and S. Rauth) Sport&Buch Strauss, Cologne, pp. 33–39.
84. Donike, M., Mareck-Engelke, U., and Rauth, S. (1994) Statistical evaluation of longitudinal studies, Part 2: The usefulness of subject-based reference ranges, in *Recent Advances in Doping Analysis* (eds M. Donike, H. Geyer, A. Gotzmann, and U. Mareck-Engelke) Sport&Buch Strauss, Cologne, pp. 157–165.
85. Sottas, P.E., Baume, N., Saudan, C., *et al.* (2007) Bayesian detection of abnormal values in longitudinal biomarkers with an application to T/E ratio. *Biostatistics*, **8**, 285–296.

86. Sottas, P.E., Saudan, C., Schweizer, C., *et al.* (2008) From population- to subject-based limits of T/E ratio to detect testosterone abuse in elite sports. *Forensic Science International*, **174**, 166–172.
87. Geyer, H. (1990) *Die gaschromatographisch/massenspektrometrische Bestimmung von Steroidprofilen im Urin von Athleten*. Institute of Biochemistry, Deutsche Sporthochschule Köln. Dissertation.
88. Dehennin, L., and Matsumoto, A.M. (1993) Long-term administration of testosterone enanthate to normal men: Alterations of the urinary profile of androgen metabolites potentially useful for detection of testosterone misuse in sport. *Journal of Steroid Biochemistry and Molecular Biology*, **44**, 179–189.
89. Franke, W.W., and Berendonk, B. (1997) Hormonal doping and androgenization of athletes: A secret program of the German Democratic Republic government. *Clinical Chemistry*, **43**, 1262–1279.
90. Conte, V. (May 15 2008) Victor Conte's letter to Dwain Chambers, in *Times Online*, London.
91. Geyer, H., Flenker, U., Mareck, U., *et al.* (2007) The detection of the misuse of testosterone gel, in *Recent Advances in Doping Analysis* (eds W. Schänzer, H. Geyer, A. Gotzmann, and U. Mareck), Sport&Buch Strauss, Cologne, pp. 133–142.
92. Donike, M., Ueki, M., Kuroda, Y., *et al.* (1995) Detection of dihydrotestosterone (DHT) doping: Alterations in the steroid profile and reference ranges for DHT and its 5 alpha-metabolites. *Journal of Sports Medicine and Physical Fitness*, **35**, 235–250.
93. Southan, G.J., Brooks, R.V., Cowan, D.A., *et al.* (1992) Possible indices for the detection of the administration of dihydrotestosterone to athletes. *Journal of Steroid Biochemistry and Molecular Biology*, **2**, 87–94.
94. Kicman, A.T., Coutts, S.B., Walker, C.J., and Cowan, D.A. (1995) Proposed confirmatory procedure for detecting 5 alpha-dihydrotestosterone doping in male athletes. *Clinical Chemistry*, **41**, 1617–1627.
95. Coutts, S.B., Kicman, A.T., Hurst, D.T., and Cowan, D.A. (1997) Intramuscular administration of 5 alpha-dihydrotestosterone heptanoate: Changes in urinary hormone profile. *Clinical Chemistry*, **43**, 2091–2098.
96. Zimmermann, J. (1986) *Untersuchung zum Nachweis von exogenen Gaben von Testosteron*. Institut für Biochemie, Deutsche Sporthochschule Köln, Dissertation.
97. Levesque, J.F., and Ayotte, C. (1999) Criteria for the detection of androstenedione oral administration, in *Recent Advances in Doping Analysis* (eds W. Schänzer, H. Geyer, A. Gotzmann, and U. Mareck), Sport&Buch Strauss, Cologne, pp. 169–179.
98. Bassindale, T., Cowan, D.A., Dale, S., *et al.* (2002) Disposition of androstenedione and testosterone following oral administration of androstenedione to healthy female volunteers; influence on the urinary T/E ratio, in

Recent Advances in Doping Analysis (eds W. Schänzer, H. Geyer, A. Gotzmann, and U. Mareck), Sport&Buch Strauss, Cologne, pp. 51–60.

99. Uralets, V.P., and Gilette, P.A. (1998) Over-the-counter anabolic steroids 4-androstene-3,17-dione, 4-androstene-3β,17β-diol, and 19-nor-4-androstene-3,17-dione: Excretion studies in men, in *Recent Advances in Doping Analysis* (eds W. Schänzer, H. Geyer, A. Gotzmann, and U. Mareck), Sport&Buch Strauss, Cologne, pp. 147–169.
100. Catlin, D.H., Leder, B.Z., Ahrens, B.D., *et al.* (2002) Effects of androstenedione administration on epitestosterone metabolism in men. *Steroids*, **67**, 559–564.
101. Dehennin, L., Ferry, M., Lafarge, P., *et al.* (1998) Oral administration of dehydroepiandrosterone to healthy men: Alteration of the urinary androgen profile and consequences for the detection of abuse in sport by gas chromatography-mass spectrometry. *Steroids*, **63**, 80–87.
102. Bosy, T.Z., Moore, K.A., and Poklis, A. (1998) The effect of oral dehydroepiandrosterone (DHEA) on the urine testosterone/epitestosterone (T/E) ratio in human male volunteers. *Journal of Analytical Toxicology*, **22**, 455–459.
103. Cawley, A.T., Hine, E.R., Trout, G.J., *et al.* (2004) Searching for new markers of endogenous steroid administration in athletes: "looking outside the metabolic box". *Forensic Science International*, **143**, 103–114.
104. Mareck, U., Geyer, H., Flenker, U., et al. (2007) Detection of dehydroepiandrosterone misuse by means of gas chromatography- combustion-isotope ratio mass spectrometry. *European Journal of Mass Spectrometry*, **13**, 419–426.
105. Wilson, W., 3rd, Pardo-Manuel de Villena, F., Lyn-Cook, B.D., *et al.* (2004) Characterization of a common deletion polymorphism of the UGT2B17 gene linked to UGT2B15. *Genomics*, **84**, 707–714.
106. Jakobsson, J., Ekstrom, L., Inotsume, N., *et al.* (2006) Large differences in testosterone excretion in Korean and Swedish men are strongly associated with a UDP-glucuronosyl transferase 2B17 polymorphism. *Journal of Clinical Endocrinology and Metabolism*, **91**, 687–693.
107. Falk, O., Palonek, E., and Bjorkhem, I. (1988) Effect of ethanol on the ratio between testosterone and epitestosterone in urine. *Clinical Chemistry*, **34**, 1462–1464.
108. Karila, T., Kosunen, V., Leinonen, A., *et al.* (1996) High doses of alcohol increase urinary testosterone-to-epitestosterone ratio in females. *Journal of Chromatography B, Biomedical Applications*, **687**, 109–116.
109. Mareck-Engelke, U., Geyer, H., Schindler, U., *et al.* (1996) Influence of ethanol on steroid profile parameters, in *Recent Advances in Doping Analysis* (eds M. Donike, H. Geyer, A. Gotzmann, and U. Mareck-Engelke), Sport&Buch Strauss, Cologne, pp. 143–148.

110. Grosse, J., Anielski, P., Sachs, H., and Thieme, D. (in press) Ethylglucuronide as a potential marker for alcohol-induced elevation of urinary testosterone/epitestosterone ratios. *Drug Testing and Analysis*, **1**.

111. Thevis, M., Geyer, H., Mareck, U., *et al.* (2007) Doping-control analysis of the 5alpha-reductase inhibitor finasteride: Determination of its influence on urinary steroid profiles and detection of its major urinary metabolite. *Therapeutic Drug Monitoring*, **29**, 236–247.

112. Reznik, Y., Herrou, M., Dehennin, L., *et al.* (1987) Rising plasma levels of 19-nortestosterone throughout pregnancy: Determination by radioimmunoassay and validation by gas chromatography-mass spectrometry. *Journal of Clinical Endocrinology and Metabolism*, **64**, 1086–1088.

113. Mareck-Engelke, U., Schultze, G., Geyer, H., and Schänzer, W. (2000) 19-norandrosterone in pregnant women, in *Recent Advances in Doping Analysis* (eds M. Donike, H. Geyer, A. Gotzmann, and U. Mareck-Engelke), Sport&Buch Strauss, Cologne, pp. 145–154.

114. Grosse, J., Anielski, P., Hemmersbach, P., *et al.* (2005) Formation of 19-norsteroids by in situ demethylation of endogenous steroids in stored urine samples. *Steroids*, **70**, 499–506.

115. Ueki, M., and Okano, M. (1999) Analysis of exogenous dehydroepiandrosterone excretion in urine by gas chromatography/combustion/isotope ratio mass spectrometry. *Rapid Communications in Mass Spectrometry*, **13**, 2237–2243.

116. de la Torre, X., Gonzalez, J.C., Pichini, S., *et al.* (2001) 13C/12C isotope ratio MS analysis of testosterone, in chemicals and pharmaceutical preparations. *Journal of Pharmaceutical and Biomedical Analysis*, **24**, 645–650.

117. World Anti-Doping Agency (2009) Endogenous Anabolic Androgenic Steroids: Testing, Reporting and Interpretive Guidance. Available at http://www.wada-ama.org/rtecontent/document/WADA_TD2009EAAS_Endogenous_Anabolic_Androgenic_Steroids_Oct2009.pdf. Accessed 10-27-2009.

118. Piper, T., Mareck, U., Geyer, H., *et al.* (2008) Determination of (13)C/(12)C ratios of endogenous urinary steroids: Method validation, reference population and application to doping control purposes. *Rapid Communications in Mass Spectrometry*, **22**, 2161–2175.

119. Flenker, U., Güntner, U., and Schänzer, W. (2008) delta13C-values of endogenous urinary steroids. *Steroids*, **73**, 408–416.

120. Morra, V., Davit, P., Capra, P., *et al.* (2006) Fast gas chromatographic/mass spectrometric determination of diuretics and masking agents in human urine: Development and validation of a productive screening protocol for antidoping analysis. *Journal of Chromatography A*, **1135**, 219–229.

121. Ventura, R., and Segura, J. (1996) Detection of diuretic agents in doping control. *Journal of Chromatography B*, **687**, 127–144.

122. Park, S.J., Pyo, H.S., Kim, Y.J., et al. (1990) Systematic analysis of diuretic doping agents by HPLC screening and GC/MS confirmation. *Journal of Analytical Toxicology*, **14**, 84–90.
123. Ventura, R., Nadal, T., Alcalde, P., *et al.* (1993) Fast screening method for diuretics, probenecid and other compounds of doping interest. *Journal of Chromatography A* **655**, 233–242.
124. Schänzer, W. (1987) The problem of diuretics in doping control, in *International Athletic Foundation World Symposium on Doping in Sport* (eds P. Bellotti, G. Benzi, and A. Ljungqvist), FIDAL Centro Studi&Richerche, Florence, pp. 89–106.
125. Carreras, C.W., Costi, P.M., and Santi, D.V. (1994) Heterodimeric thymidylate synthases with C-terminal deletion on one subunit. *Journal of Biological Chemistry*, **269**, 12444–12446.
126. Vandenheuvel, W.J., Gruber, V.F., Walker, R.W., and Wolf, F.J. (1975) GLC analysis of hydrochlorothiazide in blood and plasma. *Journal of Pharmaceutical Sciences*, **64**, 1309–1312.
127. Wallace, S.M., Shah, V.P., and Riegelman, S. (1977) GLC analysis of acetazolamide in blood, plasma, and saliva following oral administration to normal subjects. *Journal of Pharmaceutical Sciences*, **66**, 527–530.
128. Li, H., Johnston, M.M., and Mufson, D. (1977) GLC determination of urinary chlorthalidone levels. *Journal of Pharmaceutical Sciences*, **66**, 1732–1734.
129. Dünges, W. (1973) Alkylation of acidic organic compounds for gas chromatographic analysis. *Chromatographia*, **6**, 196–197.
130. Lisi, A.M., Kazlauskas, R., and Trout, G.J. (1992) Diuretic screening in human urine by gas chromatography-mass spectrometry: Use of a macroreticular acrylic copolymer for the efficient removal of the coextracted phase-transfer reagent after derivatization by direct extractive alkylation. *Journal of Chromatography*, **581**, 57–63.
131. Lisi, A.M., Trout, G.J., and Kazlauskas, R. (1991) Screening for diuretics in human urine by gas chromatography-mass spectrometry with derivatisation by direct extractive alkylation. *Journal of Chromatography*, **563**, 257–270.
132. Fagerlund, C., Hartvig, P., and Lindstrom, B. (1979) Extractive alkylation of sulphonamide diuretics and their determination by electron-capture gas chromatography. *Journal of Chromatography*, **168**, 107–116.
133. Kraft, M. (1990) *Gleichzeitiger gas-chromatographisch/massenspektrometrischer Nachweis von Hydroxy- und Phenolalkylaminen (Stimulanzien, β-Blocker, Narkotika) sowie synthetischen und endogenen anabol-androgenen Steroiden nach extraktiver Acylierung und gezielter Trimethylsilylierung*. Dissertation, Institut für Biochemie, Deutsche Sporthochschule Köln.
134. Leloux, M.S., and Maes, R.A. (1990) The use of electron impact and positive chemical ionization mass spectrometry in the screening of beta

blockers and their metabolites in human urine. *Biomedical and Environmental Mass Spectrometry*, **19**, 137–142.

135. Polettini, A. (1996) Bioanalysis of β_2-agonists by hyphenated chromatographic and mass spectrometric techniques. *Journal of Chromatography B*, **687**, 27–42.
136. Henze, M.K., Opfermann, G., Spahn-Langguth, H., and Schänzer, W. (2001) Screening of beta-2 agonists and confirmation of fenoterol, orciprenaline, reproterol and terbutaline with gas chromatography-mass spectrometry as tetrahydroisoquinoline derivatives. *Journal of Chromatography B*, **751**, 93–105.
137. Amendola, L., Colamonici, C., Rossi, F., and Botre, F. (2002) Determination of clenbuterol in human urine by GC-MS-MS-MS: Confirmation analysis in antidoping control. *Journal of Chromatography B*, **773**, 7–16.
138. Georgakopoulos, C.G., Tsitsimpikou, C., and Spyridaki, M.H. (1999) Excretion study of the beta2-agonist reproterol in human urine. *Journal of Chromatography B*, **726**, 141–148.
139. Van Eenoo, P., and Delbeke, F.T. (2002) Detection of inhaled salbutamol in equine urine by ELISA and GC/MS2. *Biomedical Chromatography*, **16**, 513–516.
140. Van Eenoo, P., Delbeke, F.T., and Deprez, P. (2002) Detection of inhaled clenbuterol in horse urine by GC/MS2. *Biomedical Chromatography*, **16**, 475–481.
141. Dumasia, M.C., and Houghton, E. (1991) Screening and confirmatory analysis of beta-agonists, beta-antagonists and their metabolites in horse urine by capillary gas chromatography-mass spectrometry. *Journal of Chromatography*, **564**, 503–513.
142. Ventura, R., Damasceno, L., Farre, M., *et al.* (2000) Analytical methodology for the detection of beta(2)-agonists in urine by gas chromatography-mass spectrometry for application in doping control. *Analytica Chimica Acta*, **418**, 79–92.
143. Thevis, M., Guddat, S., Flenker, U., and Schänzer, W. (2008) Quantitative analysis of urinary glycerol levels for doping control purposes using gas chromatography-mass spectrometry. *European Journal of Mass Spectrometry*, **14**, 117–125.
144. Guddat, S., Thevis, M., and Schänzer, W. (2008) Identification and quantification of the osmodiuretic mannitol in urine for sports drug testing using gas chromatography-mass spectrometry. *European Journal of Mass Spectrometry*, **14**, 127–133.
145. Kulicke, W-M., Roessner, D., and Kull, W. (1993) Characterization of hydroxyethyl starch by polymer analysis for use as a plasma volume expander. *Starch/Stärke*, **45**, 445–450.
146. Mishler, J.M., Borberg, H., Emerson, P.M., and Gross, R. (1977) Hydroxyethyl starch. An agent for hypovolaemic shock treatment II.

Urinary excretion in normal volunteers following three consecutive daily infusions. *British Journal of Clinical Pharmacology*, **4**, 591–595.
147. Yacobi, A., Stoll, R.G., Sum, C.Y., *et al.* Pharmacokinetics of Hydroxyethyl Starch in Normal Subjects. *Journal of Clinical Pharmacology*, **22**, 206–212.
148. Anumula, K.A., and Taylor, P.B. (1992) A comprehensive procedure for preparation of partially methylated alditol acetates from glycoprotein carbohydrates. *Analytical Biochemistry*, **203**, 101–108.
149. Lowe, M.E., and Nilsson, B. (1983) A method of purification of partially methylated alditol acetates in the methylation analysis of glycoproteins and glycopeptides. *Analytical Biochemistry*, **136**, 187–191.
150. Thevis, M., Opfermann, G., and Schänzer, W. (2000) Detection of the plasma volume expander hydroxyethyl starch in human urine. *Journal of Chromatography B*, **744**, 345–350.
151. Thevis, M., Opfermann, G., and Schänzer, W. (2000) Mass spectrometry of partially methylated alditol acetates derived from hydroxyethyl starch. *Journal of Mass Spectrometry*, **35**, 77–84.
152. Thevis, M., Opfermann, G., and Schänzer, W. (2001) Nachweis des Plasmavolumenexpanders Hydroxyethylstärke in Humanurin. *Deutsche Zeitschrift für Sportmedizin*, **52**, 316–320.
153. Thevis, M., and Schänzer, W. (2007) Current role of LC-MS(/MS) in doping control. *Analytical and Bioanalytical Chemistry*, **388**, 1351–1358.
154. Ventura, R., Roig, M., Montfort, N., *et al.* (2008) High-throughput and sensitive screening by ultra-performance liquid chromatography tandem mass spectrometry of diuretics and other doping agents. *European Journal of Mass Spectrometry*, **14**, 191–200.
155. Virus, E.D., Sobolevsky, T.G., Rodchenkov, G.M. (2008) Introduction of HPLC/orbitrap mass spectrometry as screening method for doping control. *Journal of Mass Spectrometry*, **43**, 949–957.
156. Thörngren, J.O., Östervall, F., and Garle, M. (2008) A high-throughput multicomponent screening method for diuretics, masking agents, central nervous system (CNS) stimulants and opiates in human urine by UPLC-MS/MS. *Journal of Mass Spectrometry*, **43**, 980–992.
157. Touber, M.E., van Engelen, M.C., Georgakopoulus, C., *et al.* (2007) Multi-detection of corticosteroids in sports doping and veterinary control using high-resolution liquid chromatography/time-of-flight mass spectrometry. *Analytica Chimica Acta*, **586**, 137–146.
158. Nielen, M.W., Bovee, T.F., van Engelen, M.C., *et al.* (2006) Urine testing for designer steroids by liquid chromatography with androgen bioassay detection and electrospray quadrupole time-of-flight mass spectrometry identification. *Analytical Chemistry*, **78**, 424–431.
159. Georgakopoulos, C.G., Vonaparti, A., Stamou, M., *et al.* (2007) Preventive doping control analysis: Liquid and gas chromatography time-of-flight mass spectrometry for detection of designer steroids. *Rapid Communications in Mass Spectrometry*, **21**, 2439–2446.

160. Mazzarino, M., de la Torre X., and Botre, F. (2008) A screening method for the simultaneous detection of glucocorticoids, diuretics, stimulants, anti-oestrogens, beta-adrenergic drugs and anabolic steroids in human urine by LC-ESI-MS/MS. *Analytical and Bioanalytical Chemistry*, **392**, 681–698.

161. Guddat, S., Thevis, M., Thomas, A., and Schänzer, W. (2008) Rapid screening of polysaccharide-based plasma volume expanders dextran and hydroxyethyl starch in human urine by liquid chromatography-tandem mass spectrometry. *Biomedical Chromatography*, **22**, 695–701.

162. Deventer, K., Van Eenoo, P., and Delbeke, F.T. (2005) Simultaneous determination of beta-blocking agents and diuretics in doping analysis by liquid chromatography/mass spectrometry with scan-to-scan polarity switching. *Rapid Communications in Mass Spectrometry*, **19**, 90–98.

163. Kolmonen, M., Leinonen, A., Pelander, A., and Ojanpera, I. (2007) A general screening method for doping agents in human urine by solid phase extraction and liquid chromatography/time-of-flight mass spectrometry. *Analytica Chimica Acta*, **585**, 94–102.

164. Kolmonen, M., Leinonen, A., Kuuranne, T., *et al.* (2009) Generic sample preparation and dual polarity liquid chromatography—time-of-flight mass spectrometry for high-throughput screening in doping analysis. *Drug Testing and Analysis*, **1**, 250–266.

165. Badoud, F., Grata, E., Perrenoud, L., *et al.* (2009) Fast analysis of doping agents in urine by ultra-high-pressure liquid chromatography-quadrupole time-of-flight mass spectrometry I. Screening analysis. *Journal of Chromatography A*, **1216**, 4423–4433.

166. Deventer, K., Van Eenoo, P., Delbeke, F.T. (2006) Screening for amphetamine and amphetamine-type drugs in doping analysis by liquid chromatography/mass spectrometry. *Rapid Communications in Mass Spectrometry*, **20**, 877–882.

167. Thomas A., Sigmund, G., Guddat. S., *et al.* (2008) Determination of selected stimulants in urine for sports drug analysis by solid phase extraction via cation exchange and means of liquid chromatography-tandem mass spectrometry. *European Journal of Mass Spectrometry*, **14**, 135–143.

168. Deventer, K., Pozo, O.J., Van Eenoo, P., and Delbeke, F.T. (2007) Development of a qualitative liquid chromatography/tandem mass spectrometric method for the detection of narcotics in urine relevant to doping analysis. *Rapid Communications in Mass Spectrometry*, **21**, 3015–3023.

169. Thevis, M., Geyer, H., Bahr, D., and Schänzer, W. (2005) Identification of fentanyl, alfentanil, sufentanil, remifentanil and their major metabolites in human urine by liquid chromatography/tandem mass spectrometry for doping control purposes. *European Journal of Mass Spectrometry*, **11**, 419–427.

170. Thevis, M., Opfermann, G., and Schänzer, W. (2003) Liquid chromatography/electrospray ionization tandem mass spectrometric screening and confirmation methods for β_2-agonists in human or equine urine. *Journal of Mass Spectrometry*, **38**, 1197–1206.
171. Kang, M.J., Hwang, Y.H., Lee, W., and Kim, D.H. (2007) Validation and application of a screening method for beta2-agonists, anti-estrogenic substances and mesocarb in human urine using liquid chromatography/tandem mass spectrometry. *Rapid Communications in Mass Spectrometry*, **21**, 252–264.
172. Thevis, M., Opfermann, G., and Schänzer, W. (2001) High speed determination of beta-receptor blocking agents in human urine by liquid chromatography/tandem mass spectrometry. *Biomedical Chromatography*, **15**, 393–402.
173. Mazzarino, M., and Botre, F. (2006) A fast liquid chromatographic/mass spectrometric screening method for the simultaneous detection of synthetic glucocorticoids, some stimulants, anti-oestrogen drugs and synthetic anabolic steroids. *Rapid Communications in Mass Spectrometry*, **20**, 3465–3476.
174. Pozo, O.J., Van Eenoo, P., Deventer, K., and Delbeke, F.T. (2007) Development and validation of a qualitative screening method for the detection of exogenous anabolic steroids in urine by liquid chromatography-tandem mass spectrometry. *Analytical and Bioanalytical Chemistry*, **389**, 1209–1224.
175. Mareck, U., Thevis, M., Guddat, S., *et al.* (2004) Comprehensive sample preparation for anabolic steroids, glucocorticosteroids, beta-receptor blocking agents, selected anabolic androgenic steroids and buprenorphine in human urine, in *Recent Advances in Doping Analysis* (eds W. Schänzer, H. Geyer, A. Gotzmann, and U. Mareck) Sport und Buch Strauss, Cologne, pp. 65–68.
176. Deventer, K., and Delbeke, F.T. (2003) Validation of a screening method for corticosteroids in doping analysis by liquid chromatography/tandem mass spectrometry. *Rapid Communications in Mass Spectrometry*, **17**, 2107–2114.
177. Mazzarino, M., Turi, S., and Botre, F. (2008) A screening method for the detection of synthetic glucocorticosteroids in human urine by liquid chromatography-mass spectrometry based on class-characteristic fragmentation pathways. *Analytical and Bioanalytical Chemistry*, **390**, 1389–1402.
178. Mareck, U., Geyer, H., Guddat, S., *et al.* Identification of the aromatase inhibitors anastrozole and exemestane in human urine using liquid chromatography/tandem mass spectrometry. *Rapid Communications in Mass Spectrometry*, **20**, 1954–1962.
179. Borges, C.R., Miller, N., Shelby, M., *et al.* (2007) Analysis of a challenging subset of World Anti-Doping Agency-banned steroids and antiestrogens by LC-MS-MS. *Journal of Analytical Toxicology*, **31**, 125–131.

180. Thieme, D., Grosse, J., Lang, R., *et al.* Screening, confirmation and quantification of diuretics in urine for doping control analysis by high-performance liquid chromatography-atmospheric pressure ionisation tandem mass spectrometry. *Journal of Chromatography B*, **757**, 49–57.
181. Deventer, K., Delbeke, F.T., Roels, K., and Van Eenoo, P. (2002) Screening for 18 diuretics and probenecid in doping analysis by liquid chromatography-tandem mass spectrometry. *Biomedical Chromatography*, **16**, 529–535.
182. Thevis, M., and Schänzer, W. (2005) Examples of doping control analysis by liquid chromatography-tandem mass spectrometry: Ephedrines, beta-receptor blocking agents, diuretics, sympathomimetics, and cross-linked hemoglobins. *Journal of Chromatographic Science*, **43**, 22–31.
183. Thevis, M., Krug, O., and Schänzer, W. (2006) Mass spectrometric characterization of efaproxiral (RSR13) and its implementation into doping controls using liquid chromatography-atmospheric pressure ionization-tandem mass spectrometry. *Journal of Mass Spectrometry*, **41**, 332–338.
184. Thevis, M., Kamber, M., and Schänzer, W. (2006) Screening for metabolically stable aryl-propionamide-derived selective androgen receptor modulators for doping control purposes. *Rapid Communications in Mass Spectrometry*, **20**, 870–876.
185. Politi, L., Morini, L., and Polettini, A. (2007) A direct screening procedure for diuretics in human urine by liquid chromatography-tandem mass spectrometry with information dependent acquisition. *Clinica Chimica Acta*, **386**, 46–52.
186. Thevis, M., Thomas, A., Delahaut, P., *et al.* (2006) Doping control analysis of intact rapid-acting insulin analogues in human urine by liquid chromatography-tandem mass spectrometry. *Analytical Chemistry*, **78**, 1897–1903.
187. Thevis, M., Thomas, A., Delahaut, P., *et al.* (2005) Qualitative determination of synthetic analogues of insulin in human plasma by immunoaffinity purification and liquid chromatography-tandem mass spectrometry for doping control purposes. *Analytical Chemistry*, **77**, 3579–3585.
188. Thevis, M., Thomas, A., and Schänzer, W. (2008) Mass spectrometric determination of insulins and their degradation products in sports drug testing. *Mass Spectrometry Reviews*, **27**, 35–50.
189. Thomas, A., Geyer, H., Kamber, M., *et al.* Mass spectrometric determination of gonadotrophin-releasing hormone (GnRH) in human urine for doping control purposes by means of LC-ESI-MS/MS. *Journal of Mass Spectrometry*, **43**, 908–915.
190. Thomas, A., Schänzer, W., Delahaut, P., and Thevis, M. (2009) Sensitive and fast identification of urinary human, synthetic and animal insulin by means of nano-UPLC coupled with high resolution/high accuracy mass spectrometry. *Drug Testing and Analysis*, **1**, 219–227.

191. Thomas, A., Thevis, M., Delahaut, P., *et al.* (2007) Mass spectrometric identification of degradation products of insulin and its long-acting analogues in human urine for doping control purposes. *Analytical Chemistry*, **79**, 2518–2524.
192. Bredehöft, M., Schänzer, W., and Thevis, M. (2008) Quantification of human insulin-like growth factor-1 and qualitative detection of its analogues in plasma using liquid chromatography/electrospray ionisation tandem mass spectrometry. *Rapid Communications in Mass Spectrometry*, **22**, 477–485.
193. Kay, R.G., Barton, C., Velloso, C.P., *et al.* (2009) High-throughput ultra-high-performance liquid chromatography/tandem mass spectrometry quantitation of insulin-like growth factor-I and leucine-rich alpha-2-glycoprotein in serum as biomarkers of recombinant human growth hormone administration. *Rapid Communications in Mass Spectrometry*, **23**, 3173–3182.
194. Popot, M.A., Woolfitt, A.R., Garcia, P., and Tabet, J.C. (2008) Determination of IGF-I in horse plasma by LC electrospray ionisation mass spectrometry. *Analytical and Bioanalytical Chemistry*, **390**, 1843–1852.
195. Thomas, A., Kohler, M., Delahaut, P., Schänzer, W., and Thevis, M. (2009) Determination of Synacthen in urine for sports drug testing and preliminary results for a peptide screening method by means of LC-MS/MS, in *Recent Advances in Doping Analysis* (eds W. Schänzer, H. Geyer, A. Gotzmann, and U. Mareck), Sport&Buch Strauss, Cologne, in press.
196. Thevis, M., and Schänzer, W. (2005) Identification and characterization of peptides and proteins in doping control analysis. *Current Proteomics*, **2**, 191–208.
197. Lasne, F., Crepin, N., Ashenden, M., *et al.* (2004) Detection of hemoglobin-based oxygen carriers in human serum for doping analysis: Screening by electrophoresis. *Clinical Chemistry*, **50**, 410–415.
198. Varlet-Marie, E., Ashenden, M., Lasne, F., *et al.* (2004) Detection of hemoglobin-based oxygen carriers in human serum for doping analysis: Confirmation by size-exclusion HPLC. *Clinical Chemistry*, **50**, 723–731.
199. Goebel, C., Alma, C., Howe, C., *et al.* (2005) Methodologies for detection of hemoglobin-based oxygen carriers. *Journal of Chromatographic Science*, **43**: 39–46.
200. Gotzmann, A., Voss, S., Machnik, M., and Schänzer, W. (2002) Detection of hemopure and hemoglobin in human plasma by HPLC/UV and different types of columns—An approach to screen for this substance in doping analysis, in *Recent Advances in Doping Analysis* (eds W. Schänzer, H. Geyer, A. Gotzmann, and U. Mareck), Sport und Buch Strauß, Cologne, pp. 179–187.
201. Guan, F., Uboh, C., Soma, L., *et al.* (2004) Unique tryptic peptides specific for bovine and human hemoglobin in the detection and confirmation of hemoglobin-based oxygen carriers. *Analytical Chemistry*, **76**, 5118–5126.

202. Guan, F., Uboh, C.E., Soma, L.R., *et al.* (2004) Confirmation and quantification of hemoglobin-based oxygen carriers in equine and human plasma by hyphenated liquid chromatography tandem mass spectrometry. *Analytical Chemistry*, **76**, 5127–5135.
203. Simitsek, P.D., Giannikopoulou, P., Katsoulas, H., *et al.* (2007) Electrophoretic, size-exclusion high-performance liquid chromatography and liquid chromatography-electrospray ionization ion trap mass spectrometric detection of hemoglobin-based oxygen carriers. *Analytica Chimica Acta*, **583**, 223–230.
204. Thevis, M., Ogorzalek-Loo, R.R., Loo, J.A., and Schänzer, W. (2003) Doping control analysis of bovine hemoglobin-based oxygen therapeutics in human plasma by LC-electrospray ionization-MS/MS. *Analytical Chemistry*, **75**, 3287–3293.
205. Delbeke, F.T., Van Eenoo, P., and De Backer, P. (1998) Detection of human chorionic gonadotrophin misuse in sports. *International Journal of Sports Medicine*, **19**, 287–290.
206. Kicman, A.T., Brooks, R.V., and Cowan, D.A. (1991) Human chorionic gonadotrophin and sport. *British Journal of Sports Medicine*, **25**, 73–80.
207. Stenman, U.H., Unkila-Kallio, L., Korhonen, J., and Alfthan, H. (1997) Immunoprocedures for detecting human chorionic gonadotropin: Clinical aspects and doping control. *Clinical Chemistry*, **43**, 1293–1298.
208. Gam, L.H., Tham, S.Y., and Latiff, A. (2003) Immunoaffinity extraction and tandem mass spectrometric analysis of human chorionic gonadotropin in doping analysis. *Journal of Chromatography B*, **792**, 187–196.
209. Kicman, A.T., Parkin, M.C., and Iles, R.K. (2007) An introduction to mass spectrometry based proteomics-detection and characterization of gonadotropins and related molecules. *Molecular and Cellular Endocrinology*, **260–262**, 212–227.
210. Schänzer, W., and Donike, M. (1993) Metabolism of anabolic steroids in man: Synthesis and use of reference substances for identification of anabolic steroid metabolites. *Analytica Chimica Acta*, **275**, 23–48.
211. Kazlauskas, R. (2008) WADA List, in *Recent Advances in Doping Analysis* (eds W. Schänzer, H. Geyer, A. Gotzmann, and U. Mareck), Sport&Buch Strauss, Cologne, pp. 21–32.

7 Limitations and Perspectives of Mass Spectrometry-Based Procedures in Doping Control Analysis

7.1 RECOMBINANT BIOMOLECULES

Although mass spectrometry has demonstrated utmost utility for sports drug testing purposes and is predominantly employed in this arena for the rapid and unequivocal detection and quantification of hundreds of prohibited substances,[1] several compounds are yet not amenable to MS-based analytical approaches. These compounds are mostly recombinant peptides or proteins designed to have identical structures with natural, endogenously produced peptide hormones and are commonly targeted using electrophoretic and/or immunological assays.

7.1.1 Erythropoietin

Numerous studies have been undertaken in the past to characterize in-depth the structures of recombinantly derived and naturally produced EPOs. Comprehensive information was obtained outlining distinct differences[2–13] that enable the separation of biotechnologically produced from natural urinary or plasma EPO by means of electrophoretic strategies;[14–21] however, the limited amounts of EPO excreted intact into urine as well as the considerable heterogeneity of the glycosylation pattern of EPOs have complicated and impeded the development of MS-based detection methods for recombinant EPO in human sports drug testing so far. While significantly improved sample preparation techniques (e.g., using magnetic nanoparticles for immu-

Mass Spectrometry in Sports Drug Testing: Characterization of Prohibited Substances and Doping Control Analytical Assays, By Mario Thevis
Copyright © 2010 John Wiley & Sons, Inc.

nological purification) combined with considerably increased analytical sensitivities of LC-MS(/MS) systems allow for the isolation and identification of recombinant human EPO in animal blood samples (e.g., horses),[22,23] such procedures are options in human doping control programs only for erythropoietins comprising modified primary structures such as ARANESP.[24] In principle, the sensitivity and specificity of state-of-the-art MS instruments is capable of measuring EPO in blood and/or urine; however, the unambiguous differentiation of natural EPO from recombinantly produced versions by mass spectrometry represents a challenge that has not been accomplished yet and might not be the task to perform in the near future either. Because most EPO products differ from natural urinary EPO only in the composition of the carbohydrate moiety, glycomics might be the preferred strategy to provide adequate analytical tools allowing to accept the dare of reproducibly and reliably identifying recombinantly produced EPO in human doping control samples. Until now, MS has supported the characterization of urinary and recombinant EPO, but their differentiation on a routine basis in sports drug testing samples remains to be demonstrated.

7.1.2 Human Growth Hormone

A comparable though slightly different issue represents the detection of the illicit application of recombinant human growth hormone. Immunoassays measuring different isoforms of hGH in human serum provide a means to determine the unnaturally elevated 22 kDa isoform, thus proving the exogenous source of the measured hGH.[25–29] Complementary, surface plasmon resonance,[30] and 2D-gel electrophoretic approaches[31] utilizing also the quantity of different splice variants in comparison to the major 22 kDa isoform were reported; however, also here MS is not included in the suggested doping control procedure. This is primarily due to the limited amounts of the minor though informative splice variants or glycosylated isoforms (e.g., 23 kDa, 20 kDa, 17 kDa, 12 kDa, and 9 kDa) and the necessity to obtain quantitative data on each of these. Bottom-up analytical approaches would be considered more sensitive but require an enzymatic hydrolysis and the generation of prototypical peptides. This, in turn, would require appropriate (i.e., stable isotope-labeled) internal standards of each target compound. Nevertheless, as in case of EPO, the principle capability of modern LC-MS/MS systems to measure hGH at physiological concentrations offers room and prospect for future attempts that might focus on, e.g., the minor percentage of erroneous primary structures of several recombinant hGH products (see Chapter 5) as well as the quantitation of different isoforms of hGH.

7.1.3 Human Insulin and IGF-1

In contrast to EPO and hGH, human insulin and IGF-1 represent a different challenge to MS-based doping control methods, as these compounds are entirely identical to their natural counterparts. Both compounds are measured by current MS-based procedures at physiological concentrations in blood and urine,[32–37] but these assays lack (as any other assay also) the capability of proving the source of the analyte, i.e., endogenous or exogenous. For human insulin, mass spectrometric studies yielded preliminary results that indicate differences in urinary metabolites depending on whether insulin was secreted from pancreatic β-cells or adipocytes (after subcutaneous administration);[38] hence, there might be potential for future doping control assays relying on an insulin metabolite profile. The considerable amount of IGF-1 that is naturally present in human blood samples seems, however, to exclude any mass spectrometric approach until now. A metabolite profile might be an option also here, but no data are currently available that support this idea.

7.2 UNKNOWN COMPOUNDS

Most assays described in this book represent targeted mass spectrometric approaches, i.e., the analytes of interest are known and characterized. These test methods have been designed for ample sensitivity and specificity but are mostly limited to a preselected group of compounds. This fact was maliciously exploited by cheating athletes in the past, who were using designer substances (e.g., designer steroids)[39–41] with structures unknown to the doping control community and, thus, not covered by routine initial testing methods and below the analytical radar. Consequently, alternatives to solely targeted procedures were established using MS,[42–44] bioassays,[45,46] or a combination of both[47] to overcome this issue. Product ions characterizing particular features of classes of substances as well as neutrally lost species provide valuable information on the presence of unknown compounds, and bioassays comprising adequate receptors indicate the potential bioactivity of a substance. The latter presumably yield higher sensitivities but do not provide utmost specificity; hence, the combined use of a bioassay screening and MS-based drug testing is considered particularly interesting.

7.3 PROFILING OF URINE AND/OR BLOOD

One of the constantly growing fields of medicinal, biological, and biochemical analysis are profiling projects commonly referred to as—omics arenas such as proteomics, peptidomics, lipidomics, metabolomics, etc. Mass spectrometry plays a key role in these sciences, and some might be of great assistance also to future sports drug testing programs, especially to the so-called "athlete biological passport." The knowledge concerning "natural" variations of selected plasma, blood, or urine parameters is continuously expanding, and use has been made of several low and high molecular weight markers with regard to various doping control issues including the above mentioned challenges of EPO and hGH misuse as well as autologous blood doping, etc. Instead of direct detection of a prohibited substance or method the combined consideration of concentrations and ratios of different variables has been suggested as alternatives to either indicate or even prove a doping offence by outlining the fact that "unnatural changes" are observed in the longitudinal profile of an individual athlete.

For instance, the use of two approaches referred to as ON- and OFF-models (being sensitive to either accelerated or decelerated erythropoiesis, respectively) was proposed for the detection of EPO abuse. These models utilize various blood parameters such as hemoglobin, erythropoietin, serum transferrin receptor, and reticulocytes,[48–50] and longitudinal profiling of these allow the comparison of new test results to historical baseline values.[51] These parameters are measured solely by means of immunoassays; however, urinary indicators for an EPO administration such as elevated concentrations of asymmetrical dimethylarginine, symmetrical dimethylarginine, citrulline, and arginine have been determined using MS-based approaches,[52] and additional targets might follow to be included in urinary profiling and monitoring procedures.

In a comparable manner, the plasma concentrations of IGF-1 and type 3 pro-collagen (P-III-P) have been studied to serve as indirect markers for hGH abuse in sports,[53–56] predominantly aiming the extension of the detection window. Being currently limited to less than 48 hours, altered IGF-1 and P-III-P concentrations might prove the administration of hGH for more than 4 days. Also here, the measurands are currently determined by immunological methods, but much effort is made to transfer the analysis of IGF-1 to LC-MS/MS.[36,37]

7.4 ALTERNATIVE SPECIMENS

Besides the search for new target analytes or analytical methods, the question about whether alternative specimens are suitable for doping control purposes has been raised.[57–65] Hair and oral fluid, which both are successfully applied outside the field of doping controls, e.g., in drugs-of-abuse testing, have been frequently discussed in this regard, and advantages as well as disadvantages compared to urine and blood were found.[66]

The advantageous properties of hair testing are (a) the principle option of retrospective analyses, and (b) the non-invasive nature of sampling; however, constitutional as well as technical issues have complicated the introduction of hair testing as a means for routine doping controls. Scalp hair grows approximately 1 cm / month; hence, the sectional analysis of a 6 cm strand of hair would principally allow the detection of drug (ab)use within the last 6 months, which represents an extremely large detection window and has made hair analysis a valuable tool in forensics and toxicology. Nevertheless, in routine sports drug testing, hair has not been taken into account yet due to major drawbacks resulting from (a) considerably different incorporation rates of prohibited substances depending on the color of the hair, and (b) the necessity that athletes *have* scalp hair. Here, concerns regarding the individuals' equality against the controls have been mentioned and hair analysis has been used only in selected occasions of doping rule violations.[67]

In contrast to the desirably long detection window of hair analysis, the use of oral fluid for doping control purposes was suggested to complement sports drug testing programs from the opposite end and support the determination of a timely drug administration. Stimulants and cannabinoids in particular represent classes of drugs that are prohibited in-competition only. Urine analysis allows the detection of minute amounts of these compounds and respective metabolites, and threshold values have been established for selected analytes such as ephedrine, methylephedrine, cathine, and the carboxy-metabolite of tetrahydrocannabinol. However, the distinction whether a detected drug such as tetrahydrocannabinol was administered shortly before competition or represents residuals from much earlier (and thus irrelevant) applications proved complicated using the data of urine analysis only.[68–70] The same issue applies to a variety of stimulants that are not provided with a threshold value.[71] A solution to these problems might be oral fluid analysis, which reflects plasma concentrations without requiring invasive blood sampling. Only pharmacologically relevant oral fluid and, thus, plasma concentrations of banned compounds in combination with the detection of the prohibited drug (or its metabolites) in urine were suggested to be classified as adverse analytical finding.

REFERENCES

1. Thevis, M., and Schänzer, W. (2007) Mass spectrometry in sports drug testing: Structure characterization and analytical assays. *Mass Spectrometry Reviews*, **26**, 79–107.
2. Hokke, C.H., Bergwerff, A.A., Van Dedem, G.W.K., *et al.* (1995) Structural analysis of the sialylated N- and O-linked carbohydrate chains of recombinant human erythropoietin expressed in Chinese hamster ovary cells. Sialylation patterns and branch location of dimeric N-acetyllactosamine units. *European Journal of Biochemistry*, **228**, 981–1008.
3. Kawasaki, N., Haishima, Y., Ohta, M., *et al.* (2001) Structural analysis of sulfated N-linked oligosaccharides in erythropoietin. *Glycobiology*, **11**, 1043–1049.
4. Kawasaki, N., Ohta, M., Hyuga, S., *et al.* (2000) Application of liquid chromatography/mass spectrometry and liquid chromatography with tandem mass spectrometry to the analysis of the site-specific carbohydrate heterogeneity in erythropoietin. *Analytical Biochemistry*, **285**, 82–91.
5. Nimtz, M., Martin, W., Wray, V., *et al.* (1993) Structures of sialylated oligosaccharides of human erythropoietin expressed in recombinant BHK—21 cells. *European Journal of Biochemistry*, **213**, 39–56.
6. Recny, M.A., Scoble, H.A., and Kim. Y. (1987) Structural characterization of natural human urinary and recombinant DNA-derived erythropoietin. Identification of des-arginine 166 erythropoietin. *Journal of Biological Chemistry*, **62**, 17156–17163.
7. Sasaki, H., Ochi, N., Dell, A., and Fukuda, M. (1988) Site-specific glycosylation of human recombinant erythropoietin: Analysis of glycopeptides or peptides at each glycosylation site by fast atom bombardment mass spectrometry. *Biochemistry*, **27**: 8618–8626.
8. Skibeli, V., Nissen-Lie, G., and Torjesen, P. (2001) Sugar profiling proves that human serum erythropoietin differs from recombinant human erythropoietin. *Blood*, **98**, 3626–3634.
9. Stübiger, G., Marchetti, M., Nagano, M., *et al.* (2005) Characterization of N- and O-glycopeptides of recombinant human erythropoietins as

potential biomarkers for doping analysis by means of microscale sample purification combined with MALDI-TOF and quadrupole IT/RTOF mass spectrometry. *Journal of Separation Sciences*, **28**, 1764–1778.

10. Stübiger, G., Marchetti, M., Nagano, M., *et al.* (2005) Characterisation of intact recombinant human erythropoietins applied in doping by means of planar gel electrophoretic techniques and matrix-assisted laser desorption/ionisation linear time-of-flight mass spectrometry. *Rapid Communications in Mass Spectrometry*, **19**, 728–742.
11. Zhou, G.H., Luo, G.A., Zhou, Y., *et al.* (1998) Application of capillary electrophoresis, liquid chromatography, electrospray-mass spectrometry and matrix-assisted laser desorption/ionization—time of flight—mass spectrometry to the characterization of recombinant human erythropoietin. *Electrophoresis*, **19**, 2348–2355.
12. Groleau, P.E., Desharnais, P., Cote, L., and Ayotte, C. (2008) Low LC-MS/MS detection of glycopeptides released from pmol levels of recombinant erythropoietin using nanoflow HPLC-chip electrospray ionization. *Journal of Mass Spectrometry*, **43**, 924–935.
13. Nagano, M., Stubiger, G., Marchetti, M., *et al.* (2005) Detection of isoforms of recombinant human erythropoietin by various plant lectins after isoelectric focusing. *Electrophoresis*, **26**, 1633–1645.
14. Lasne, F., and de Ceaurriz, J. (2000) Recombinant erythropoietin in urine. *Nature*, **405**, 635.
15. Lasne, F., Martin, L., Crepin, N., and de Ceaurriz, J. (2002) Detection of isoelectric profiles of erythropoietin in urine: Differentiation of natural and administered recombinant hormones. *Analytical Biochemistry*, **311**, 119–126.
16. Lasne, F., Martin, L., Martin, J.A., and de Ceaurriz, J. (2007) Isoelectric profiles of human erythropoietin are different in serum and urine. *International Journal of Biological Macromolecules*, **41**, 354–357.
17. Reichel, C., Kulovics, R., Jordan, V., *et al.* (2009) SDS-PAGE of recombinant and endogenous erythropoietins: Benefits and limitations of the method for application in doping control. *Drug Testing and Analysis*, **1**, 43–50.
18. Kohler, M., Ayotte, C., Desharnais, P., *et al.* (2008) Discrimination of recombinant and endogenous urinary erythropoietin by calculating relative mobility values from SDS gels. *International Journal of Sports Medicine*, **29**, 1–6.
19. Pascual, J., Belalcazar, V., de Bolos, C., *et al.* (2004) Recombinant erythropoietin and analogues: A challenge for doping control. *Therapeutic Drug Monitoring*, **26**, 175–179.
20. Segura, J., Pascual, J.A., and Gutiérrez-Gallego, R. (2007) Procedures for monitoring recombinant erythropoietin and analogues in doping control. *Analytical and Bioanalytical Chemistry*, **388**, 1521–1529.

21. Lasne, F., Martin, L., Martin, J.A., and de Ceaurriz, J. (2009) Detection of continuous erythropoietin receptor activator in blood and urine in anti-doping control. *Haematologica*, **94**, 888–890.
22. Guan, F., Uboh, C.E., Soma, L.R., *et al.* (2007) LC-MS/MS method for confirmation of recombinant human erythropoietin and darbepoetin alpha in equine plasma. *Analytical Chemistry*, **79**, 4627–4635.
23. Guan, F., Uboh, C.E., Soma, L.R., *et al.* (2008) Differentiation and identification of recombinant human erythropoietin and darbepoetin Alfa in equine plasma by LC-MS/MS for doping control. *Analytical Chemistry*, **80**, 3811–3817.
24. Guan, F., Uboh, C.E., Soma, L.R., *et al.* (2009) Identification of darbepoetin alfa in human plasma by liquid chromatography coupled to mass spectrometry for doping control. *International Journal of Sports Medicine*, **30**, 80–86.
25. Bidlingmaier, M., and Strasburger, C.J. (2007) Technology insight: Detecting growth hormone abuse in athletes. *Nature Clinical Practice. Endocrinology & Metabolism*, **3**, 769–777.
26. Bidlingmaier, M., Wu, Z., and Strasburger, C. (2003) Problems with GH doping in sports. *Journal of Endocrinological Investigation*, **26**, 924–931.
27. Bidlingmaier, M., Wu, Z., and Strasburger, C.J. (2000) Test method: GH. *Baillière's Clinical Endocrinology and Metabolism*, **14**, 99–109.
28. Bidlingmaier, M., Wu, Z., and Strasburger, C.J. (2001) Doping with growth hormone. *Journal of Pediatric Endocrinology and Metabolism*, **14**, 1077–1084.
29. Bidlingmaier, M., Suhr, J., Ernst, A., *et al.* (2009) High-sensitivity chemiluminescence immunoassays for detection of growth hormone doping in sports. *Clinical Chemistry*, **55**; 445–453.
30. Gutierrez-Gallego, R., Bosch, J., Such-Sanmartin, G., and Segura, J. (2009) Surface plasmon resonance immuno assays: A perspective. *Growth Hormone and IGF Research*, **19**, 388–398.
31. Kohler, M., Püschel, K., Sakharov, D., *et al.* (2008) Detection of recombinant growth hormone in human plasma by a 2D-PAGE method. *Electrophoresis*, **29**, 4495–4502.
32. Thevis, M., Thomas, A., Delahaut, P., *et al.* (2006) Doping control analysis of intact rapid-acting insulin analogues in human urine by liquid chromatography–tandem mass spectrometry. *Analytical Chemistry*, **78**, 1897–1903.
33. Thevis, M., Thomas, A., Delahaut, P., *et al.* (2005) Qualitative determination of synthetic analogues of insulin in human plasma by immunoaffinity purification and liquid chromatography-tandem mass spectrometry for doping control purposes. *Analytical Chemistry*, **77**, 3579–3585.
34. Thevis, M., Thomas, A., and Schänzer, W. (2008) Mass spectrometric determination of insulins and their degradation products in sports drug testing. *Mass Spectrometry Reviews*, **27**, 35–50.

35. Thomas, A., Schänzer, W., Delahaut, P., and Thevis, M. (2009) Sensitive and fast identification of urinary human, synthetic and animal insulin by means of nano-UPLC coupled with high resolution/high accuracy mass spectrometry. *Drug Testing and Analysis*, **1**, 219–227.
36. Bredehöft, M., Schänzer, W., and Thevis, M. (2008) Quantification of human insulin-like growth factor-1 and qualitative detection of its analogues in plasma using liquid chromatography/electrospray ionisation tandem mass spectrometry. *Rapid Communications in Mass Spectrometry*, **22**, 477–485.
37. Kay, R.G., Barton, C., Velloso, C.P., *et al.* (2009) High-throughput ultra-high-performance liquid chromatography/tandem mass spectrometry quantitation of insulin-like growth factor-I and leucine-rich alpha-2-glycoprotein in serum as biomarkers of recombinant human growth hormone administration. *Rapid Communications in Mass Spectrometry*, **23**, 3173–3182.
38. Thomas, A., Thevis, M., Delahaut, P., *et al.* (2007) Mass spectrometric identification of degradation products of insulin and its long-acting analogues in human urine for doping control purposes. *Analytical Chemistry*, **79**, 2518–2524.
39. Catlin, D.H., Ahrens, B.D., and Kucherova, Y. (2002) Detection of norbolethone, an anabolic steroid never marketed, in athletes' urine. *Rapid Communications in Mass Spectrometry*, **16**, 1273–1275.
40. Catlin, D.H., Sekera, M.H., Ahrens, B.D., *et al.* (2004) Tetrahydrogestrinone: Discovery, synthesis, and detection in urine. *Rapid Communications in Mass Spectrometry*, **18**, 1245–1249.
41. Sekera, M.H., Ahrens, B.D., Chang, Y.C., *et al.* (2005) Another designer steroid: Discovery, synthesis, and detection of "madol" in urine. *Rapid Communications in Mass Spectrometry*, **19**, 781–784.
42. Thevis, M., Geyer, H., Mareck, U., and Schänzer, W. (2005) Screening for unknown synthetic steroids in human urine by liquid chromatography-tandem mass spectrometry. *Journal of Mass Spectrometry*, **40**, 955–962.
43. Pozo, O.J., Deventer, K., Eenoo, P.V., and Delbeke, F.T. (2008) Efficient approach for the comprehensive detection of unknown anabolic steroids and metabolites in human urine by liquid chromatography-electrospray-tandem mass spectrometry. *Analytical Chemistry*, **80**, 1709–1720.
44. Mazzarino, M., Turi, S., and Botre, F. (2008) A screening method for the detection of synthetic glucocorticosteroids in human urine by liquid chromatography-mass spectrometry based on class-characteristic fragmentation pathways. *Analytical and Bioanalytical Chemistry*, **90**, 1389–1402.
45. Zierau, O., Lehmann, S., Vollmer, G., *et al.* (2008) Detection of anabolic steroid abuse using a yeast transactivation system. *Steroids*, **73**, 1143–1147.
46. Houtman, C.J., Sterk, S.S., van de Heijning, M.P., *et al.* (2009) Detection of anabolic androgenic steroid abuse in doping control using mammalian reporter gene bioassays. *Analytica Chimica Acta*, **637**, 247–258.

47. Nielen, M.W., Bovee, T.F., van Engelen, M.C., *et al.* (2006) Urine testing for designer steroids by liquid chromatography with androgen bioassay detection and electrospray quadrupole time-of-flight mass spectrometry identification. *Analytical Chemistry*, **78**, 424–431.
48. Parisotto, R., Gore, C., Emslie, K., *et al.* (2000) A novel method utilising markers of altered erythropoiesis for the detection of recombinant human erythropoietin abuse in athletes. *Haematologica*, **85**, 564–572.
49. Parisotto, R., Wu, M., Ashenden, M.J., *et al.* (2001) Detection of recombinant human erythropoietin abuse in athletes utilizing markers of altered erythropoiesis. *Haematologica*, **86**, 128–137.
50. Gore, C.J., Parisotto, R., Ashenden, M.J., *et al.* (2003) Second-generation blood tests to detect erythropoietin abuse by athletes. *Haematologica*, **88**, 333–344.
51. Sharpe, K., Ashenden, M.J., and Schumacher, Y.O. (2006) A third generation approach to detect erythropoietin abuse in athletes. *Haematologica*, **91**, 356–363.
52. Appolonova, S.A., Dikunets, M.A., and Rodchenkov, G.M. (2008) Possible indirect detection of rHuEPO administration in human urine by high-performance liquid chromatography tandem mass spectrometry. *European Journal of Mass Spectrometry*, **14**, 201–209.
53. Holt, R.I., and Sonksen, P.H. (2008) Growth hormone, IGF-I and insulin and their abuse in sport. *British Journal of Pharmacology*, **154**, 542–556.
54. Holt, R.I. (2007) Meeting reports: Beyond reasonable doubt: Catching the growth hormone cheats. *Pediatric Endocrinology Reviews*, **4**, 228–232.
55. Holt, R.I., Erotokritou-Mulligan, I., and Sönksen, P.H. (2009) The history of doping and growth hormone abuse in sport. *Growth Hormone and IGF Research*, **19**, 320–326.
56. Holt, R.I.G. (2009) Detecting growth hormone abuse in athletes. *Drug Testing and Analysis*, **1**, DOI: 10.1002/dta.1059.
57. Rivier, L. (2000) Is there a place for hair analysis in doping controls? *Forensic Science International*, **107**, 309–323.
58. Sachs, H. (1997) History of hair analysis. *Forensic Science International*, **84**, 7–16.
59. Thieme, D., Grosse, J., Sachs, H., and Mueller, R.K. (2000) Analytical strategy for detecting doping agents in hair. *Forensic Science International*, **107**, 335–345.
60. Kintz, P., and Samyn, N. (2002) Use of alternative specimens: Drugs of abuse in saliva and doping agents in hair. *Therapeutic Drug Monitoring*, **24**, 239–246.
61. Gaillard, Y., and Pepin, G. (1997) Hair testing for pharmaceuticals and drugs of abuse: Forensic and clinical applications. *American Clinical Laboratory*, **16**, 18–22.
62. Gaillard, Y., and Pepin, G. (1999) Testing hair for pharmaceuticals. *Journal of Chromatography B, Biomedical Sciences and Applications*, **733**, 231–246.

63. Kintz, P. (1996) *Drug Testing in Hair*. CRC Press, Boca Raton.
64. Kintz, P. (1998) Hair testing and doping control in sport. *Toxicology Letters*, **102–103**, 109–113.
65. Kintz, P. (2003) Testing for anabolic steroids in hair: A review. *Legal Medicine (Tokyo)*, **5** Suppl 1, S29–33.
66. Gallardo, E., and Queiroz, J.A. (2008) The role of alternative specimens in toxicological analysis. *Biomedical Chromatography*, **22**, 795–821.
67. Segura, J. (2009) Is anti-doping analysis so far from clinical, legal or forensic targets?: The added value of close relationships between related disciplines. *Drug Testing and Analysis*, **1**, DOI: 10.1002/dta.1055.
68. Huestis, M.A., and Cone, E.J. (1998) Differentiating new marijuana use from residual drug excretion in occasional marijuana users. *Journal of Analytical Toxicology*, **22**, 445–454.
69. Gustafson, R.A., Levine, B., Stout, P.R., *et al.* (2003) Urinary cannabinoid detection times after controlled oral administration of delta9-tetrahydrocannabinol to humans. *Clinical Chemistry*, **49**, 1114–1124.
70. Goodwin, R.S., Darwin, W.D., Chiang, C.N., *et al.* (2008) Urinary elimination of 11-nor-9-carboxy-delta9-tetrahydrocannnabinol in cannabis users during continuously monitored abstinence. *Journal of Analytical Toxicology*, **32**, 562–569.
71. Strano-Rossi, S., Colamonici, C., and Botre, F. (2008) Parallel analysis of stimulants in saliva and urine by gas chromatography/mass spectrometry: Perspectives for "in competition" anti-doping analysis. *Analytica Chimica Acta*, **606**, 217–222.

INDEX

Note: Page numbers in *italics* refer to figures; those in **bold** to tables.

Mass Spectrometry in Sports Drug Testing: Characterization of Prohibited Substances and Doping Control Analytical Assays, By Mario Thevis
Copyright © 2010 John Wiley & Sons, Inc.

WILEY-INTERSCIENCE SERIES IN MASS SPECTROMETRY

Series Editors

Dominic M. Desiderio
Departments of Neurology and Biochemistry
University of Tennessee Health Science Center

Nico M. M. Nibbering
Vrije Universiteit Amsterdam, The Netherlands

John R. de Laeter • *Applications of Inorganic Mass Spectrometry*
Michael Kinter and Nicholas E. Sherman • *Protein Sequencing and Identification Using Tandem Mass Spectrometry*
Chhabil Dass • *Principles and Practice of Biological Mass Spectrometry*
Mike S. Lee • *LC/MS Applications in Drug Development*
Jerzy Silberring and Rolf Eckman • *Mass Spectrometry and Hyphenated Techniques in Neuropeptide Research*
J. Wayne Rabalais • *Principles and Applications of Ion Scattering Spectrometry: Surface Chemical and Structural Analysis*
Mahmoud Hamdan and Pier Giorgio Righetti • *Proteomics Today: Protein Assessment and Biomarkers Using Mass Spectrometry, 2D Electrophoresis, and Microarray Technology*
Igor A. Kaltashov and Stephen J. Eyles • *Mass Spectrometry in Biophysics: Confirmation and Dynamics of Biomolecules*
Isabella Dalle-Donne, Andrea Scaloni, and D. Allan Butterfield • *Redox Proteomics: From Protein Modifications to Cellular Dysfunction and Diseases*
Silas G. Villas-Boas, Ute Roessner, Michael A.E. Hansen, Jorn Smedsgaard, and Jens Nielsen • *Metabolome Analysis: An Introduction*
Mahmoud H. Hamdan • *Cancer Biomarkers: Analytical Techniques for Discovery*
Chabbil Dass • *Fundamentals of Contemporary Mass Spectrometry*
Kevin M. Downard (Editor) • *Mass Spectrometry of Protein Interactions*
Nobuhiro Takahashi and Toshiaki Isobe • *Proteomic Biology Using LC-MS: Large Scale Analysis of Cellular Dynamics and Function*
Agnieszka Kraj and Jerzy Silberring (Editors) • *Proteomics: Introduction to Methods and Applications*
Ganesh Kumar Agrawal and Randeep Rakwal (Editors) • *Plant Proteomics: Technologies, Strategies, and Applications*
Rolf Ekman, Jerzy Silberring, Ann M. Westman-Brinkmalm, and Agnieszka Kraj (Editors) • *Mass Spectrometry: Instrumentation, Interpretation, and Applications*
Christoph A. Schalley and Andreas Springer • *Mass Spectrometry and Gas-Phase Chemistry of Non-Covalent Complexes*
Riccardo Flamini and Pietro Traldi • *Mass Spectrometry in Grape and Wine Chemistry*
Mario Thevis • *Mass Spectrometry in Sports Drug Testing: Characterization of Prohibited Substances and Doping Control Analytical Assays*

Printed in the United States
By Bookmasters